Technology for Smart Futures

Mohammad Dastbaz • Hamid Arabnia
Babak Akhgar
Editors

Technology for Smart Futures

Springer

Editors
Mohammad Dastbaz
Deputy Vice Chancellor
University of Suffolk
Suffolk, United Kingdom

Hamid Arabnia
Department of Computer Science
The University of Georgia
Athens, GA, USA

Babak Akhgar
Director of CENTRIC
Sheffield Hallam University
Sheffield, United Kingdom

ISBN 978-3-319-60136-6 ISBN 978-3-319-60137-3 (eBook)
DOI 10.1007/978-3-319-60137-3

Library of Congress Control Number: 2017949263

Printed on acid-free paper

This Springer imprint is published by Springer Nature
The registered company is Springer International Publishing AG
The registered company address is: Gewerbestrasse 11, 6330 Cham, Switzerland

Preface

First we thought the PC was a calculator. Then we found out how to turn numbers into letters with ASCII — and we thought it was a typewriter. Then we discovered graphics, and we thought it was a television. With the World Wide Web, we've realized it's a brochure.

—Douglas Adams

The technological changes over the last four decades have perhaps rightly been termed as "the third industrial revolution." It was some 36 years ago on 12 August 1981, when IBM launched its first PC. The first IBM PC had an Intel 8088 processor that ran at the speed of 4.77 MHz, it had a base memory of 16 kB expandable to 256 kB, and it carried no hard disk drive and only benefited from two 5–1/4 in, 160 kB capacity "floppy disk" drives.

Comparing what was on offer then and now, where significant computing power is packed in our smart devices, demonstrates the amazing journey of technological advances over the past four decades. Devices capable of easily and seamlessly exchanging "terabytes" of data, processors capable of over 4.5 GHz of processing speed, and a global connected community over a billion strong point to a different world and possibilities that would have been very hard to imagine in 1981. As historical evidence shows, the development of human society has always been associated with significant social and economic costs. It is therefore not surprising that the current phenomenal growth also comes with significant costs in terms of energy that is required to keep a 24/7365 virtual world running and for us to access as we wish.

The current volume, second in our four-part series on sustainable futures, attempts to shed some light on some of the key concepts surrounding the impact of technology on our future.

Section one of this volume is dedicated to "Internet of things" (IoT) and "smart living," where subjects and concepts such as cloud computing, big data, "fog" computing, cognitive middleware, and context aware interaction in a smart living environment are explored.

Section two of the volume is dedicated to a number of very interesting case studies related to "smart living." Case studies around data science applications for independent and healthy living and a comprehensive framework for elderly healthcare monitoring in a smart environment are presented.

The final section of the book deals with the "technological" as well as some of the "environmental" challenges we face. Issues around environmental responsibility, energy consumption by network infrastructure and how we can monitor and reduce it, the ethical dimension of designing health technologies, and virtual vulnerability are discussed.

Eric Schmidt, executive chairman of Google, "In Tech We Trust and The Future of the Digital Economy" at Davos 2015, stated that: "Everyone gets smarter because of this technology… and the empowerment of people is the secret to technological progress. [In the future], the Internet will disappear… you won't even sense it, it will be part of your presence all the time…" While the development of the past 40 years certainly provides the ground for such predictions, we should also be very mindful that this future is not marred by developing further divide between the north and the south and the "haves" and "have-nots." There is no doubt in my mind that we can and we should use technology as a force for good and for developing a bright future for humanity, but this can only happen if we are aware of the shortcomings as well as the strength of the emerging trends and technologies and ultimately who controls them. The current volume goes some way to provide the bases for some of these discussions.

Suffolk, UK Mohammad Dastbaz

Acknowledgments

We would like to thank all the contributing authors for their tireless work and for providing us with their valuable research work which has made this edited volume possible. We would also like to thank the Springer editorial and production team, Amanda Quinn, Brian Halm, and Brinda Megasyamalan for their patience and valuable advice and support.

Special thanks go to Prof. Hissam Tawfik, Laura Messer, and Fiona Scarth who worked tirelessly in organizing our schedule, getting all the necessary forms done, and sending numerous e-mails and gentle reminders when necessary.

Finally, our thanks go to all our colleagues at University of Suffolk, Leeds Beckett University, Sheffield Hallam University, and University of Georgia whose work has made a significant contribution to our sustainable development agenda and has informed some of the ideas and core discussions, which are presented in this edited volume.

Mohammad Dastbaz
Hamid Arabnia
Babak Ahgkar

Contents

Contributors

Babak Akhgar is professor of informatics and director of CENTRIC (Centre of Excellence in Terrorism, Resilience, Intelligence and Organised Crime Research) at Sheffield Hallam University (UK) and fellow of the British Computer Society. He has more than 100 refereed publications in international journals and conferences on information systems with specific focus on knowledge management (KM). He is member of editorial boards of several international journals and has acted as chair and programme committee member for numerous international conferences.

M. Alahmad (senior IEEE member) received his B.Sc. in electrical engineering from Birzeit University in Ramallah/West Bank in 1999 and both M.Sc. and the Dr.-Ing. degrees in microwave engineering from Technische Universitat Munchen, Munich, Germany, in 2002 and 2006, respectively. Currently, he is faculty member at the Electrical Department at the United Arab Emirates University, and his research interest is involved in the design and fabrication of self-powered low-powered nano-based electronic devices and systems along with applied electromagnetics for bio-medical applications. Dr. Al Ahmad is senior member of IEEE. He has 10 years of electronic materials and device fabrication research, with universities, national laboratories and industrial experiences. He has managed several research projects and teams with annual budgets of up to US$1 million. He is also a principal (lead) author of around 55 published articles in journals and international conferences and has over 40 presentations at international conferences wherein several were invited. Dr. Al Ahmad has conducted a research in energy harvesting technologies and frequency agile circuits at the Siemens AG/CNRS and King Abdullah University of Science and Technology (KAUST) and has published in this domain more than 20 journal and 35 conference papers and more under review.

M. Al Hemairy is a Ph.D. candidate at Coventry University, UK. He obtained his master's degree in data and knowledge management from the BUiD and University of Edinburgh in 2010. His research interest area is information security and networking, smart technologies and pervasive healthcare.

Mohamed has published more than ten research papers in peer-reviewed journals and reputable conferences, and he has also published book chapters and lecture notes. Mohamed has recently filed two patents in the US Patent Office [USPTO] from his novel research work.

Besides his research excellence, Mohamed has a vast experience in managing research and technology transfer activities and is currently working at UAE University as head of Intellectual Property and Commercialization Office at the Research and Innovation Division. During the last few years, he helped over a hundred researchers to file more than 100 inventions and managed more than 210 patent applications with 45 patents granted and many patents successfully commercialised to the industry. Moreover, Mohamed has organised more than ten national and international conferences and many research symposiums.

Dhiya Al-Jumeily is the associate dean of External Engagement for the Faculty of Engineering and Technology at Liverpool John Moores University, UK. He has extensive research interests covering a wide variety of interdisciplinary perspectives concerning the theory and practice of applied computing in medicine, human biology and healthcare. He has published well over 170 peer-reviewed scientific publications, six books and three book chapters, in multidisciplinary research areas. Dhiya has successfully supervised 16 Ph.D. students' studies and has been an external examiner to various UK and overseas universities for undergraduate, postgraduate and research degrees. He has been actively involved as a member of the editorial board and review committee for a number of peer-reviewed international journals and is on programme committees as a general chair for a number of international conferences. Dhiya is also a successful entrepreneur. He is the head of enterprise for the Faculty of Engineering and Technology. He has been awarded various commercial and research grants, nationally and internationally, over £3 M from overseas research and educational partners, from the British Council, UK, and directly from industry with portfolio of various knowledge transfer programmes between academia and industry. Dhiya has extensive leadership experience including head of enterprise for the Faculty of Engineering and Technology, leading the Applied Computing Research Group and developing and managing the professional doctorate programme in engineering and technology for the university, and is a founder and chair of the IEEE International Conference Series on Developments in eSystems Engineering DeSE (www.dese.org.uk) since 2007. He has a large number of international contacts and leads or participates in several international committees in his research fields. Dhiya has one patent and coordinated over ten projects at national and international levels.

Dr. Al-Jumeily is a senior member of the IEEE and a chartered IT professional. He is also a fellow of the UK Higher Education Academy.

S. Amin is the postgraduate programme director and principal lecturer in network computing in the Faculty of Engineering and Computing. His main area of research is health informatics.

One of his research interests centres on the design of biomedical image processing algorithms and multimedia applications on parallel computers and distributed computing and their implementation on cluster of workstations (CoW). He is also involved in the area of context-aware multimedia information systems for e-health decision support. Previously, he was involved with a European community project at the Parallel Algorithms Research Centre, Loughborough University.

Dr. Amin is the principal investigator of several funded projects. He has published more than 90 papers in peer-reviewed journals, reputable conferences and book chapters on many aspects of this research. He has organised international conferences and workshops, as well as acted as session chair on programme committees for many international conferences.

He served on the IEEE Committee for Signal Processing Chapter for the UK and Ireland and as vice-chairman of the British Computer Society–Middle East Section.

Obinna Anya is a research staff member at IBM Research–Almaden. His research interest lies broadly in the area of enterprise collaborative systems and spans from computer-supported co-operative work to e-health, user experience analytics and context-aware design. In his current research, Obinna combines machine learning and in-depth analysis of customer engagement data in mobile and cloud services for generating actionable insights for user behaviour prediction, personalised service recommendation and improved customer engagement. Obinna holds a Ph.D. in computer science, an M.Sc. in distributed systems – both from the University of Liverpool – and a B.Sc. in computer science from the University of Nigeria. Before coming to IBM, Obinna worked as a research scientist at Liverpool Hope University, where he was the lead researcher on a British Council-sponsored project on context-aware collaborative environments for e-health decision support. Dr. Anya's work has been published in several high-quality peer-reviewed journals and academic conferences.

Hamid Arabnia is professor of computer science at the University of Georgia, USA. He is editor-in-chief of *The Journal of Supercomputing* (one of the oldest journals in computer science) published by Springer and has been associate editor of *IEEE Transactions on Information Technology in Biomedicine* (2008–2011). He is the book series editor-in-chief of *Transactions of Computational Science and Computational Intelligence* (Springer) and of *Emerging Trends in Computer Science and Applied Computing* (Elsevier).

M. Arzamendia was born in Asuncion, Paraguay, in 1977. He received a B.E. in electrical engineering from the University of Brasilia, Brazil, in 2002 and an M.S. in electronic engineering from Mie University, Japan, in 2009. Currently, he is a researcher at the Laboratory of Distributed Systems (LSD) in the Faculty of Engineering at the National University of Asuncion (FIUNA), Paraguay, and is working towards his Ph.D. at the Department of Electronic Engineering at the University of Seville, Spain. His main interests include wireless sensor networks, embedded systems applications and evolutionary algorithms.

Thar Baker is a senior lecturer in software systems, director of Computer Science Research Group and a member of Applied Computing Research Group in Liverpool John Moores University, UK. Thar has published numerous refereed research papers in multidisciplinary research areas including the Internet of things, cloud computing, distributed software systems, algorithm design and green computing. He has been actively involved as a member of the editorial board and review committees for a number of international journals and conferences.

Anar Bazarhanova is a first year Ph.D. candidate in the DINE research project – Digital Infrastructures and Enterprises Integration at Aalto University – Department of Computer Science. Before coming to Aalto, she completed the Erasmus Mundus Joint Degree programme PERCCOM – Pervasive Computing and Communications for Sustainable Development – and obtained the following degrees: master in complex systems engineering (University of Lorraine, France), master in pervasive computing and communications for sustainable development (Lulea University of Technology, Sweden) and master of science in technology (Lappeenranta University of Technology, Finland). This book chapter is based on her master's thesis work "Belief Rule-Based Environmentally Responsibility Assessment System for SMEs" which she did at Leeds Beckett University, UK, under the supervision of Dr. Ah-Lian Kor and Prof. Colin Pattinson.

Adie Blanchard is a sport and exercise science graduate from the University of Portsmouth. She joined Rescon in January 2014, researching, developing and delivering human performance technologies and services. Adie is operations lead for the Lincus software as a self-service and shared care management platform, where she has led training and support for several Lincus deployments into health and social organisations, from initial implementation to national rollout. She supports day-to-day project management, is experienced in complex system testing and has supported the continued development and refinement of the Lincus platform. Her research focus is on the barriers to technology implementation and methods to overcome them, in which she has published and presented.

Emma Bond is an associate professor at the University of Suffolk. She has extensive social science teaching and research experience, and her work on virtual environments, mobile technologies and risk has attracted national and international acclaim.

Paul Chamberlain is professor of design, head of the Art and Design Research Centre and codirector of Lab4Living at Sheffield Hallam University in the UK. He has led major design projects engaging diverse disciplinary specialists and commercial partners that have played a significant role in supporting regional industrial reconstruction. Paul explores the multisensory aspects of design and adopts an inclusive approach to co-creation and applies this with a focus on healthcare, disability and ageing.

Claire Craig is a reader in design and creative practice in health at Sheffield Hallam University and is codirector of Lab4Living, an interdisciplinary research group which brings together designers, healthcare practitioners, engineers and psychologists with the aim of designing products, services and environments to improve the quality of life and well-being for individuals. Much of her research focuses on the role of technology in supporting well-being in older people and people with dementia.

Mohammad Dastbaz is a professor of informatics and deputy vice chancellor at the University of Suffolk. Professor Dastbaz remains research active and has published over 60 refereed journals and conference papers, books and book chapters. He is on a number of editorial boards of international journals, has been chair of a number of international conferences (including IEEE's Information Visualisation) and remains a member on a number of international conference committees.

His latest publications include a number of edited volumes such as *Green Information Technologies* (Elsevier Publishers, 2015) and *Building Sustainable Futures: Design and the Built Environment*, published by Springer (2016) – a collaboration with the University of California–Berkeley – and a series of four edited volumes on "technology and sustainable futures" published by Springer. The first book in the series titled *Building Information Modelling, Building Performance, Design and Smart Construction*, a collaboration with the University of Cambridge, is already published (2017).

Professor Mohammad Dastbaz is a fellow of the BCS, The Chartered Institute for IT; of the Higher Education Academy; and of the Royal Society of Arts (RSA).

Tom Dawson is a medical practitioner, human performance specialist and entrepreneur with over 25 years of practical and commercial experience. He has a doctorate in physiology, anatomy and genetics from the University of Oxford where he was previously director of the Science and Medicine of Athletic Performance programme. He also has medical and clinical biochemistry degrees from Otago University and multiple postgraduate qualifications. He has led the commercial and scientific direction of Rescon since 2011 and continues to actively contribute to scientific and patent literature. Tom has led diverse research projects for Rescon including development of a unique wearable sensor system and the development and system design of the Lincus evaluation and management platform.

Jiamei Deng is professor of artificial intelligence and energy in Leeds Sustainability Institute at Leeds Beckett University (UK). His research interests and expertise lie in the areas of big data, intelligent systems, power plant safety, and control systems design. Professor Deng holds a Ph.D. in cybernetics from University of Reading in 2005. He has published more than 65 refereed publications in reputable international journals and conferences such as IEEE Transactions on Neural Networks and Learning Systems.

Mohammed Dighriri has a B.Sc. and M.Sc. degree in computing and information systems from Liverpool John Moores University (LJMU) in 2013 and 2014, respectively. Mr. Dighriri is currently doing a Ph.D. in computer science in the Department of Computer Science at LJMU. He is a lecturer in management information system (MIS) at the University of Business and Technology (UBT) in the Kingdom of Saudi Arabia since 2014. His research expertise includes the Internet of things, mobile network, cloud computing and business application.

Paul Fergus is a senior lecturer in the Faculty of Engineering and Technology at Liverpool John Moores University and a visiting professor at Supelec University in France. Paul has over 5 years' experience as a senior software engineer in industry and has worked on several development projects for the Prison Service, Ericsson, Nokia, NMC, Nissan, and Pilkington Glass. He has been an active researcher for more than 10 years and has published over 150 international peer-reviewed papers on topics that include artificial intelligence, Semantic Web, signal processing, bioinformatics, and data science.

Jean-Philippe Georges He received his Ph.D. in computer science from the University of Lorraine (France) in 2005 about performance evaluation of switched Ethernet architectures based on network calculus modelling. This work was applied in the framework of the project FP6 STREP NeCST Networked Control System Tolerant to faults (2004–2007) and continued in 2006 as researcher in the Laboratory of Process Control and Automation at the Aalto University (Finland). From 2006 as associate professor, he is researcher at the Centre de Recherche en Automatique de Nancy. His interests include performance evaluation, green IT, quality of service metrics for networked control systems, real-time networking as well as wireless communications.

Laura Gilbert leads Rescon's information technology team and has a strong background in data analysis and statistics, with a focus on creating novel, informative and scientifically robust ways to analyse variable datasets. She holds a doctorate in particle physics from the University of Oxford and bachelor's degree and master's in natural sciences from the University of Cambridge. Laura has worked as an academic researcher in particle physics, where she also held a number of lectureships. Laura previously worked for the UK Ministry of Defence as an analyst and after that in three financial institutions as a quantitative analyst, developing predictive analytics tools. Laura has a wide experience of data visualisation, computer virtualisation, GRID computing, management and analysis of large datasets and web development. Laura is fluent in multiple languages, including python, Java, R, php and JavaScript.

D. Gregor is professor of distributed systems, Engineering Faculty of the National University of Asuncion (FIUNA), Paraguay. He received a bachelor's degree in systems analysis and computer engineering from the American University, Asuncion, Paraguay, in 2007. He received M.Sc. and Ph.D. degrees in electronic, signal processing and communications from the University of Seville, Spain, in 2009 and 2013, respectively. In 2014, he received an honourable mention at the National Science Award, Paraguay. He is a researcher categorised Level 1 by the National Council of Science and Technology (CONACYT), Paraguay. He is currently head of the Laboratory of Distributed Systems (LSD), Engineering Faculty of the National University of Asuncion, Paraguay. His research focuses on the application of intelligent transportation systems (ITS); interoperability in sensor networks, embedded systems and instrumentation systems; distributed systems applied to renewable energy; and development and management of open source software project for embedded systems.

Khan Mohammad Habibullah is lecturer of Computer Science and Engineering Department, East West University, Bangladesh.

His areas of interest include sustainable IT, cloud computing and software development with sustainability aspects.

T. Hamid is an academic with rich experience in research, teaching, professional training and curriculum development. He received his first Ph.D. degree in MIS in 2004 and the second Ph.D. degree in cybersecurity in 2014 from Bedfordshire University, UK.

In 2012, he joined the Bedfordshire University as research fellow in cyberstalking and security and in 2014 became a director of study/research fellow at Cardiff Metropolitan University/LSC. Since December 1990, he has successfully supervised to completion nine Ph.D. students as leading supervisor and 14 M.Sc. students. He is an internal/external examiner/referee for M.Phil./Ph.D. research students and a trainer in Java programming, Oracle 12C, Informatica PowerCenter, CCNA, CISSP, CIMP and ethical hacking. His current research interests include cybersecurity, cyberstalking, networking and information security. He was the reviewer and session chair for International Conference on Developments in eSystems Engineering (2015, 2016) and International Conference on Systems, Signals and Image Processing (IWSSIP) (2015); a member of the technical committee for *Central European Journal of Computer Science*; a reviewer for *Central European Journal of Computer Science*; and a founder of international conference and exhibition of Symposium on Emerging Trends in advanced computing (SETAC-2012) – 15th May 2012.

Amin Hosseinian-Far is currently a senior lecturer in the School of Computing, Creative Technologies and Engineering at Leeds Beckett University. He gained his B.Sc. (Hons) in business information systems from the University of East London, M.Sc. in satellite communication and space systems from the University of Sussex and Ph.D. from the University of East London. He has held lecturing and research positions at the University of East London and at a number of private HE institutions and strategy research firms in the capital. Amin is a member of the Institution of Engineering and Technology (IET), a fellow of the Royal Society of Arts (RSA) and a senior fellow of Higher Education Academy (HEA).

Abir Hussain is a reader (associate professor) in image and signal processing and she is the head of the Applied Computing Research Group at the Faculty of Engineering and Technology. She completed her Ph.D. study at the University of Manchester, UK, in 2000 with a thesis entitled "Polynomial Neural Networks for Image and Signal Processing". She has published numerous refereed research papers in conferences and journal in the research areas of neural networks, signal prediction, telecommunication fraud detection and image compression. She has worked with higher order and recurrent neural networks and their applications to financial, physical, e-health and image compression techniques. She has developed with her research students a number of recurrent neural network architectures. Her research has been published in a number of high esteemed and high impact journals such as the *Expert Systems with Applications*, *PLOS ONE*, *Electronics Letters*, *Neurocomputing* and *Neural Computing and Applications*. She is a Ph.D. supervisor and an external examiner for research degrees including Ph.D. and M.Phil. She is one of the initiators and chairs of the Developments in eSystems Engineering (DeSE) series, most notably illustrated by the IEEE technically sponsored DeSE International Conference Series.

Robert Keight is a Ph.D. candidate at Liverpool John Moores University, where he is currently investigating the development of artificial intelligence and advanced computational solutions, directed towards the realisation of next generation frameworks in human healthcare and proactive wellness. He gained his master's in information communication technology from Liverpool John Moores University in 2014, UK, prior to which he obtained a B.A. (Hons) degree in graphic design from NEWI, Wrexham, 2005, in the UK. His research interests include the areas of data science, machine learning and artificial intelligence, ranging from fundamental questions about the nature of intelligence to computationally effective solutions. He is especially interested in the domains of healthcare, including related fields such as bioinformatics, in addition to the consideration of emerging paradigms in finance and advanced manufacturing.

Vyacheslav Kharchenko is head of Computer Systems and Networks Department and an honoured inventor of Ukraine. He has authored more than 250 papers, 12 national standards, 604 inventions and 34 books and has supervised more than 40 Ph.D. and D.Scs. students. His expertise includes dependability of instrumentation and control systems and system on chip and diversity, safety and sustainability of IT systems and infrastructure. He has experience in project management (nine projects).

Alexandra Klimova was born in Artem, Russia, in 1985. She received Ph.D. in mechanical engineering in 2011. Alexandra has extensive research experience in vehicle dynamics and FEA. She is currently working as an international project coordinator and assistant professor at ITMO University, Saint Petersburg, Russia. She is highly experienced in development and coordination of international educational projects and programmes. Her areas of interest include internalisation of education, project management and knowledge management.

Ah-Lian Kor is a course leader of Leeds Beckett University. She received M.Sc. in sustainable computing, specialising in software development, web applications, data analytics and AI. She is active in sustainable IT, intelligent systems, decision support systems and data centre research. She forges an industrial collaboration with Drax Power Station which aims to develop and evaluate a virtual environment (in the form of an expert system) to support decision making in fault diagnosis for the centrifuge system for Drax Power Station. She is an intelligent system academic expert for the Knowledge Transfer Partnership Project with Premier Farnell (a global electronics engineering company). Ah-Lian is a Project Team Committee member for several EU projects (e.g. ALIOT, GreenCo, PERCCOM, etc.). She has chaired sessions in reputable international conferences (e.g. FTC 2016), a publicity chair for CyberSciTech 2017, and sits in Technical Programme Committee for journals and conferences (e.g. ICAART, SEEDS, IARIA Intelligent Systems, IEEE Computing Conference, etc.).

Gyu Myoung Lee works in the Department of Computer Science at Liverpool John Moores University (LJMU) as a senior lecturer since 2014 and, with KAIST, Korea, as an adjunct professor since 2012. His research interests include the Internet of things, future networks, multimedia services and energy-saving technologies including smart grids. He has been actively working for standardisation in ITU-T, IETF and oneM2M. He has been actively working for standardisation in ITU-T, IETF, oneM2M, etc., and currently serves as a rapporteur of Q11/13, Q16/13 and Q4/20 as well as an editor in ITU-T. He is a senior member of IEEE.

Dimitar Minovski From 2016 he is a Ph.D. student at Luleå University of Technology with a focus on areas such as future 5G networks and IoT. More specifically, his research includes the integration of IoT in the future networks with emphasis on QoS and QoE. He completed his M.Sc. in computer science at the University of Lorraine (France) in 2016 about pervasive computing and sustainable development.

Reza Montasari is a lecturer at the Birmingham City University, where he lectures on the topics of digital forensics and cybersecurity. Reza was awarded his Ph.D. in digital forensics at the University of Derby, having previously completed his M.Sc. in computer forensics and B.Sc. (Hons) in multimedia computing at the University of South Wales. Reza has a sound knowledge of the principles and technology relating to computer forensic science, including AccessData, EnCase and Cellebrite products as well as other open source tools. His research interests include digital forensic investigation, computer forensics and anti-forensics, data recovery, image analysis, cybercrime and cyber war, cybercrime detection and prevention and digital forensic investigation process models.

J. Mustafina is the head of Foreign Languages Department in Kazan Federal University. She received her Ph.D. in 2007 and full professorship in 2012. Her scientific interests cover the sociolinguistic studies, particularly protection of minor languages in the globalised world. A special area of her research is interdisciplinary approach towards improvement of the social spheres of human life using technology and applied computing. Jamila has published over 70 peer-reviewed scientific articles, four books and two book chapters. She is currently supervising five Ph.D. students and two postdocs. She is also acting as a deputy editor-in-chief of the peer-reviewed scientific journal *Education and Self-Development*. Jamila has been awarded a number of research grants nationally and internationally.

Colin Pattinson is dean of the School of Computing, Creative Technologies and Engineering. His research expertise includes the relationship between power-saving network designs and management systems and low-cost, low-energy networking solutions. He is a member of the EU e-InfraNet Green ICT Network which aims to identify and promote an environmentally sustainable e-infrastructure for research and higher education in the ERA. He is leading projects on green computing technologies and energy-efficient data centres. He has recently completed a review for Cisco Systems of their energy monitoring as well as control software, and he is a Project Team Committee member for several EU projects (e.g. ALIOT, GreenCo, PERCCOM, etc.).

Faye Prior is a researcher at Rescon Technologies. She received a B.Sc. in sport and exercise science at Aberystwyth University, followed by a master's in clinical exercise physiology from Liverpool John Moores University. Her research interests are physical activity for health, prevention and management of long-term conditions and smart technology for health monitoring. She has been involved in a number of national research projects and is currently engaged in a public health Ph.D. research project.

Muthu Ramachandran is currently a principal lecturer in the School of Computing, Creative Technologies and Engineering as part of the Faculty of Arts, Environment and Technology at Leeds Beckett University in the UK. Previously, he spent nearly 8 years in industrial research (Philips Research Labs and Volantis Systems Ltd., Surrey, UK) where he worked on software architecture, reuse and testing. His first career started as a research scientist where he worked on real-time systems development projects. Muthu is an author and co-author of books: *Software Components: Guidelines and Applications* (Nova Publishers, NY, USA, 2008) and *Software Security Engineering: Design and Applications* (Nova Publishers, NY, USA, 2011). He has also widely authored and published nine books, over 100 journal articles, over 50 book chapters and over 200 conference papers on various advanced topics in software engineering, software security, cloud computing and education. Muthu has been leading conferences as chair and as keynote speaker on global safety, security and sustainability, emerging services,

IoT, Complexis 2016, big data and software engineering. Muthu is a member of various professional organisations and computer societies such as IEEE and ACM, a fellow of BCS and a senior fellow of HEA. He was also invited as keynote speaker on several international conferences. Muthu had worked on several research projects including all aspects of software engineering, SPI for SMEs (known as a PRISM model), emergency and disaster management systems, software components and architectures, good practice guidelines on software developments, software security engineering and service and cloud computing. Project details can be accessed at soft-research.com.

Vanessa Rawlings is a lecturer and researcher in the Children, Young People and Education Department at the University of Suffolk. She is a qualified teacher with over 20 years' experience teaching children and adults in both the private and public sector and in formal and informal education provision.

D.G. Reina was born in Seville, Spain, in 1983. He received his bachelor's degree in electronic engineering and M.S. and Ph.D. degrees at the University of Seville in 2009, 2011 and 2015, respectively. He is currently working as an assistant professor at the University of Seville. He has been a visitor researcher at Liverpool John Moores University, Freie Universität Berlin, Colorado School of Mines and Leeds Becket University. His research interests include multi-hop ad hoc networks such as MANETs, VANETs, DTNs and FANETs and the application of metaheuristics in optimisation problems related to multi-hop ad hoc networks.

Eric Rondeau He is professor at the University of Lorraine (France) and is working at CRAN (Research Centre for Automatic Control in Nancy) – UMR 7039 CNRS. His research activities are green networking, the Internet of things, smart building and industrial networks. He is the director of the Erasmus Mundus Joint Master Degree in Pervasive Computing and Communications for Sustainable Development (PERCCOM). He was the coordinator of FP6 European project entitled Networked Control System Tolerant to faults (NeCST). He is member of IFAC TC8.3 on Modelling and Control of Environmental Systems. He published more than 150 papers in conferences, journals and books.

T.V. Santosh joined Reactor Safety Division of Bhabha Atomic Research Centre as scientific officer in 2001 after successful completion of postgraduate diploma in nuclear science and engineering. He received his B.E. in electrical and electronics from Bangalore University, M.Tech. in reliability engineering from IIT Bombay and Ph.D. in engineering from Homi Bhabha National Institute, Mumbai. He is faculty of BARC Training School and member of Atomic Energy Regulatory Board Committees for Power Plant Safety Review.

He has more than 50 publications in refereed journals and conferences. He is referee to several Elsevier and IEEE Transaction journals. His areas of research include artificial intelligence, development of operator support system, ageing and reliability studies. He has several years of experience in the field of risk and reliability assessment of nuclear and other industrial plants. He has developed the diagnostic system for 220 MWe Indian PHWRs and remaining life prediction models for I&C cables for PSA applications.

M. Serhani is currently an associate professor, College of Information Technology, U.A.E University, Al Ain, U.A.E. He is also an adjunct faculty in CIISE, Concordia University, Canada. He holds a Ph.D. in computer engineering from Concordia University in 2006 and M.Sc. in software engineering from University of Montreal, Canada, in 2002. Serhani's main research interests include cloud data centres for data-intensive e-health applications and services; SLA enforcement in cloud data centres, big data value chain, cloud federation and monitoring, non-invasive smart health monitoring, Web services architecture, management and QoS monitoring; Web services in mobile and ad hoc networks; management of communities of Web services; and Web services applications and security. He has a large experience earned throughout his involvement and management of different R&D projects. He served on several organising and Technical Programme Committees, including the IEEE WiMob, IEEE DEST, IEEE IWCMC, WMNC, NDT and IIT. He was the programme co-chair of the IEEE Conference on Innovations in Information Technology (IIT'13), chair of IEEE Workshop on Web Service (IWCMC'13), chair of IEEE Workshop on Web, Mobile and Cloud Services (IWCMC'12), and co-chair of International Workshop on Wireless Sensor Networks and Their Applications (NDT'12). He has published more than 90 refereed publications including conferences, journals, a book and book chapters.

Charlotte Lilly Slack has received her B.Sc. (Hons) in business information technology from Leeds Beckett University. She is currently studying a master's in information and technology within the same institution.

Hissam Tawfik is professor of computer science in the School of Computing, Creative Technologies and Engineering at Leeds Beckett University (UK). His research interests and expertise lie in the areas of biologically inspired computing, intelligent systems and simulation applications and e-health. Professor Tawfik holds a Ph.D. in computer engineering from the University of Manchester in 2000. He has been publishing since 1998 and currently has more than 100 refereed publications in reputable international conferences and journals such as *IEEE Transactions on IT in Biomedicine*, *Advanced Engineering Informatics*, *Ad Hoc Networks* and *IET Communications* and the proceedings of the IEEE World Congress on Computational Intelligence. Professor Tawfik is editor of the international journal on *Future Generation Computer Systems* and the international journal of *Neural Computing and Applications*, guest editor of the *International Journal of Medical Informatics* and chair of the International Conference Series on Developments in eSystems Engineering.

S.L. Toral was born in Rabat, Morocco, in 1972. He received the M.S. and Ph.D. degrees in electrical and electronic engineering from the University of Seville, Spain, in 1995 and 1999, respectively. He is currently a full professor with the Department of Electronic Engineering, US. His main research interests include ad hoc networks and their routing protocols, deployment of wireless sensor networks, flying ad hoc networks, real-time and distributed systems, intelligent transportation systems and embedded operating systems. He is actually an author or co-author of 73 papers in major international peer-reviewed journals (with JCR impact factor) and of over 100 papers in well-stablished international conferences and workshops.

P.K. Vijayan is former director of Reactor Design and Development Group, Bhabha Atomic Research Centre, and is a chemical engineer from Calicut University. He joined Reactor Engineering Division, BARC, in 1977. He obtained his Ph.D. from Indian Institute of Technology, Bombay. His domain of work includes design and detailed engineering of advanced heavy water reactor, thermal hydraulics and safety analysis. He is internationally acclaimed for his contributions in the field of natural circulation thermal hydraulics. He is a senior professor of Homi Bhabha National Institute and has also served as chairman, Board of Studies in Engineering Sciences at HBNI. He is also a recipient of DAE Scientific and Technical Excellence Award and Homi Bhabha Science and Technology Award for his contributions in the field of thermal hydraulic design validation of advanced reactors.

Gopika Vinod joined Reactor Safety Division of Bhabha Atomic Research Centre as a scientific officer from the 37th batch of training school after completing her graduation in computer engineering. She received her doctoral degree in reliability engineering from Indian Institute of Technology, Bombay, and also is postdoctoral fellow at Steinbeis Advanced Risk Technologies, Germany. She is a recipient of DAE Young Engineer Award 2007. She has been the visiting scientist at Brookhaven National Laboratory. Currently she is heading the Probabilistic Safety Section of Reactor Safety Division of BARC. She also holds the position of faculty with Homi Bhabha National Institute.

She has been actively involved in reliability, safety and risk analysis of Indian nuclear power plants. She has worked on the development of reliability-based operator support systems, such as risk monitor, and symptom-based diagnostic system, for Indian nuclear power plants. Her other areas of research activities include risk informed in-service inspection, reliability of computer-based systems, dynamic reliability analysis, etc.

She is in the editorial board of *International Journal of System Assurance Engineering and Management*, Springer, and journal referee to *IEEE Transactions on Reliability*, *Reliability Engineering and System Safety*, *Risk Analysis*, *Annals of Nuclear Energy*, *Nuclear Engineering and Design*, etc.

She is also member of technical committee for various conferences such as PSA 2011, ISSRE 2012, PSA 2013, ICRESH 2005, ICRESH 2010, ICRESH 2015, etc.

Max Yanovsky is associate professor at the Computer Systems and Network Department. He is an experienced developer of mobile applications and large-scale distributed systems and an administrator of the department's ICT infrastructure. His Ph.D. thesis was dedicated to research and development of green Wi-Fi networks by adapting its parameters as a function of the number and location of subscribers.

The original version of this book was revised. An erratum to this book can be found at DOI 10.1007/978-3-319-60137-3_18

Part I
Internet of Things (IoT), and "Smart Living"

Chapter 1
IoT-Enabled Smart Living

Ah-Lian Kor, Colin Pattinson, Max Yanovsky, and Vyacheslav Kharchenko

1.1 Introduction

By 2020, a quarter of Europeans will be over 60 years of age, and this will impact on our health care, economy, and social security systems.[1] According to EC, Europe is already spending nearly 10% GDP on health care due to EU aging population.[2] Consequently, the EU Health2020[3] aims to "significantly *improve* the *health* and *well-being* of *populations, reduce health inequalities, strengthen public health* and *ensure people-centered health systems* that are *universal, equitable, sustainable* and of *high quality.*" In order to support action plan for healthy aging,[4] this proposal aims to provide an inclusive, integrated, and user-centric IoT-ecosystem and services to promote smart health care, smart quality of life, and smart social inclusion.

---

[1] https://ec.europa.eu/research/social-sciences/pdf/policy_reviews/kina26426enc.pdf

[2] http://ec.europa.eu/research/participants/portal/doc/call/h2020/common/1587763-08._health_wp2014-2015_en.pdf

[3] http://www.euro.who.int/en/health-topics/health-policy/health-2020-the-european-policy-for-health-and-well-being/about-health-2020

[4] http://www.euro.who.int/__data/assets/pdf_file/0008/175544/RC62wd10Rev1-Eng.pdf

A.-L. Kor (✉) • C. Pattinson
School of Computing, Creative Technologies, and Engineering, Leeds Beckett University, Leeds, UK
e-mail: A.Kor@leedsbeckett.ac.uk

M. Yanovsky • V. Kharchenko
Department of Computer Systems and Networks, National Aerospace University "KhAI", Kharkiv, Ukraine

M. Dastbaz et al. (eds.), *Technology for Smart Futures*,
DOI 10.1007/978-3-319-60137-3_1

1.1.1 Project Objectives

The following list of concrete objectives will help achieve the aforementioned goals of SMART-ITEM system and services:

Objective 1 Provide a user-centered service where users are co-creators of good health. In order to ensure user-centeredness and co-creation, users are extensively involved throughout SMART-ITEM iterative development life cycle (i.e., requirements phase, development phase, evaluation phase, and integration). Users will be empowered to make health-related decisions (with or without clinicians) and responsibly act on information provided in their respective secured e-health passports. Access to SMART-ITEM services is facilitated via multimodal (and lingual) facility and devices (PDAs, mobile phones, tablets, laptops, desktops, etc.).

Objective 2 Provide an integrated one-stop shop portal for IoT-enabled services for smart living within a secured living environment. The rationale for such a portal is to coordinate the multiple services provision and afford a business model which offers users easy access to multiple services in one "location."

Objective 3 Provide real-time user health information and analysis. User's self-monitoring health data (e.g., through multi-sensor physiological monitoring – heart rate, respiration, temperature, oxygenation, etc.) will be securely transmitted (via wireless technologies), analyzed in real time (by an intelligent data system), and fed back via the e-health passport application. Permission will be granted by the user to a third party so that they could view the e-health passport information (e.g., clinicians, family members, carers, etc.).

Objective 4 Provide synchronous and asynchronous user activity information and analysis. Wearable activity tracker could be employed to track physical and health activities (e.g., steps taken, stairs climbed, sleep hours and quality logged, distance travel, etc.).

Objective 5 Provide an IoT-based smart, safe, and secured living environment (i.e., smart homes). This will help manage temperature, air quality, etc. and detect falls, intrusion, gas leaks, etc.

Objective 6 Build an online community for users. The community will comprise smart living experts and users (that are stratified based on needs, demographics, and languages). The primary communication channel will be via social media. It aims to facilitate knowledge transfer (from experts to novices) and knowledge/experience sharing (among users with similar needs); facilitate social interaction, engagement, and integration; and provide emotional support.

Objective 7 Bring about benefits to users and stakeholders. A range of impact analyses will be conducted to determine the benefits of SMART-ITEM (e.g., cost-effective analysis, economic impact analyses, user satisfaction, environmental impact analysis, health economics analysis, financial performance analysis of health economy, etc.).

In summary, in this project, we shall seize a list of opportunities: (i) empower elderly population to manage their own health and stay active, healthy, and independent as long as possible; (ii) develop IoT-enabled systems which are people-centered and home-based integrated care; and (iii) support an economy which adapts to new categories of clients and their needs (note: with ICT-driven innovation we can tap into the growing Silver economy of 85 million consumers over 65 and € 3000 billion).

1.2 Related Work

IBM's vision of a smarter home enabled by cloud technology[5] has the following characteristics: *Instrumented,* the ability to sense and monitor changing conditions; *Interconnected,* the ability to communicate and interact, with people, systems, and other objects; and *Intelligent,* the ability to make decisions based on data, leading to better outcomes. According to IBM, the categories of smarter home services are as follows:

Entertainment and convenience – the Internet-enabled television will be processed to personalized entertainment content via a portal.

Energy management – management of home electrical appliances to deliver automatic savings to home owners (e.g., automatically synchronized lighting, climate control sensors, etc.).

Safety and security – as an example, deployment of centralized alarm services using sensors and cameras and instant notification of relevant parties could enhance home security.

Health and wellness – smarter home scales and sensors monitor fitness, well-being, and consistent activities.

These home devices could collect evaluative information about current health condition for disease prevention and overall wellness. In SMART-ITEM, we shall categorize smart home services into three categories which are integrated into a one-stop web portal for ease of use: smart health and care, smart quality of life, and SMART-ITEM and community of support.

1.2.1 Smart Health and Care

The domain of e-health is extended with wearables or mobile IoT devices for home-based or on-the-go (mobile) monitoring of vital data.[6] Some examples of these devices are wirelessly connected glucometers, heart rate and blood pressure monitors,

---

[5] http://www.ibm.com/smarterplanet/global/files/uk__uk_en__cloud__a_smarter_home_enabled_by_cloud_computing.pdf

[6] http://www.itu.int/en/ITU-T/academia/kaleidoscope/2014/Pages/SpecialSessionE-health,-IoT-and-Cloud.aspx

scales, oximeter, etc. Tracking health and well-being information is vital for the elderly population. However, current e-health systems require users to manually input their vital data through smart phones or tablets (e.g., Vitrucare System[7] by Dynamic Health Systems, Wellness Layers Digital Therapeutic Platform,[8] Honeywell Lifecare Solutions[9] (e.g., Honeywell HomMed), Philips eTraC program and eCare Companion,[10] etc.). Garmin[11] and Fitbit activity and sleep trackers with their fitness apps afford the following facilities[12]: progress and performance of all day activity, sleep goals and trends, stay connected with family and friends, provides guidance and tools for nutrition and weight goals, and access through smartphones or desktops. In the same vein, Garmin activity trackers[13] and smart scales and apps[14] provide the following functionalities: personalized activity goals, track progress and performance, analysis of sleep trends, wellness and weight management, heart rate analysis, and smart notifications. A synergy of IoT and cloud-based technologies will provide the means for smart devices to automatically transmit and platforms to analyze as well as store data. This will automatically remove the limitations of manual data collection. In August 2013, Cooking Hacks launched a new version of the first biometric shield for Arduino and Raspberry Pi: the e-Health Sensor Platform V2.0[15] which allows users to run biometric and medical applications for body monitoring using ten different sensors connected to the shield via cables: pulse, oxygen in blood (SPO2), airflow (breathing), body temperature, electrocardiogram (ECG), glucometer, galvanic skin response (GSR – sweating), blood pressure (sphygmomanometer), patient position (accelerometer), and muscle/electromyography sensor (EMG). Biometric information gathered is then wirelessly transmitted to the cloud using one of the following six connectivity options: Wi-Fi, 3G, GPRS, Bluetooth, 802.15.4, and ZigBee depending on the type of application. The cloud will provide the following facilities: permanent storage and processing and analysis platform. Analyzed data could be visualized in real time and sent back to a laptop or smartphone (with iPhone or Android OS).

Senno Pro[16] is a smart insole gait analysis system[17] which consists of smart insoles that can be placed in shoes and a data analyzer, which are connected via

[7] http://www.dynamichealthsystems.co.uk/vitrucare-supported-selfcare

[8] http://www.wellnesslayers.com/solution-3-2/

[9] https://www.linkedin.com/company/honeywell-hommed

[10] http://www.usa.philips.com/b-dam/b2bhc/master/Products/Category/enterprise-telehealth/etrac/eTrAC.pdf

[11] http://sites.garmin.com/en-GB/wellness/

[12] http://www.fitbit.com/uk/apphttp://www.fitbit.com/uk/app

[13] https://buy.garmin.com/en-US/US/wearables/activity-tracking/c10002-atFILTER_USERPROFILE_ACTIVITYTRACKING_01-p1.html

[14] https://buy.garmin.com/en-US/US/prod530464.html

[15] https://www.cooking-hacks.com/documentation/tutorials/ehealth-biometric-sensor-platform-arduino-raspberry-pi-medical/

[16] http://www.gait-analysis-smart-insole.com/

[17] http://www.cse.buffalo.edu/~wenyaoxu/papers/conference/xu-petra2012a.pdf

wireless communication protocols so that users' activities can be monitored in real time. Digitsole[18] smart shoes are connected, interactive, heated, and shock absorbent with automatic tightening that can be controlled via smartphones. This innovative footwear product is designed for health and comfort. 3 L Labs' FootLogger[19] affords the following health-care facilities: activity tracking (calories, distance, and time), early prediction of accidents from falling, early prediction of dementia and spinal disease, monitors rehabilitation or recovery after operations, and monitors safety and activity of senior citizens.

SMART-ITEM e-health and activity differentiation is as follows:

(i) *Low-cost sensing, gateway, and cloud-based platform system* (e.g., for SMART-ITEM estimated total cost for weight and SPO2 monitoring is £205.98 – Raspberry Pi 3 Model B (£26 per unit); iHealth PO3 Wireless Pulse Oximeter (£79.99 per unit); and iHealth HS6 Body Analysis Wireless Scale (£99.99 per unit). On the other hand, the price of MASIMO Radical 7 Rainbow RAD-7[20],[21] is US$1795 (equivalent to £1268). Honeywell HomMed[22] estimated total cost is US$26,350 (equivalent to £18,680) – Genesis DM Pro BP, wireless scale and SPO2 Kit (US$4500 per unit), Genesis Touch Kit (US$1000 per unit), Genesis DM peripherals and cables (GPRS – US$450; Software for LifeStream Base Platform US$5000; and LifeStream Connect US$10000; Training (Clinical Strategic Planning Meeting Onsite US$1800; Clinical Strategic Planning Meeting via teleconference US$600; Clinical Training Initial Day-Onsite US$1800; Additional Consecutive Day US$1200).

(ii) *Automated wireless data transfer* (from devices to gateway and finally the cloud) to reduce the risk of errors.

(iii) *Allows a single point of access to a range of services and apps.*

(iv) *Interoperability* with a variety of medical and fitness devices including their natively available apps.

(v) *HIPAA compliance.*[23]

(vi) *Use cloud-based service-oriented platform* that adopts the service-oriented architecture (SOA) approach which is a paradigm for organizing and utilizing distributed capabilities that is under the control of different ownership domains and implemented using various technology stacks.[24]

---

[18] http://www.digitsole.com/smartshoes

[19] http://www.footlogger.com/

[20] http://masimo.co.uk/rainbow/radical7.htm

[21] http://www.ebay.com/itm/MASIMO-Radical-7-Rainbow-RAD-7-Factory-Refurbished-in-the-box-/152023656222

[22] http://www.vnaa.org/Files/CorporatePartnerships/MVPDocuments/Honeywell-2013-pricing.pdf

[23] http://technologyadvice.com/blog/healthcare/the-importance-of-hipaa-compliance-and-health-app-security/

[24] http://www.adobe.com/enterprise/pdfs/Services_Oriented_Architecture_from_Adobe.pdf

Such architecture is scalable and modular where its services are accessible from external applications via application programming interfaces (APIs).

In summary, SMART-ITEM will be an e-health and well-being innovation because it will offer a low-cost solution which harnesses the power of wireless data transfer, synergy of private, public, and hybrid cloud-based computational power, aggregation, and storage, as well as display of analyzed active living-related data which is not accessible before. Users' vital and activity data will be collected over time and analyzed using a range of complex algorithms. The analyzed data could be sent to caregiver (note: this is an optional functionality) for further analysis and review. This could be exploited to enable preventive care and allow prompt diagnosis of possible complications. *SMART-ITEM decision support system* will help users make appropriate decisions based on recommendations given by the system. SMART-ITEM will support local council's promotion for active aging for the elderly population (e.g., Age Friendly Leeds,[25] Leeds Older and Active,[26] etc.).

1.2.2 Smart Quality of Life

The category will encompass the following: smart home automation and safety and security. Smart devices can be used to collect relevant data about the home. As an example, smart thermostats could collect data about occupancy and homeowners' temperature preferences followed by appropriate adjustments. Samsung has launched the Smart Home Ecosystem[27] which connects home devices (e.g., televisions, home appliances, and smartphones) using a single integrated platform. These devices can be controlled by users through an application which connects the devices in a home. *Samsung SmartThings Starter Kit*[28] provides everything needed to create a smart home. It is iOS, Android, and Windows phone compatible. SmartThings Starter Kit provides the following functionalities: (i) home monitoring from anywhere, (ii) home control with an app, (iii) home security and protection from damage and danger, and (iv) scalability by increasing hundreds of compatible smart devices to enhance home automation. Its components are hub to connect smart sensors, lights, locks, cameras, etc.; motion sensor to monitor movement in the home; multi-sensor – to monitor whether doors, windows, drawers, or garage are open or closed; presence sensor to know when people, pets, and cars arrive or leave home; and a power outlet to control lights, electronics, and small appliances. The *Smart Home Cloud API*[29] provides methods to control and monitor *Samsung Smart Home* devices. Through this *Smart Home Control Service*, a partner's application can connect with various devices and

[25] https://betterlivesleeds.wordpress.com/age-friendly-leeds/

[26] http://www.altogetherbetter.org.uk/leeds-older-and-active

[27] http://www.computerweekly.com/news/2240212008/Samsung-launches-Smart-Home-ecosystem-to-connect-home-devices

[28] http://www.samsung.com/uk/consumer/smartthings/kit/kit/F-STR-KIT-UK/.

[29] http://developer.samsung.com/smart-home

provide enhanced services to their customer. This service operates through cloud-to-cloud integration between the Partner Cloud and Smart Home Cloud. This is facilitated by the provision of several REST API by Samsung for partners so they are able to integrate their system to the *Samsung Smart Home Cloud*. The body of the REST API uses a standard JSON document (called *Smart Home Data*), and this means that partner developers will need to understand the JSON document.

Apple Homekit[30] is Apple's vision for the future smart home with a set of interconnecting apps and devices (via Bluetooth or Wi-Fi technologies) that control lighting, heating, and the cycles of your washing machine. HomeKit is a framework[31] for communicating with and controlling connected accessories in a user's home. Users can configure HomeKit accessories in the home or create actions to control those devices, group actions together, and trigger them using an iOS device[32] or Siri (an Apple personal voice assistant).[33] Thus, it is a technology platform which cannot be bought because it is aimed at app developers and manufacturers but consumers can only buy HomeKit-compatible gadgets. The focal point of the HomeKit framework is a unifying protocol (i.e., standardization via a singular SDK) to act as the conduit for devices or gadgets to communicate with iOS[34] and to make a public API for the configuration and communication with these devices.[35] The object orientation (OO) or ontological model has been used to organize the hierarchy of classes and subclasses of objects, and thus it is essential for developers to understand the relationships among the objects in the model. Finally, HomeKit is privacy compliant because HomeKit data is always encrypted so that only people in the household the settings of the lights and thermostats. A list of accessories that are Homekit-enabled is found here.[36]

Google has launched its new IoT platform and language, *Brillo and Weave.*[37] Brillo is an Android operating system for IoTs, while Weave provides a common language to facilitate devices connection via Wi-Fi, Bluetooth, or Thread (an IPv6-based wireless network protocol).[38] Brillo[39] is optimized to run on devices with a small computing footprint. It consists of a kernel, for managing input and output commands to the device, as well as a hardware abstraction layer, which will make the operating system interoperable across disparate hardware devices. The hardware abstraction layer then connects to the wireless networking protocol (such as Wi-Fi, Bluetooth, or Thread) that the device uses to access a gateway or the Internet. Weave[40]

[30] http://www.wareable.com/smart-home/apple-homekit-essential-guide-2016

[31] https://developer.apple.com/homekit/

[32] http://www.apple.com/uk/ios/homekit/

[33] http://www.apple.com/uk/ios/siri/

[34] http://www.programmableweb.com/news/how-to-get-started-apple-homekit/how-to/2014/11/28

[35] https://developer.apple.com/library/ios/documentation/HomeKit/Reference/HomeKit_Framework/

[36] http://www.apple.com/uk/shop/accessories/all-accessories/home-automation?page=1

[37] http://9to5google.com/2016/01/05/brillo-weave-devices-ces/

[38] http://www.iotjournal.com/articles/view?13089/

[39] https://developers.google.com/brillo/

[40] https://developers.google.com/weave/

will ensure that the IoT devices talk to each other, talk to the cloud, and talk to the Android phone. Standardized schemas are used to help devices "understand" the programming language that refers to a command (e.g., a command for locking and unlocking an electronic door lock will be understood by the following relevant devices that work with that lock: gateway, alarm system, and user's Android phone). Google has also launched a certification program that enables developers who employ Weave application programming interfaces (APIs) that will be able to make sure that connected devices interoperate as they should. Weave can work in conjunction with Brillo, or it can run on another software stack. *Nest Automation*[41] is Google's plan for a smart home, and some of the Nest products are *Learning Thermostat* (smartphone controllable thermostat, capability of learning heating preferences, and creating custom heating schedules), *Nest Protect* (a reinvention of the standard smoke detector), *Nest Cam* (aims to keep the home secured with 24/7 live streaming, advanced night vision, and motion and sound alerts), and *Nest Weave* (involves developer-friendly Nest IoT protocol to connect more and more Nest-compatible products; offers a high level of security with predefined application planes and application-specific keys, low latency interaction, low power requirements, and zero reliance on Wi-Fi). The new Nest cloud API[42] or the Nest Developer Program provides developers of other products to make even more smart home connections.

OpenHAB Home Automation[43] uses the Raspberry Pi, and it is a mature, open-source, and integrated home automation platform that runs on a variety of hardware with an interoperable protocol which is a common language that allows the different varieties of hardware to talk to each other so as to create a really automated and smart environment within the home. OpenHAB is a piece of software that integrates different home automation systems and technologies into one single solution and is thus a solution for the most flexible smart home hub. In summary, OpenHAB has the following characteristics[44],[45]: is absolutely vendor-neutral as well as hardware or protocol agnostic, integrates a range of home automation technologies into one, has a powerful rule engine to fulfill any automation needs, comes with different web-based UIs as well as native UIs for iOS and Android, is extensible to integrate with new systems and devices, and provides APIs that can be integrated into other systems in order to provide uniform user interfaces.

Homebridge[46] is a lightweight NodeJS server that runs on a home network and emulates the iOS HomeKit API. *Raspberry Pis* running Homebridge can create *HomeKit* smart home[47] which controls Samsung SmartThings, Philips Hue lighting,

[41] http://www.appcessories.co.uk/nest-automation-googles-plan-for-your-smart-home/

[42] https://developer.nest.com/

[43] http://www.makeuseof.com/tag/getting-started-openhab-home-automation-raspberry-pi/

[44] http://www.openhab.org/

[45] http://www.openhab.org/features/introduction.html

[46] https://github.com/nfarina/homebridge

[47] http://www.automatedhome.co.uk/apple/smartthings-raspberry-pi-homebridge-siri-homekit-controlled-smart-home.html

and Sonos. *RaZberry*[48] uses the *Raspberry Pi platform* with *Z-Wave* which is the leading wireless communication technology for smart homes. The Razberry platform adds all the components needed to turn a Raspberry Pi board into a fully operational and inexpensive Z-Wave gateway. They are:

(i) Hardware and operating system – Raspberry Pi board and Rasperian OS
(ii) Z-Wave hardware – RaZberry daughter card which connects to the GPIOs of Raspberry and carry the Sigma Designs' Z-Wave transceiver
(iii) Z-Wave firmware – runs on the transceiver and is compatible to the original Sigma Designs' firmware spec
(iv) Z-Wave stack – first certified Z-Wave communication stack, handles all Z-Wave network communication, Java script automation engine, built-in web server
(v) User interface – web-based Z-Way Demo User Interface using the Z-Way JSON API and demonstrating all functions of Z-Way
(vi) User application – *Bluetooth's Gateway Smart Starter Kit*[49] provides guidance on how to connect Bluetooth® sensors to IoT via the web using *Bluetooth RESTful APIs* by creating a simple *Bluetooth gateway* on a *Raspberry Pi* board

This kit also shows how to move data from all of *Bluetooth sensors* into the *cloud* without a mobile device while giving users the ability to communicate and control all of them from one place in the web.

1.2.3 IoT and Security Management

Growth in IoTs brings increased security risks. According to Symantec,[50] IoT systems are often highly complex, requiring end-to-end security solutions that span cloud and connectivity layers. Resource-constrained IoT devices are not sufficiently powerful to support traditional security solutions. Symantec suggests that comprehensive IoT security should encompass four cornerstones:

(i) Communications protection – requires encryption and authentication for devices (e.g., elliptic curve cryptography in resource constrained chips like 8 bit, 8 MHz chips of IoT, and embedded "device certificate" keys into IoT devices (e.g., embedment of X.509 certificates[51] into a hardware device during manufacturing, which allows service providers to perform strong authentication to prevent unauthorized devices from obtaining access to services), helping mutually authenticate a wide range of devices including cellular base stations, televisions, etc.

[48] http://razberry.z-wave.me/index.php?id=1

[49] https://www.bluetooth.com/develop-with-bluetooth/developer-resources-tools/gateway

[50] https://www.symantec.com/content/en/us/enterprise/white_papers/iot-security-reference-architecture-wp-en.pdf

[51] https://www.symantec.com/en/uk/device-certificate-service/

(ii) Devices protection – requires code signing, ensures all code is authorized to run, and runtime protection. Code signing cryptographically ensures that codes are not overwritten and host protection in order to protect devices after code has begun running. This entails the verification that all codes running on each IoT device have been authorized for that device and protected by a strong root of trust.
(iii) Devices management – over-the-air (OTA) manageability must be built into the devices before they are shipped.
(iv) Understanding of the system – it is imperative to have an IoT Security Analytics capability to understand one's network by flagging anomalies that might be suspicious, dangerous, or malicious.

1.3 Underlying Concept and Methodology

1.3.1 Overall Underpinning Concept

The SMART-ITEM project will be anchored on an adaptation of biomimicry basic life's principles[52] (see Fig. 1.2).

1.3.1.1 Adapted Biomimicry Life's Principles

Biomimicry is an emerging approach to innovation that seeks sustainable solutions to human challenges by emulating nature's time-tested patterns and strategies and its goal is to create sustainable products, or processes.[53] The ultimate goal for SMART-ITEM is to *create conducive conditions* (based on biomimicry life's principles depicted in Fig. 1.2) for the development and deployment of an IoT-enabled ecosystem system and services which is sustainable, efficient, effective and beneficial to target users and the elderly population. The principles are as follows:

(i) *Evolve to survive*: replicate strategies that work (note: SMART-ITEM will be built on Vitrucare and Wellness Layers's success stories on smart health care and also other work listed in *Scope 4: Exploitation of previous work*)
(ii) *Be resource efficient*: use multifunctional design (as depicted in Fig. 1.1, SMART-ITEM will afford multifunctionalities to support users in their self-empowered independent journey)
(iii) *User-centered design* will be employed throughout the entire life cycle of SMART-ITEM IoT ecosystem and services
(iv) *Be locally attuned and responsive*: use feedback loops and leverage cyclic processes, cultivate cooperative relationships (among the users and stakeholders

[52] http://www.being-here.net/page/2531/applying-biomimicry-to-global-issues
[53] http://biomimicry.org/what-is-biomimicry/

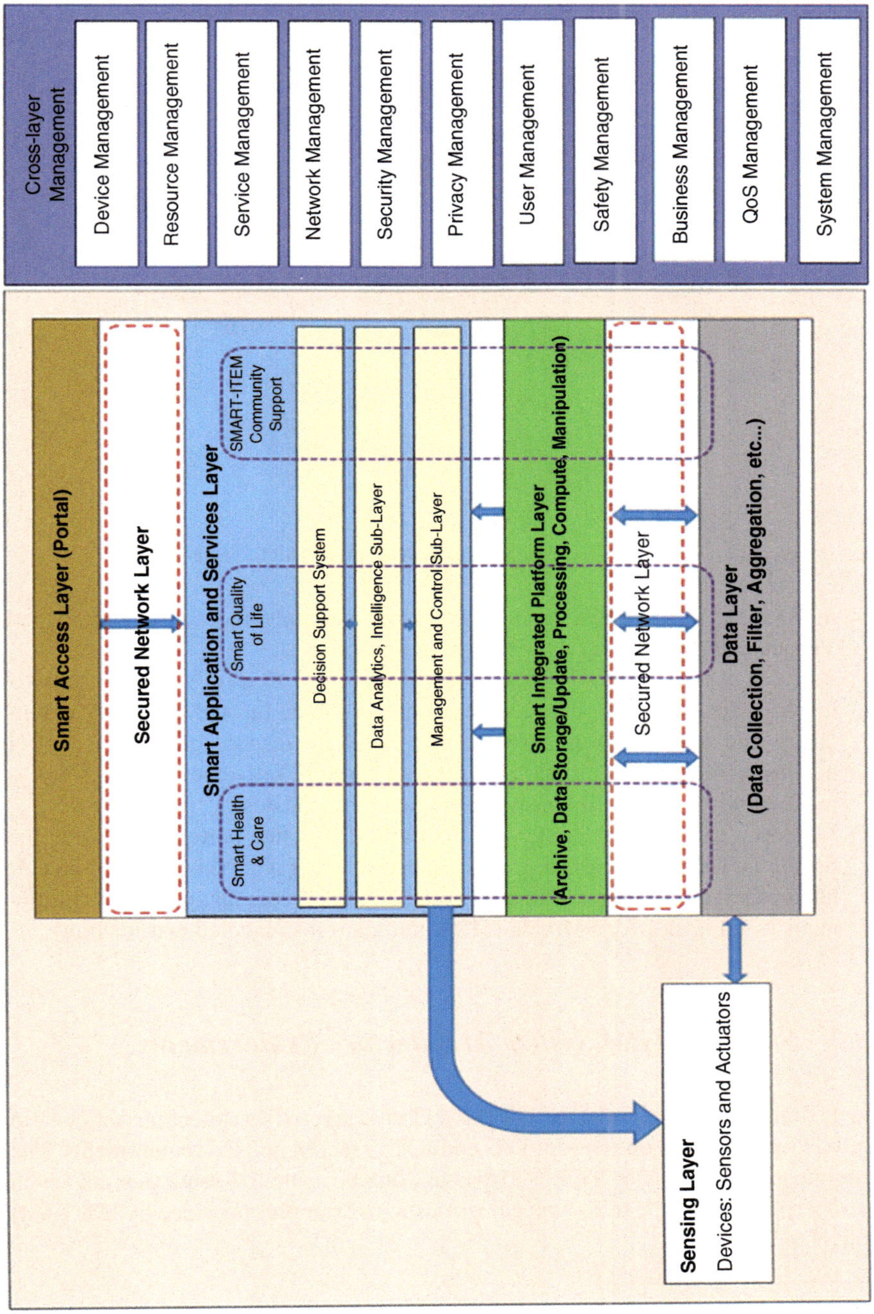

Fig. 1.1 SMART-ITEM layered architecture and targeted IoT-enabled services

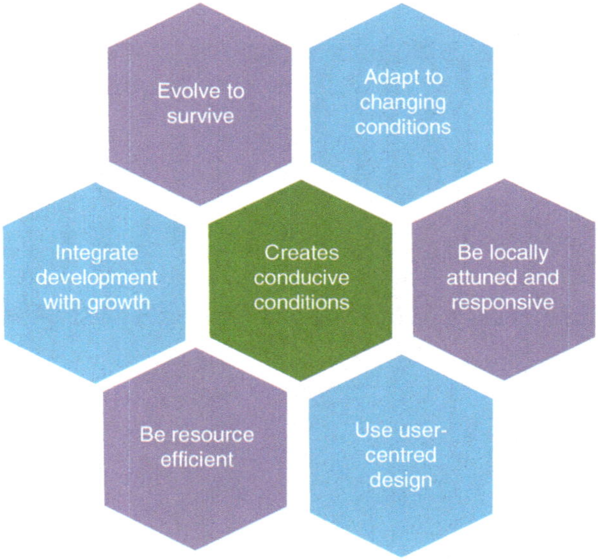

Fig. 1.2 Adaptation of biomimicry basic life's principles for SMART-ITEM

in SMART-ITEM), use readily available materials (e.g., use and extend Vitrucare and Wellness Layer's existing platform)

(v) *Adapt to changing conditions*: incorporate diversity (e.g., diversity of supporting activities shown in Fig. 1.2), embody resilience through variation redundancy and decentralization (note: users will be responsible for their own healthy aging process), maintain integrity through self-renewal (note: this is made possible via continuous improvement and updates)

(vi) *Integrate development with growth*: build from the bottom-up (note: the SMART-ITEM IoT-enabled ecosystem and services will be built based on collated user requirements and analysis) and combine modular and nested components (see Fig. 1.1 for SMART-ITEM modular cross-layered architecture)

1.3.2 SMART-ITEM System Architecture Requirements

The technical requirements for SMART-ITEM layered architecture (shown in Fig. 1.1) will be based on relevant ITU and IOT-A recommended requirements. The recommendation ITU-T Y.2068[54] (2015) has 2068 possible IoT requirements which have been coded into the following categories with examples provided by ITU (ibid) and IOT-A[55]:

[54] http://www.itu.int/rec/T-REC-Y.2068-201503-I

[55] http://www.iot-a.eu/public/requirements/copy_of_requirements

1.3.2.1 Functional Requirements

Implementation, Operability, and Integration Requirements Interoperability of technologies, applications, services, and management functionalities; autonomous collaboration between devices and services; integration with existing systems using process modeling; involves process execution engine; cross-layered architecture; self-configuration, self-optimization, self-protection, self-management, and auto-bootstrap management.

Application Support Requirements Support interworking between different application protocols. OGC (Open Geospatial Consortium) has developed the open SensorThings API[56] standard to provide an open and unified way to interconnect IoT devices, data, and applications over the web. It builds on web protocols and the OGC Sensor Web Enablement standards and applies an easy-to-use REST-like style.

Service Requirements Service description (e.g., ITU-T F.744[57] – service description and requirements for ubiquitous sensor network middleware), organization, composition, programmability, orchestration including orchestration engine, meter service reputation, service prioritization, usage tracking, service subscription according to application domain, and implementation.

Communication Requirements Autonomous communication; distributed communications; multi-cast messages; communication stacks with small data footprints; low power and lossy network communication; stateless communication methods; common addressing schemes such as IPv6; routing over heterogeneous network; network handover and handoff support; single, simple, and standardized management interfaces for all communication protocols; support for intermittent and command-based communication with devices.

Device Management Requirements Unique identification of devices, resource control, support for sleepy devices, support for multi-homing devices, translational functionality for IoT gateways, sensors will be web accessible.

Data Management Requirements Data acquisition, data filtering, handling, aggregation/fusion, data integrity, data flow modeling, information retrieval, CRUD functions, and data freshness.

1.3.2.2 Nonfunctional Requirements

Privacy, Trust, and Security Requirements Anonymity support, security of data transfer and communications, access control, authentication, user authorization, encryption, non-repudiation, device security, privacy at atomic level, location privacy, infrastructure services (e.g., resolution services, security services, manage-

---

[56] http://ogc-iot.github.io/ogc-iot-api/

[57] https://www.itu.int/rec/dologin_pub.asp?lang=e&id=T-REC-F.744-200912-I!!PDF-E&type=items

ment services) that are trustable, untraceable digital activities, unlinkability, communication confidentiality, communication integrity, pseudonymization mechanisms, unobservability, data portability, security management (e.g., secure over-the-air/over-the-network device management), and identity management (e.g., identity and access control – subject-based, role-based, or attribute-based; multiparty authentication and cryptography in the IoT; group authentication and authorization; authorization, authentication, and credentials requirements; uniquely addressable; ownership transfer; anonymous devices and blinded identities

Semantics Requirements Semantic description of physical entities and services, self-description of things (e.g., sensors, actuators, and processors) using standard encoding or model language (e.g., OGC's encoding standard, SensorML[58]).

Resilience Requirements Performance; reliability – support the appropriate level of reliability in aspects such as communication, service, and data management capabilities to meet system requirements; fault tolerance; availability; integrity; scalability; evolvability; timeliness; and modularity – support components that can be combined in different configurations to form systems as needed.

Extensibility Requirements Use of external computing resources, for example, the cloud, for information processing.

Quality of Service (QoS) Requirements Transport (a.k.a network) layer fairness).

Accessibility and Usability Requirements Support end user accessibility/usability preferences and requirements based on user-centered design and accessibility/usability based on W3C guidelines[59] and IEC/PAS 62883:2014(E)[60] which specifies a framework for adaptive handling of explicit interaction among humans and AAL spaces. This will also include usability for developers of IoT systems who will be involved in the configuration, operation, and management of the system.

Discoverability Requirements Support discovery services for sensors and sensor data.

Awareness Requirements – Location Awareness, e.g., location and distance; locating physical entities based on geographical parameters, standardized location model, location information representation; *time-awareness* – should support time synchronicity among the actions of interconnected components when using communication and service capabilities; *context awareness*: enable flexible, user customized and autonomic services based on the related context of IoT components and/or users; and *content awareness*: to facilitate services such as path selection and routing of communications based on content.

Manageability Support management capabilities to address aspects such as data management, device management, network management, risk management (e.g., end point controls, gateway controls, network controls, cloud and application controls, overall systems controls), and interface maintenance and alerts.

---

[58] http://www.opengeospatial.org/standards/sensorml.

[59] https://www.w3.org/WAI/intro/usable.

[60] https://webstore.iec.ch/webstore/webstore.nsf/publication/7577.

Energy Awareness Requirements Manage energy consumption of devices, applications, networks, limit communication range, limit local processing and storage capacity (note: this is the reason for moving these functionalities to the cloud), and support sleeping modes.

Standard and Compliance Requirements Regulation compliances, standard compliance, etc.; *OGC Standards in the Sensor Web Enablement (SWE)*[61] framework are as follows:

Observations and Measurements (O&M) – general models and XML encodings for observations and measurements

PUCK Protocol Standard – defines a protocol to retrieve a SensorML description, sensor "driver" code, and other information from the device itself, thus enabling automatic sensor installation, configuration, and operation

Sensor Model Language (SensorML) – standard models and XML schema for describing the processes within sensor and observation processing systems

Sensor Observation Service (SOS) – open interface for a web service to obtain observations and sensor and platform descriptions from one or more sensors

Sensor Planning Service (SPS) – an open interface for a web service by which a client can determine the feasibility of collecting data from one or more sensors or models and submit collection requests; SWE Common Data Model – defines low-level data models for exchanging sensor-related data between nodes of the OGC® Sensor Web Enablement (SWE) framework

SWE Service Model – defines data types for common use across OGC Sensor Web Enablement (SWE) services

oneM2M[62] develops standards for M2M and IoTs and published technical specifications which address the need for a common M2M service layer – functional architecture, technical requirements, security solutions, service layer core protocol specification, management enablement, CoAP protocol binding, HTTP protocol binding, MWTT protocol binding, common terminology, and interoperability testing. *ISO/IEC 29182* encompasses Sensor Network Reference Architecture (SNRA) for sensor networks, and it consists of seven parts, namely, ISO/IEC 29182-1: 2013[63] (general overview and requirements for sensor networks), ISO/IEC 29182-2: 2013[64] (vocabulary and terminology), ISO/IEC 29182-3[65] (reference architecture views), ISO/IEC 29182-4:2013[66] (entity models), ISO/IEC 29182-5:2013[67] (interface definitions), ISO/IEC 29182-6[68] (applications), and ISO/IEC 29182-7[69] (interoperability guidelines).

[61] http://www.opengeospatial.org/domain/swe#standards

[62] http://www.onem2m.org/about-onem2m/why-onem2m

[63] https://webstore.iec.ch/preview/info_isoiec29182-1%7Bed1.0%7Den.pdf

[64] https://webstore.iec.ch/preview/info_isoiec29182-2%7Bed1.0%7Den.pdf

[65] https://webstore.iec.ch/preview/info_isoiec29182-3%7Bed1.0%7Den.pdf

[66] https://webstore.iec.ch/preview/info_isoiec29182-4%7Bed1.0%7Den.pdf

[67] https://webstore.iec.ch/preview/info_isoiec29182-5%7Bed1.0%7Den.pdf

[68] https://webstore.iec.ch/preview/info_isoiec29182-6%7Bed1.0%7Den.pdf

[69] https://webstore.iec.ch/preview/info_isoiec29182-7%7Bed1.0%7Den.pdf

IEEE has published standards for IoT,[70] and some of examples of relevant standards are for Ethernet, Telecommunications and Information Exchange Between Systems, Air Interface for Broadband Wireless Access Systems, Scalable Storage Interface, Smart Transducer Interface for Sensors and Actuators, Interconnecting Distributed Resources with Electric Power Systems, Monitoring, Information Exchange, and Control of Distributed Resources Interconnected with Electric Power Systems, Design, Operation, and Integration of Distributed Resource Island Systems with Electric Power Systems, etc.

Impact Requirements Support components, services, and capabilities which lead to minimal environmental impact but optimal economic as well as social impact of an implementation.

1.3.3 SMART-ITEM System Architecture

The high level SMART-ITEM system layered architecture with functional entities is shown in Fig. 1.1. Physical entities in a sensor network consist of hardware, actual devices, sensor nodes, gateways, components, etc., while functional entities represent tasks that may be carried out by or on the physical entities.[71]

SMART-ITEM detailed architecture (in Fig. 1.3) is adapted from the sensor network functional architecture in ISO/IEC 29182-3.[72] The three domains in the functional architecture are sensing domain, network domain, and service domain. The core functions of the system are coded into two broad categories:

1.3.3.1 Data, Information, and Communications-Related Functions

(i) *Data acquisition* – (or collection) from the sensors.
(ii) *Data filtering* – based on selected criteria.
(iii) *Data processing* – data aggregation (i.e., raw data collected from a range of sensors) and data fusion (i.e., combining and process aggregated data).
(iv) *Data analytics and intelligence* – i.e., to extract feature from the data using data mining techniques, supervised or unsupervised machine learning algorithms, etc. which result in information or intelligence that will trigger off an appropriate event and could be visualized in various modes such as pie charts, graphs, etc.
(v) *Data storage*
(vi) *Data communication* – this involves the transmission of data (e.g., from the sensing layer to the data layer, from the data layer to the platform layer, etc.),

[70] http://standards.ieee.org/innovate/iot/stds.html

[71] https://webstore.iec.ch/preview/info_isoiec29182-4%7Bed1.0%7Den.pdf

[72] http://www.sis.se/PageFiles/15118/Study%20Report%20on%20IoT%20Reference%20Architectures%20and%20Frameworks.pdf

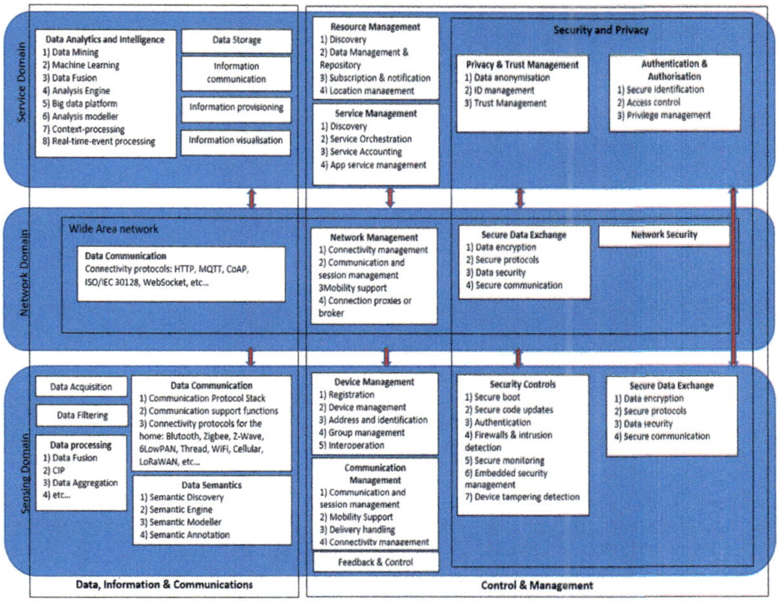

Fig. 1.3 SMART-ITEM detailed architecture

and this encompasses connectivity protocols for home devices and also for the network.

(vii) *Data semantics* – encompasses description and meaning of data.

1.3.3.2 Control and Management-Related Functions

(i) *Resource Management* – provides efficient and effective management of IoT resources (e.g., discovery of devices and services, location awareness, etc.)

(ii) *Service Management* – supports service orchestration, discovery, app service management to support users, etc.

(iii) *Device Management* – encompasses functions such as registration, device configuration, address and identification, and management of a group of devices.

(iv) *Security and Privacy* – some of the relevant functions are privacy and trust management, authentication and authorization, secure data exchange using encryption and secured protocols, security controls through firewalls, intrusion detection, runtime verification and malware detection, key management, device tampering detection, and bootstrapping.

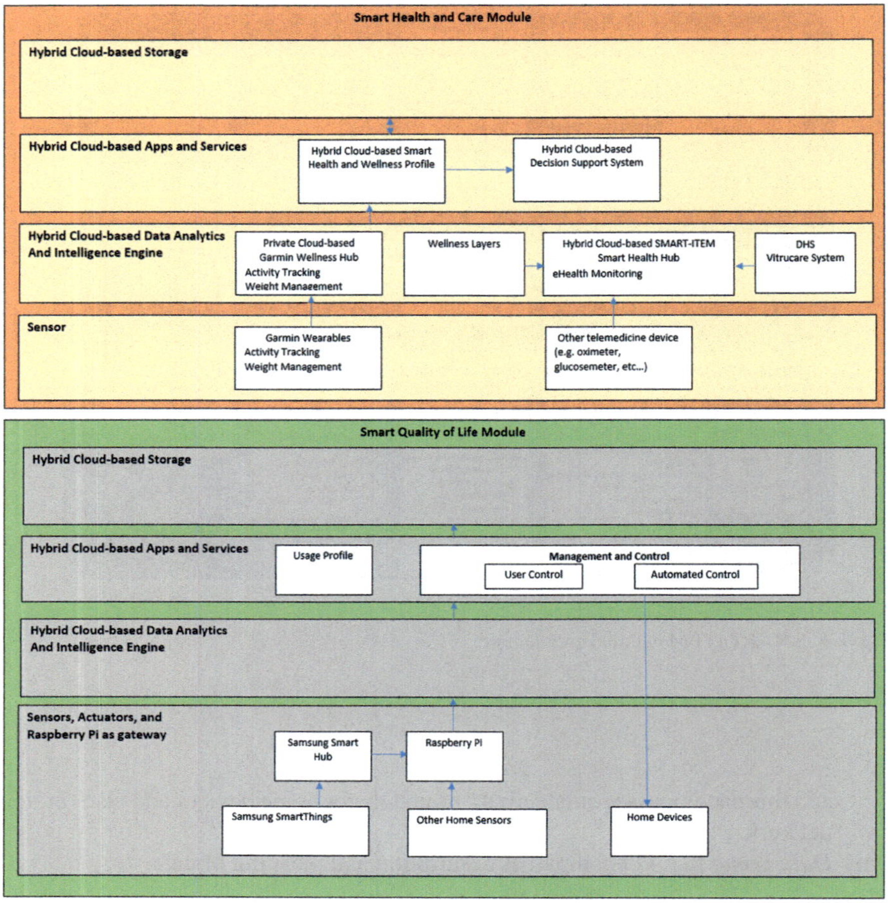

Fig. 1.4 Data flow in the smart health and care and smart living modules

Figure 1.4 depicts the data flow for the smart health and care as well as the smart living modules. The first image shows the integration of Garmin Wellness Portal, Wellness Layers Platform, DHS Vitrucare Platform, and SMART-ITEM platform. On the other hand, the second image shows the integration of Samsung SmartThings sensor data and other home automation sensor data (e.g., air quality, temperature, etc.) using a Raspberry Pi 3+ integrated gateway which will be customized for the project so that it will be interoperable for a range of communication protocols (e.g., ZigBee, Bluetooth, Z-Wave, etc.). Data processing, analytics, and storage will be executed in private clouds across eight cities (note: these are hosted in academic/ research institutions and co-shared with city councils or medical organizations) in the first instance to save costs but will be later migrated to the Amazon EC2 for scalability and sustainability purposes.

1.3.4 SMART-ITEM Integrated Software Architecture

SMART-ITEM integrated software architecture (depicted in Fig. 1.1) is modular and cross-layered. The design and development of each module in the SMART-ITEM system and services will be based on the component-based software design and development approach[73] with the following principles: (i) Components are independent (note all the SMART-ITEM modules can be developed independently at the initial stage; (ii) communication is through well-defined interfaces (note: in SMART-ITEM, the interaction among the modules are made possible via integrations; and (iii) components have shared platforms (note: in SMART-ITEM, a cloud-based platform will be used). As previously discussed, the architecture consists of six layers (i.e., sensing layer, data layer, platform layer, application and services layer, secured network layer, and access layer) and three modules (smart health and care, smart quality of life, SMART-ITEM community of support). The Microsoft .NET framework will be employed to support the access and application/services layers because it will support the software required for the development of all the modules. Additionally, this will facilitate deep customization, scalable, and integration. The flexible structures will be cost-effective and allow timely deployment (as in the agile methodology).

1.3.5 SMART-ITEM Open Platform

The SMART-ITEM open platform will be developed using security, compliances, and QoS-related tools from existing open platforms (i.e., FIRE+ and FIWARE[74]) that are currently exploited by CTI and LTU for their smart cities projects. It will also adopt existing standards (e.g., one M2 M, ITU/ISO/IEC standards for IoT). Examples of IoT services enablement[75] which will be relevant for this project are (i) *Generic Enablers* for advanced middleware and interfaces to network and devices; advanced web-based user interface; applications/services and data delivery; cloud hosting; data/context management; and security; and (ii) backend device management GE[76] and IoT data edge consolidation[77] (IDEC) GE – addresses the need to process data in real time. Implemented features include filtering, aggregating and merging real-time data from different sources. It is fully integrated with other enablers of FIWARE, especially using the Open Mobile Alliance (OMA) Next Generation Service Interface Context Enabler (NGSI 9 / NGSI 10) which is a very useful and easy format to encapsulate all data and events from RFID tags, Zigbee, or IETF devices, as many other smart things.

---

[73] http://cs.ecs.baylor.edu/~maurer/CSI5v93CLP/Papers/JAsCo.pdf

[74] https://www.fiware.org/2015/03/25/fiware-a-standard-open-platform-for-smart-cities/

[75] http://catalogue.fiware.org/chapter/internet-things-services-enablement

[76] http://catalogue.fiware.org/enablers/backend-device-management-idas

[77] http://catalogue.fiware.org/enablers/iot-data-edge-consolidation-ge-cepheus

UniversAAL[78] aims to create an open platform and standards necessary for the development of Ambient Assisted Living solutions.[79] Security for the different layers in a universAAL platform (which are relevant to SMART-ITEM system and services) are:

Platform services security – basic security such as user authentication, authorization, secure communication, and functional manifests

Application services security – provides value-added security to the universAAL platform such as document encryption and consent management

Container security – container-specific security features (e.g., provided by OSGi or Android) where OSGi security contributes to the security of the universAAL infrastructure such as OSGi service bundle signing, bundle permission management, and sandboxing. OSGi security also supports some of the security platform services such as secure communication

Java security – provides the Java security model such as the Java sandbox and basic security primitives (note: this is further enhanced with third-party cryptographic libraries)

Device and operating system security – which provides a secure environment for computing and data storage by ensuring a trustworthy execution environment, separation of applications and user accounts, etc.

UniversAAL and its tools[80] could be exploited to create, run, and publish universAAL applications.

1.4 Methodology

The SMART-ITEM project will employ the following integrated approach/methodology: (i) user-centered design approach, (ii) agile methodology for software development (integrated with PDSA (plan, do, study, act) with three iterative cycles with 3 years), and (iii) integrated software architecture.

1.4.1 User-Centered Design Approach

According to Norman et al. [1], there are two sides to a system interface, namely, system perspective and user perspective. The system perspective is changed through proper design while the user perspective is changed through training and experience. User experience will be positive if the user has a good conceptual understanding of the system which has been designed and built around the user needs and requirements (user requirement analysis). Thus, the user-centered design which focuses on the user's

[78] http://ercim-news.ercim.eu/en87/special/universaal-an-open-platform-and-reference-specification-for-building-aal-systems

[79] https://github.com/universAAL/platform/wiki

[80] https://github.com/universAAL/platform/wiki/RD-Available-Tools

Fig. 1.5 PDSA cycle

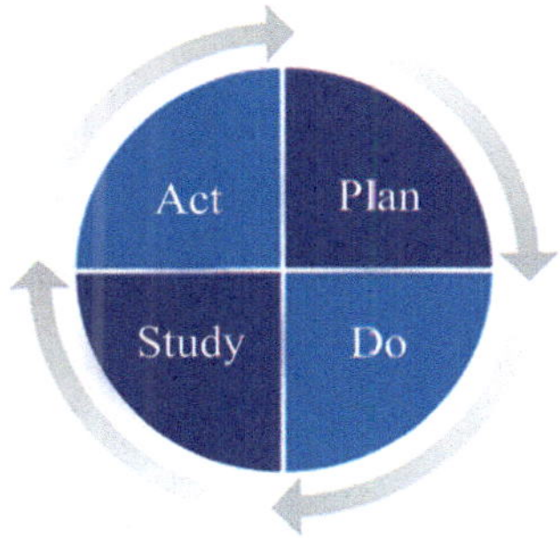

needs will entail the following [2]: user requirements analysis, activity/task analysis (necessary for the user interaction), initial testing and evaluation, and iterative designs.

1.4.2 Integrated Agile Methodology for Software Development (with PDSA)

The PDSA Cycle (also known as Deming Wheel/Cycle)[81] is a systematic series of steps for continuous improvement of the SMART-ITEM IoT-enabled ecosystem and services. The steps are (i) *Plan*: identify a goal or purpose, define success metrics, and action plan; (ii) *Do*: implement action plan (e.g., make a product); (iii) *Study*: monitor outcomes, validate the plan, and identify areas for improvement; and (iv) *Act*: revise step (i) based on results in step (iii). These four steps are iterated as part of a continuous improvement cycle (Fig. 1.5).

Agile methodology [3, 4] for software development will be adopted for this project due to its simplicity, flexibility, and adaptability. Just like PDSA, the agile methodology is iterative and incremental, with every aspect of the development life cycle (i.e., requirements, design, implementation, evaluation, etc.) being continually revisited for continuous improvement. In the agile paradigm, it will provide greater user satisfaction because it delivers a potentially consumable solution for each iteration and enables users to enhance (i.e., evolve for the better) their requirements throughout the entire project [3]. Additionally, it has a quicker delivery compared with the traditional waterfall methodology. Figure 1.6 depicts our integrated agile methodology and PDSA. The steps involved are:

(i) *PLAN*: Define requirements (*Research*) – user requirements: elderly population (greater or equal 65 years old) sample from several European cities; technical requirements; standards; legal and compliance requirements (for all the participating countries)
(ii) *PLAN*: Design models (*Design*) – user stories (i.e., begins with understanding the user, capture it in the form of a persona (a tool to help visualize the user and needs, followed by using the tasks and goals of the "personas" to support

[81] https://www.deming.org/theman/theories/pdsacycle

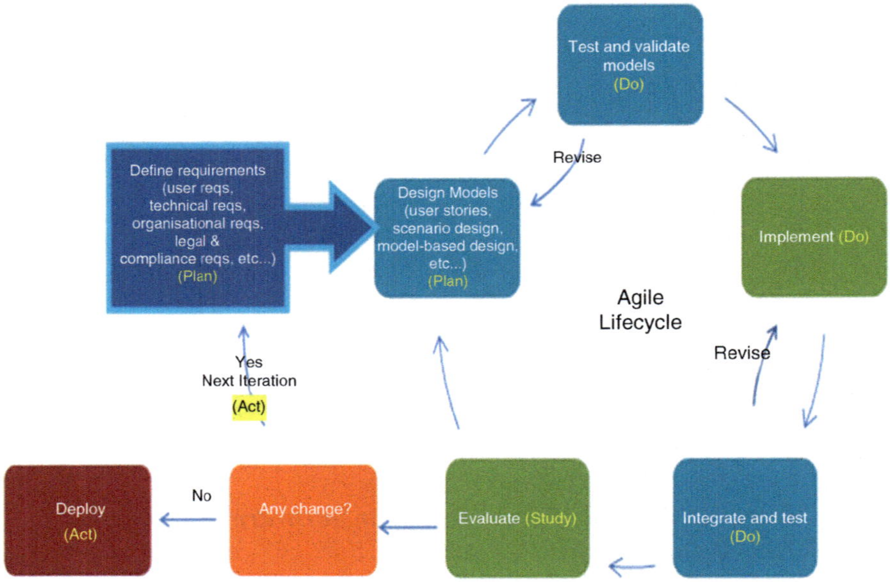

Fig. 1.6 Integrated agile life cycle and PDSA (Extracted from [6, 7])

decisions on system functionalities[82]), scenario-based designs (i.e., grounding design in an understanding and capturing the interrelationships of tasks users carry out over time),[83] model-based designs (e.g., Object, View, and Interaction Design, or OVID [5] which systematically builds models using based on a specification of the user model (*Design, Modeling, and Architecture*)

(iii) *DO*: Test and verify design models (*Design, Modeling, and Architecture*) in (ii) using simulations or experimental environment

(iv) *DO*: Implement (*Development and Integration*) – Develop system and services for the real environment with at least three complete cycles (shown in Fig. 1.7) for SMART-ITEM web portal, modules (SMART health and care (with secured individual personalized view), smart quality of life, and SMART-ITEM community of support)

(v) *DO*: Integrate and Test (*Development and Integration*) – integration within and between the six layers: sensing layer, data layer, platform layer (i.e., data access layer and cloud integration layer), application and services layer, secured network layer, and access layer)

(vi) *Study*: Evaluation of SMART-ITEM System and Services (*Evaluation: Research*) – expert appraisal, users, impact analyses (*Evaluation: Impact Analyses*)

(vii) Act: Next iteration if revision is required (note: the entire cycle will be repeated – possible evolution of requirements, revision of designs, changes in implementation, etc.).

---

[82] http://usability-bremen.de/wp-content/uploads/2013/03/Winter-Holt-et-al-2012-Persona-driven-agile-development.pdf

[83] http://www.ie.zjut.edu.cn/htm/uploadfiles/2007101052010670.pdf

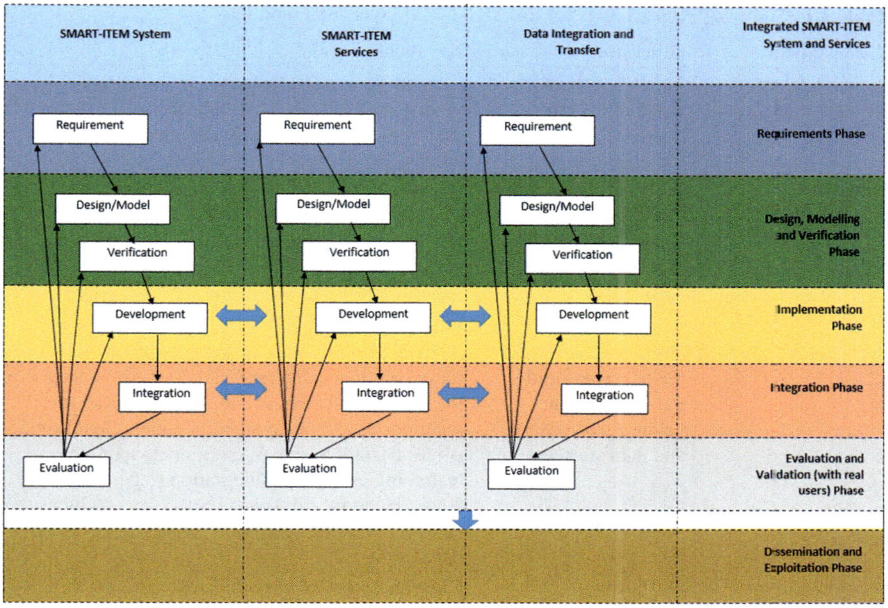

Fig. 1.7 Iterative component-based software development life cycle

On the other hand, if the evaluation results meet the SMART-ITEM system and services goals, then, it shall be deployed in the real environment.

1.4.3 Pilot Study for the Design, Development, and Deployment of SMART-ITEM System and Services

Users will be employed for the pilot tests (i.e., age must be more than 65 years of age living at home or in a care home). Initially, a small-scale pilot of 50 users will be conducted only within a private cloud environment. Three iterations of large-scale pilots are conducted (note: number of target users is 2200). During the first two iterative cycles, only the distributed private clouds (one for each city and hosted in the academic/research institutes) are used. However, during the third iterative cycle, the services are migrated to the public cloud for future scalability and sustainability.

1.5 Some Relevant Technologies for SMART-ITEM

Each SMART-ITEM system element is depicted in Fig. 1.1. Relevant technologies for SMART-ITEM ecosystem system and services are tabulated in Table 1.1. The system

Table 1.1 Some key technologies for SMART-ITEM layers and modules

Layer/others	Description	Key technology
Sensing layer	Smart health and care Telemonitoring devices for activity tracking and vital signs (with wireless capabilities)	Garmin Vivo smart (activity and heart rate) vívosmart® HR (Part Number: 010–01955-06) Weighing scale (Garmin Index smart scale – white) ONTEC CMS50EW color OLED bluetooth fingertip SPO2 monitoring pulse oximeter
	Smart quality of life Smart home automation	Raspberry Pi 3: Model B+ (Internet gateway) SAMSUNG SmartThings Starter Kit Samsung Galaxy Grand Prime DUAL SIM Simfree 5 inch Touchscreen Smartphone – Gold
Data layer	Data integration (for the telemonitored data and patient health record data) and data storage	.NET Framework for data integration of XML and relational data using ADO.NET. A list of data integration solutions in the cloud: proprietary solutions (e.g., Informatica, Pendaho, SnapLogic, Azure Biztalk Services, etc.); open-sourced technologies for data integration (e.g., JSON, XML, Hadoop and Apache Spark, etc.). Open-source web-/cloud-based MySQL will be used for data storage
Platform layer	Platform	Cloud-based platform – Microsoft Azure which supports the Microsoft .NET framework
Application and services layer	Database, access, and manipulation	PHP and MYSQL in Azure, XML, and C# (for simple and scalable data manipulation)
	Data analytics and visualization	Python for .NET (pythonnet), IronPython, R.NET (statistical language)
	Expert system	CLIPS rule-based programming language, CLIPS.NET (user interface for CLIPS), fuzzyclips
	Machine learning	Infer.NET (bayesian network), Accord.NET framework, Machine Learning Python (MLPY)
	Social community	Social Networking in .NET, ASP.NET Discussion Forum, ASP.NET Community Suite
Access layer	User interface	Jquery, Ajax (for enhanced interactivity and responsiveness), ASP.NET, C#.NET, APIs
Cloud infrastructures	Private cloud (servers and storage)	HP ProLiant DL380p G8 Storage HDD Tray Hard Drive Caddy 2.5″ Intel DC S3710 Series 200GB 2.5″ Seagate ST2000NX0273 2.5 inch 2 TB Internal Hard Drive RAM 16GB RAM Memory for HP-Compaq ProLiant DL380p G8 (DDR3-10600 - Reg)
	Public cloud and services	Amazon EC2 (Europe, Ireland) – web portal Amazon S3 (Europe, Ireland) – storage Amazon SNS (Europe, Ireland) – Simple Notification Service AWS Data Transfer Out AWS Data Transfer In AWS IoT Suite
Modeling	Simulations	Matlab and Simulink

elements are interdependent. The four types of interactions are data flow, information (processed data) flow, access, and integration (between and within layers).

1.6 Impact Analyses

Different categories of IoT technology impact assessment have been tabulated in Table 1.2.

1.7 Conclusion

The SMART-ITEM project will be a valuable project because of the scale and variety of potential users. The pilot target audience for SMART-ITEM use cases across eight cities in six countries is 2200 users. Secondly, users are co-creators of health and well-being. The project involves the development of an innovative and seamless integrated IoT-enabled ecosystem and services to support smart living. The SMART-ITEM system and services complex modular and vertically cross-layered architecture consists of three modules and six layers (see Fig. 1.1). Currently, integrating

Table 1.2 Impact assessment methods

Methods	Description
Economic impact assessment	*Benefit/costs analysis* B/C = present value of project benefit divided by present value of project cost by definition *Cost-effectiveness* This is the ratio of cost per unit of desired results (e.g., cost per person use). This test is applicable when the benefit measure cannot be reliably translated into money terms (e.g., carbon footprint reduction). It is most usefully applied when there is a clear goal (measure) for the desired level of benefit results *Net present value* NPV = present value of project benefit minus the present value of project cost (present value is at the point of decision-making) *Gross value added* GVA – measures the contribution to an economy of an individual producer, industry, sector, or region. It is used in the calculation of gross domestic product (GDP) *Return on investment* Benefits derived from investment on some resources
Environmental impact assessment	ISO 14001 – identify and evaluate environment assessment of IoT solutions The life cycle assessment of the IoT technology would be conducted according to ETSI's LCA methodology for Environmental Engineering (EE); Life Cycle Assessment (LCA) of ICT equipment, networks, and services
Health informatics impact assessment	Examples are time efficiency, user effectiveness, user behavioral change, etc.

http://www.etsi.org/deliver/etsi_ts/103100_103199/103199/01.01.01_60/ts_103199v010101p.pdf

IoT solutions from different providers is challenging because they are accompanied with proprietary apps and web access. Additionally, the interoperability between the different IoT devices poses another challenge due to the varying communication protocols. Using a customized Raspberry Pi 3+ as an integrated gateway for different sensors data will be beyond the state of the art.

References

1. Norman, D. A., et al. (1986). *User centered system design: New perspectives on human-computer interaction*. Hillsdale: Lawrence Erlbaum Associates Publishers.
2. Ritter, F. E., et al. (2014). *Foundations for designing user-centered systems*. London: Springer. doi:10.1007/978-1-4471-5134-0_2.
3. IBM. (2012). Agile for Dummies®, IBM Limited Editio. Wiley.
4. Kor, A. L., et al. (2015). *An Integrated Healthcare System and Services for Empowering Patients*. International Conference on Health Informatics and Medical Systems, 27–30 July, 2015, Las Vegas.
5. Roberts, D., Berry, D., Isensee, S., & Mullaly, J. (1998). *Designing for the user with OVID*. Indianapolis: MacMillan.
6. http://agilemethodology.org/
7. Kor, A. L., et al. (2016). *SMART-ITEM: IoT-Enabled Living*. Proceedings of IEEE Future Technologies Conference 2016, 6–8 December, 2016, San Francisco.

Chapter 2
Emerging Trends in Cloud Computing, Big Data, Fog Computing, IoT and Smart Living

Amin Hosseinian-Far, Muthu Ramachandran, and Charlotte Lilly Slack

2.1 Introduction

There can be many definitions towards the explanation and delivery that cloud computing has on the impact of the current twenty-first-century generation, too many for anyone to have achieved a rigorous meaning. According to Babcock [1], the 'cloud' is mostly specified and put into the software category labelled as a service, where a software application can be accessed online, at free hand. The main challenge to defining cloud computing is because much like other descriptive buzzwords within the technology industry, there can be many definitions to each individual or firm. Cloud computing provides 'ubiquitous', 'convenient' and 'on-demand' access to a networked and common group of configurable computing assets argued by Samani et al. [2]. 'As of now, computer networks are still in their infancy. But as they grow up and become more sophisticated, we will probably see the spread of 'computer utilities' which, like present electric and telephone utilities, will service individual homes and offices across the country' [3]. This concept by Leonard Kleinrock in 1969 was the familiar quote that inspired the development of the Internet today. Also, dating back to 1960, John McCarthy referenced that 'computation may someday be organized as a public utility', defining what is today the 'cloud' [4]. Armbrust et al. state that 'Cloud Computing refers to both the applications delivered as services over the Internet and the hardware and systems

A. Hosseinian-Far (✉)
School of Computing & Communications, The Open University, Jennie Lee Building, Walton Hall, Milton Keynes MK7 6AA, UK
e-mail: Amin.Far@open.ac.uk

M. Ramachandran • C.L. Slack
School of Computing, Creative Technologies & Engineering, Leeds Beckett University, Leeds, LS6 3QR, UK
e-mail: M.Ramachandran@leedsbeckett.ac.uk; lillyslack@hotmail.co.uk

M. Dastbaz et al. (eds.), *Technology for Smart Futures*,
DOI 10.1007/978-3-319-60137-3_2

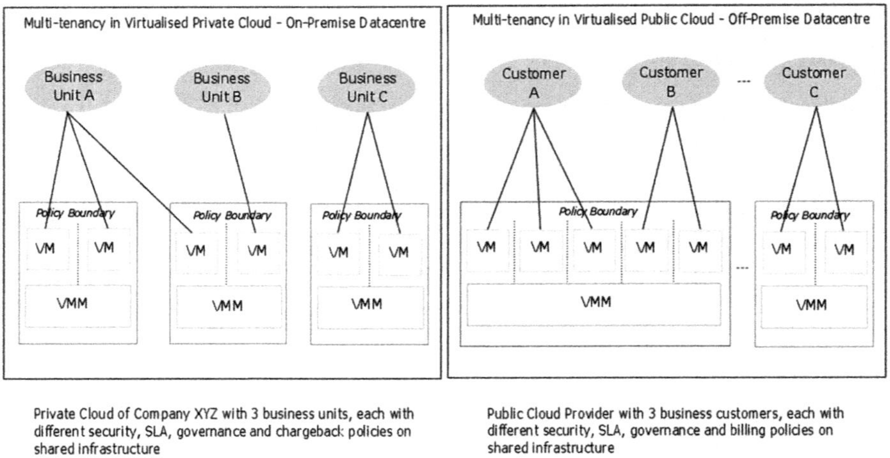

Fig. 2.1 Multitenancy (Adapted from Cloud Security Aliance [8])

software in the datacentres that provide these services' [5]. Compared to the types of services involved in cloud, and its overall commitment, Dhar (2012) gives the most defined yet infonaut definition. Dhar (2012) states, 'Cloud Computing is a style of computing in which dynamically scalable and often virtualized resources are provided as a service over web browsers and the Internet' [6]. Cloud computing has many benefits and concepts from the Web 2.0, offering infrastructure as its service [7].

Figure 2.1 outlines the concept of multitenancy within cloud computing environments [8]. The notion of multitenancy is simply referring to resource sharing within the cloud environment. In the first rectangle on the left, this resource sharing is illustrated as an example for a private cloud environment in a schematic way. On right, this resource sharing is presented in a public cloud environment. Multitencancy in public cloud can be considered as one of the major barriers for expansion of cloud computing due to existing security risks [9]. Risks such as losing privacy and/or integrity in public cloud may prevent many decision makers to authorise the implementation of digital services using cloud computing in a smart city.

2.2 Cloud Models

Cloud computing provides three types of services such as software as a service (SaaS), platform as a service (PaaS) and infrastructure as a service (IaaS). This section will look at each of these services.

2.2.1 Service Models

Due to the continuous evolution of the cloud, several cloud models have come to exist to keep up with the market demand and the Internet's continuous evolution. In the National Institute of Standards and Technology (NIST), Mell and Grance (2011) have defined three different service models [10]:

Cloud Software as a Service (SaaS) – SaaS has an offer to applications that are provided on their cloud infrastructure [11]. Overall, accountability and managerial organization is ran by SaaS, operating the applications too. The user is responsible of managing the app settings.

Cloud Platform as a Service (PaaS) – PaaS offers a structure to position apps that are already past the development stage, using certain programming supported for the PaaS. As stated before, and by Mell and Grance, the responsibility is upon the SaaS, who do not have control over the users' settings and apps. The user is again responsible for the settings and configurations [11].

Cloud Infrastructure as a Service (IaaS) – IaaS provides the correct resources and tools to suit the users system and app. The user has a responsibility of their operating system, storage and apps and network configurations [10].

The three indicated service models allow the use of multi-tenancy environment, apart from when the IaaS model is in use as the users are in total control over responsibility (see Fig. 2.2 for a detailed illustration of the details).

The software as a service category is mainly comprised of the APIs, applications, data and the required presentation platform for the software on the cloud. This would sit on the platform as service where the platform for hosting the software is offered as a service within the cloud environment. Lastly, there is the infrastructure as a service which is comprised of the necessary hardware, facilities and the associated protocols.

2.2.2 Deployment Models

There are several deployment models available in regard to the cloud today, and more and more will develop as the cloud technicalities expand [11]. The models stated below are the current models involved in cloud services so far and can be used with any of the service models listed above (SaaS, PaaS or laaS) [11].

Public Cloud – When discussing the topic of cloud, it is often specified and referred to as 'the public cloud'. This is due to the most well-known and popular cloud services being open to the public eye [2]. These service examples include e-mail services, such as Hotmail, Apple iCloud or the popular storage services, such as Dropbox. All of these cloud services are made available to the public, who can access the services via the Internet. Whilst many of the general public customers from a particular firm may use a 'public cloud' service, the nature of the public model will mean that it is available to any individual customer [2].

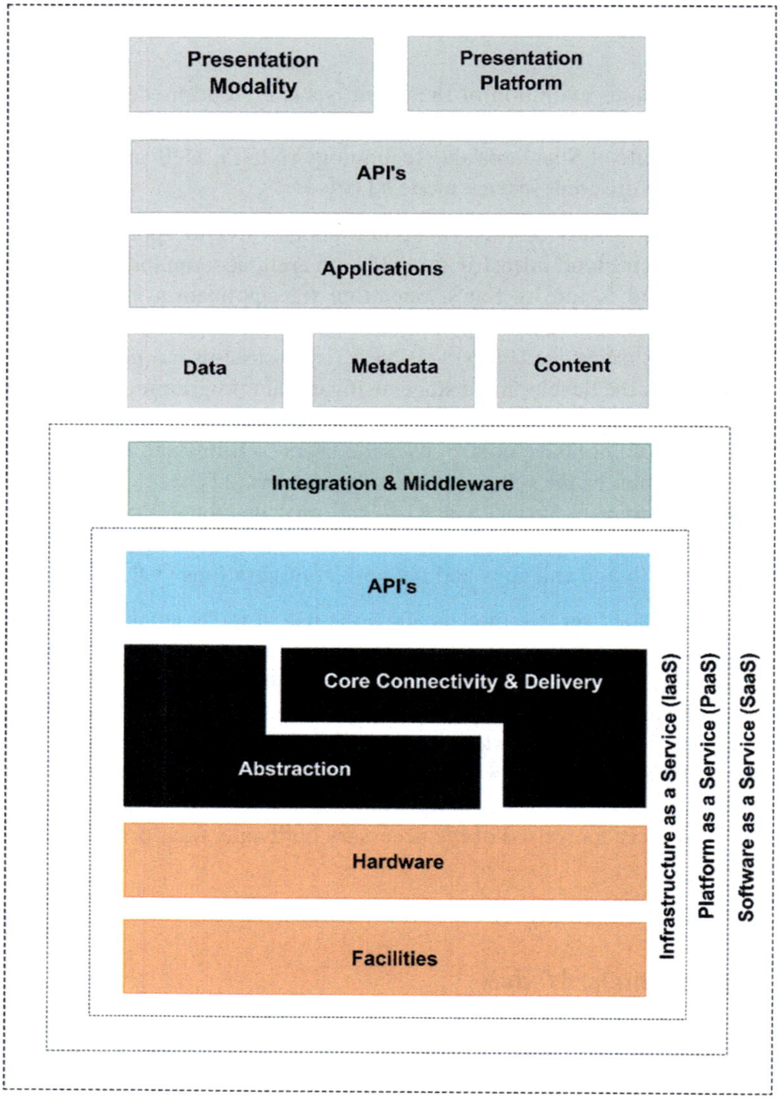

Fig. 2.2 Service models (Adapted from Cloud Security Aliance [8])

Private Cloud – Whilst the public cloud uses a substructure for several customers, the defined 'private cloud' is reserved to one individual person, a single customer. Whether the infrastructure of the firm is on sight, or off sight, the individual client will be the only one with access and control to its location and placing [2]. Only one organization constructs within their personal cloud [11]. Operated in the 'private cloud' is only information that will be or is regulated and controlled, such as personal information and data that should not be duplicated or repeated outside of this cloud.

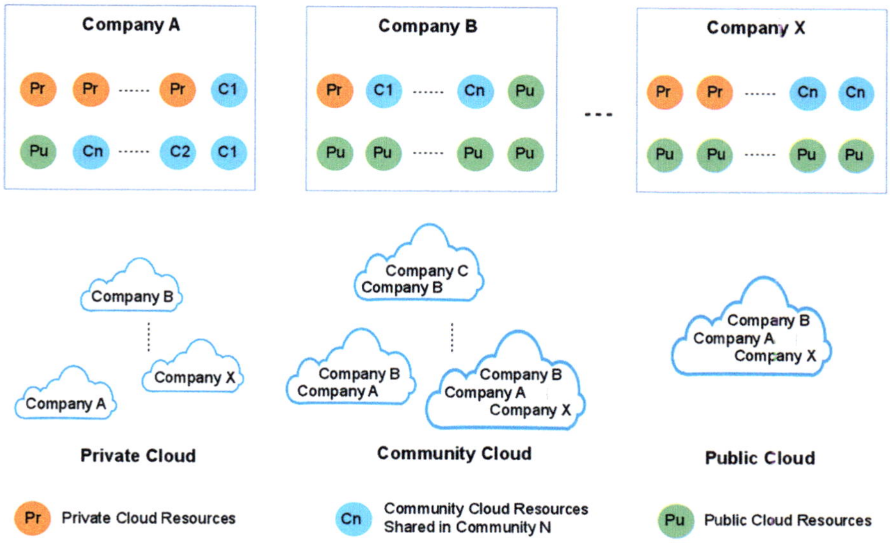

Fig. 2.3 Disparate resource-sharing requirements in different companies (Adapted from Lu et al. [15])

It is a preferred model for hosting private data where restrictions and boundaries can be applied. In comparison to the 'public cloud' and 'community cloud' (see below and Fig. 2.3), a private cloud can be 'trusted' with limited access [2].

Community Cloud – A community cloud extends the concept of a private cloud, to allow multiple known members to incorporate into shared concerns in the private cloud. The community of this particular cloud is multiple stakeholders, with similar goals or the same preferences for security, and through the whole party, these stakeholders are provided and granted the same data access [2]. The members in the cloud may wish to review those looking and seeking entry in their cloud community [12]. The community cloud may be organized, and managed by the firms, or outsourced to a third party [11]. In a busy cloud market position, many cloud service providers can differentiate themselves better using 'community cloud' [13] (see Fig. 2.3).

Hybrid Cloud – In between the original 'private' and 'public' cloud, without contrast to the 'community' cloud, is the 'hybrid' cloud. The 'hybrid cloud' is designed to prepare for the implementations that are in between the public and private cloud [2]. The hybrid is an integrated cloud service, which can utilize the private and public to perform the different functions that are needed within the community of the firm [14]. The hybrid cloud involves several different cloud infrastructures, with different deployment models combined [11].

As discussed earlier in this chapter, resource sharing is one of the objectives of cloud computing as it would benefit storage, processing and manipulation of data across shared resources. However, security, privacy and integrity of the data in a public cloud are often a concern for many decision makers. The public cloud is the desired and relevant deployment model that can be used in a smart city, nonetheless appropriate strategies should be arranged for maximising data security, privacy and integrity.

2.3 Cloud Services

The cloud services today have many providers, which are familiar with the common public, including Dropbox, iCloud and Google. All these cloud services involve various services, where files and information are kept on their servers, which are connected to the Internet. Therefore, instead of having to keep them on a single computer, you can use any online device to access the same files. This is useful for the day-to-day business concept of having your important data backed up and having easy access anywhere to 'the cloud'. With the security aspect for the product, the security of these services plays an important role in the development of software systems, such as the cloud. Security requirement is often considered after the design of the system, which is shown by an analysis of software development processes [16].

2.3.1 Security and Compliance

Some of the most important concerns in cloud computing are security and privacy issues [17]. Throughout the use of the cloud, there is a constant need for large amounts of personal data, and sensitive data is also managed in the process. The security concepts and the privacy image of the cloud are among the primary reasonings for the cloud's existence, as several surveys state. For different firms, it is essential that the analysis and concept of the security and privacy issues that adopt the cloud-computing infrastructure are understood, before its adoption into business culture [17].

The cloud service is based on the storage of personal and sensitive information, and the safety of its data, which raises concerns as to whether the cloud context can be trusted. Many industries for solutions have made sure to take into account the understanding of organizational structures and laws and regulations that are bound around the social aspects of the situation. When a firm participates into the cloud IT infrastructure, their data and important documentation is then stored in a different environment, which is managed and maintained externally to their organization. From this there is a sudden feel of lack of control within the organization. The organization is appropriately giving up their administrative control and processes. Therefore, with respect to the security of the cloud, the consideration of elements including data integrity, data transfer and recovery should be revised [17].

Considering the cloud security landscape, if overall there becomes a failure to ensure appropriate security and safety when the cloud services are in use, this could result in higher costs and potential loss of business [18], thus disregarding any potential benefits that cloud computing can have on these specific measures for the firm [18]. If the consideration in a firm is to move to the cloud computing process, the stakeholders, including customers, must have a clear mentality and approach [18]. They also need to understand the probable security benefits and risks associated with cloud computing and set realistic expectations with their cloud provider [18].

2.3.2 Cloud Security Controls

Cloud security and its procedures can only be effective if the right security measures are in place and implemented correctly. The security management of the cloud service should be in place according to the cloud architecture that provides the control of the service. With the precise security management in place, the security controls can then be used and identified to prevent any issues from occurring. These individual controls can be put into place to avoid any flaws and reduce the chance of an attack on the systems. An example would be a logon warning, which ensures the person is aware of the right to enter the information storage and the cloud storage site [19]. The cloud security controls are the key pillars of cloud security [20].

Deterrent Controls – Deterrent controls are measures to diminish any sort of attacks that may occur when on cloud architecture [21]. An example of this could be a warning sign on a housing property, or a pop up on an IT structure, suggesting further action should be proposed when using software. These are the same as deterrent controls, stating there will be opposing concerns and consequences if they were to proceed [19].

Preventive Controls – Preventive controls ensure that any security issues that will intendedly arise will have already been approached. With the managing of certain issues, damage can then be limited [19]. An example of this would be the proxy server. The preventative acts as a 'bouncer' personality between your information and the person wanting to access it. On request for this information, it can be granted and passed through servers, reducing the risk of a security crack [19]. With the correct preventive controls, damages on the cloud system will be minimized [21].

Corrective Controls – Corrective controls are exactly what its name says and are security issues that can be resolved quickly and efficiently after an attack or an attempt on damage has been made; this real-time security is important [19]. An example of a corrective control in place may be a limitation on how much time is spent on your cloud storage service, thus minimizing potential breaches. Compared to preventive controls, the correctives ones take control as an attack is taking place [21].

Detective Controls – Detective controls are the 'detectives' of the cloud controls. They detect any illegal users, who are working against the preventative and corrective controls to make sure there is minimal damage. Usually with detective controls, they can stop an attack before they even begin [19] by detecting the potential attack before it happens [21].

Furthermore, having the four storage cloud controls when accessing, or using a Cloud service, can maintain its quality and storage [19]. The potential damage that can be limited due to the use of the controls can prove extremely useful by the end of the online use to the cloud; the security of sensitive information can be completely secure if used properly. To make sure that the security of the cloud is secure, up to date and attack free, the security has to be in line with the standard security guidelines [21] with these security controls in place.

2.4 Cloud Users and Organisations

Many users today have an idea of the cloud being a glamourous security to their data and private documents, which delivers sufficient services based on their desires [22]. From a cloud user's view, a good cloud service will accommodate copious amounts of resources for its technicality. This means a user will always feel the need to be able to request more resources according to his/her need [22]. From an organisational perspective, the service is always providing elasticity, where the cloud resources are adaptive to the business and user needs [22]. There are also several cloud security principles, which summarise the essential security principles when there is an evaluation towards the cloud services and why they are important to an organization [23]. Some organisations wish to satisfy many of the security principles whilst many subsidize.

2.5 Big Data Notion

Since the emergence of databases, businesses have immensely benefited from the data organisation capabilities that the databases offer. The history of relational databases traces back to 1974 when IBM initiated a project called IBM System R aiming at developing a database system for research purposes [24]. Though, the first commercial relational database was developed by Oracle in 1979 [25]. Relational databases are using tables (relations) to organise data records. Since then, there have been various techniques and architectures for relational database development. One of these initial established well-known techniques is the Cobb's normalisation technique for redundancy reduction in the database [26]. Further extensions were added to Codd's notion of normalisation. There were other novel techniques for developing object-oriented relational databases. Attaran and Hosseinian far proposed a hierarchical novel algorithm that derives the database from the relations of the system's classes [27]. Big data was initially defined by Magoulas in 2005 as 'big data is when the size of the data becomes part of the problem' [28]. Goes associated big data with the four V's attributes [29]. Volume is the first characteristic by which manipulation of large volumes of data is often considered within the realm of bi data science. The second characteristic is variety. The data types used for analytics in data science are varied and may belong to different industries and sectors. They are also often presented in different patterns and structures [30]. Most of this large volume of data is needed for real-time analytics, and therefore the speed of transactions and data creation is often fast. Examples can be found in many sectors, e.g. financial sector and banking in which transactions are to be updated in real time or the stock market prices are to be analysed and informed rapidly [31]. The last V stands for veracity by which the reliability of the gathered data is assessed. The big data veracity is very context dependent; for instance the data produced within the social networking context may contain irrelevant details including spams which would make the analytics task very challenging [32].

2.6 Fog Computing

Fog computing has emerged to address the need for low latency, location awareness, widespread geographical distribution, strong mobility, strong wireless access, strong presence of streaming and real-time applications and a support for the needed heterogeneity. All these characteristics of fog computing coupled with the advancement for wireless sensor networks (WSNs) and Internet of things (IoT) devices make it suitable for the future of smart living. However there are research challenges that remain active on security and privacy of cloud computing, fog computing and IoT [9]. Our earlier work on cloud security has developed a generic framework known as cloud computing adoption framework [33]. However, privacy issues have not been addressed with respect to emerging IoT sensors, devices, cars, home appliances, drones and other applications that are expected to reach and use magnitude of location data as well as personal data.

The following section aims to forecast how these four emerging technologies, i.e. cloud computing, big data analytics, fog computing and Internet of things, will work together for smart living.

2.7 Smart Living

The notion of smart living and cities is often described as a governmental initiative with a view to improve citizens' lives and living standards. Jimenez et al. believe that 'interoperability' between different organisations and departments is the main challenge to achieve smarter city perception [34]. There are a number of enabling technologies which some are mentioned within this paper. There are numerous applications to transform cities to smart cities, from education, communications, to infrastructure and energy and beyond. Considering the application, an area of technology is required to enable the transformation process. The Internet of things with its various technologies such as wireless sensor networks (WSNs) is often considered to be an indispensable necessity. The overall idea is to reduce the energy consumption and costs, increase efficiency and enable faster data transfer from point A to B. Smart grid is also considered as essential enabling technologies which would reduce energy consumption intelligently. Narrow-band power-line communication (PLC) is an established technology which allows data transfer over the low-noise internal electricity wiring, whilst the broadband PLC intends to transfer data over the city electricity grid. The main challenge in implementation of broadband PLC would be to overcome the noise [35]; nonetheless, currently, electricity providers across the UK are moving towards smart metering (Fig. 2.4).

Real-time monitoring and reduction of CO2 footprints is an essential step to ensuring the environmental friendliness of a smart city. A good example of this can be explained within the perspective of the London Plan. The London Plan authorised by the Greater London Authority (GLA) is a longitudinal development plan

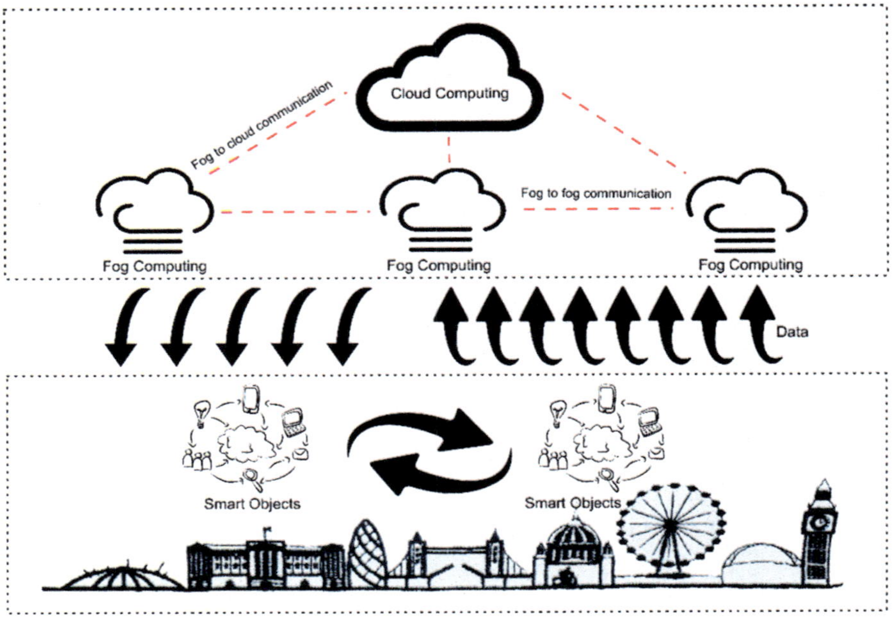

Fig. 2.4 Cloud and fog computing, IoT and data generation (own elaboration)

for the city of London, UK, which outlines different policy principles for improving the quality of life and the city. The plan has undergone many modifications since its introduction in 2004. One of the sub policies of this spatial development plan is to reduce the CO2 footprint by 2050 [36]. Such a policy disposes the CO2 reduction through the use of solar panels, combined heat and power (CHP) and micro CHB and insulation of the residential and industrial buildings (energy efficiency initiatives) as the applied approach to the policy [37]. These are accompanied by a set of executive targets; for instance, within the initial plan, the CO2 reduction is to be monitored every 5 years. This plan was followed by the 'Smart London Plan' [38] in which the exploitation of big data, IoT, cloud computing and in general digital technology could facilitate the initial longitudinal strategy. Big data technologies together with IoT would facilitate many of the monitoring and analysis tasks discussed within the initial proposal. The sensor networks could offer real-time monitoring of the smart city's polluters, road traffic and any application in which the real-time monitoring is essential.

Cloud computing would improve 'data storage', 'handling' and 'caching' in the application of smart grid [39]. Rimal et al. have developed a novel resource management framework which utilises cloud computing for infrastructural networking [39]. Their research results assess the use of cloud for broadband access traffic without affecting the performance in traditional architecture.

2.8 Conclusions

Cloud computing, big data analytics, fog computing and Internet of things (IoT) are emerging and disruptive technologies that can enable smart living. This chapter aimed to outline an overview of these technologies and highlight the factors that would be necessary to consider before strategizing resources and applying the technologies within the context. Although the technology may seem not to be the bottleneck anymore for implementing digital systems in smart cities, nonetheless security, privacy and integrity are some of the notions that are alerting many policy makers and strategists before realising smart living through digital technologies.

References

1. Babcock, C. (2010). *Management strategies for the cloud revolution*. New York: McGraw Hill.
2. Samani, R., Reavis, J., & Honan, B. (2015). *CSA guide to cloud computing: Implementing cloud privacy and security* (2nd ed.). Syngress. https://www.elsevier.com/books/csa-guide-to-cloud-computing/samani/978-0-12-420125-5.
3. Kleinrock, L. (2005). A vision of the internet. *ST Journal of Research, 2*, 4–5.
4. Sood, S. K. (2012). A combined approach to ensure data security in cloud computing. *Journal of Network and Computer Applications, 35*(6), 1831–1838.
5. Armbrust, M., Fox, A., Griffith, R., Joseph, A. D., Katz, R. H., Konwinski, A., Lee, G., Patterson, D. A., Rabkin, A., Stoica, I., & Zaharia, M. (2009). *Above the clouds: A Berkeley view of cloud computing*. Berkeley: University of California at Berkeley.
6. Dhar, S. (2012). From outsourcing to cloud computing: Evolution of IT services. *Management Research Review, 35*(8), 664–675.
7. Chang, V. (2015). An overview, examples and impacts offered by emerging services and analytics in cloud computing. *International Journal of Information Management*, 1–11.
8. Cloud Security Aliance. (2011). Security guidance for critical areas of focus in Cloud Computing V2.1. December 2011. [Online]. Available: https://cloudsecurityalliance.org/guidance/csaguide.v3.0.pdf. Accessed 2016.
9. Jahankhani, H., & Hosseinian-Far, A. (2016). Challenges of Cloud Forensics. in Working Paper.
10. Mell, P., & Grance, T. (2011). *The NIST definition of cloud*. Gaithersburg: US National Institute of Standards and Technology.
11. Ros, J. (2012). *Security in the cloud: The threat of coexist with an unknown tenant on a public environment*. London: Royal Holloway.
12. Gartner. (2016). Community cloud. [Online]. Available: http://www.gartner.com/it-glossary/community-cloud. Accessed 2016.
13. Joseph, A. O., Kathrine, J. W., & Vijayan, R. (2014). Cloud security mechanisms for data protection: A survey. *International Journal of Multimedia and Ubiquitous Engineering, 9*(9), 81–90.
14. Interoute. (2016). What is a hybrid cloud? [Online]. Available: http://www.interoute.com/cloud-article/what-hybrid-cloud. Accessed 2016.
15. Lu, Y., Xu, X., & Xu, J. (2014). Development of a hybrid manufacturing cloud. *Journal of Manufacturing Systems, 33*(4), 551–566.
16. Mouratidis, H., Argyropoulos, N., & Shei, S. (2016). Security requirements Engineering for cloud computing: The secure Tropos approach. In D. Karagiannis, H. C. Mayr, & J. Mylopoulos (Eds.), *Domain-specific conceptual modeling* (pp. 357–380). Cham: Springer International Publishing.

17. Kalloniatis, C., Mouratidis, H., & Islam, S. (2013). Evaluating cloud deployment scenarios based on security and privacy requirements. *Requirements Engineering, 18*(4), 299–319.
18. Cloud Council. (2015). Security for cloud computing; Ten steps to ensure success. Cloud Standards Customer Council.
19. CloudTweaks. (2013). Cloud Storage Security Controls. [Online]. Available: http://cloudtweaks.com/2013/03/cloud-storage-security-controls/. Accessed 2016.
20. Rawle, D. (2011). Riding the cloud storm – Responding to cloud risks. [Online]. Available: http://www.bytes.co.uk/files/4514/0500/5660/david_rawle_-_cloud_security_challenges_presentation.pdf. Accessed 2016.
21. Chakraborty, N., & Patel, R. S. (2014). Security challenges in cloud computing: A comprehensive study. *International Journal of Computer Science Engineering and Technology(IJCSET), 4*(1), 1–4.
22. Park, S. C., & Ryoo, S. Y. (2013). An empirical investigation of end-users' switching toward cloud computing: A two factor theory perspective. *Computers in Human Behavior, 29*(1), 160–170.
23. CESG. (2014). Summary of cloud security principles. [Online]. Available: https://www.gov.uk/government/publications/cloud-service-security-principles/cloud-service-security-principles. Accessed 2016.
24. Wade, B. W. (2012). IBM relational database systems: The early years. *IEEE Annals of the History of Computing, 34*(4), 38–48.
25. Ferreira, P. (2015). "The History of Oracle," [Online]. Available: http://www.dba-oracle.com/t_history_oracle.htm. [Accessed 2016].
26. Codd, E. F. (1970). A relational model of data for large shared data banks. *Communications of the ACM, 13*(6), 377–387.
27. Attaran, H., & Hosseinian-Far, A. (2011). A novel technique for object oriented relational database design. *IEEE 10th International Conference on Cybernetic Intelligent Systems (CIS)* (pp. 128–132). IEEE.
28. Loukides, M. (2013). Big data is dead, long live big data: Thoughts heading to Strata. [Online]. Available: http://radar.oreilly.com/2013/02/big-data-hype-and-longevity.html. Accessed 2016.
29. Goes, P. B. (2014). Big data and IS research. *MIS Quarterly, 38*(3), 3–8.
30. Abbasi, A., Sarker, S., & Chiang, R. H. (2016). Big data research in information systems: Toward an inclusive research agenda. *Journal of Association for Information Systems, 17*(2), 2–31.
31. Srivastava, U., & Gopalkrishnan, S. (2015). Impact of big data analytics on banking sector: Learning for Indian banks. *Procedia Computer Science, 50*, 643–652.
32. Abbasi, A., & Adjeroh, D. (2014). Social media analytics for smart health. *IEEE Intelligent Systems, 29*(2), 60–64.
33. Chang, V., & Ramachandran, M. (2016). Towards achieving cloud data security with the cloud computing adoption framework. *IEEE Transaction on Service Computing, 9*(1), 138–151.
34. Jimenez, C. E., Solanas, A., & Falcone, F. (2014). E-government interoperability: Linking open and smart government. *IEEE Computer Society, 47*(10), 22–24.
35. Hosseinpournajarkolaei, A., Jahankhani, H., & Hosseinian-Far, A. (2014). Vulnerability considerations for power line communication's supervisory control and data acquisition. *International Journal of Electronic Security and Digital Forensics, 6*(2), 104–114.
36. Hosseinian-Far, A., Pimenidis, E., & Jahankhani, H. (2011). Financial assessment of London plan policy 4A. 2 by probabilistic inference and influence diagrams. In *Artificial intelligence applications and innovations* (Vol. 364, pp. 51–60). Berlin Heidelberg: Springer.
37. Hosseinian-Far, A., Jahankhani, H., & Pimenidis, E. (2011). Using probabilistic networks for the London plan knowledge representation. In *10th IEEE International Conference On Cybernetic Intelligent Systems*, London.
38. London Mayor's Office. (2013). Smart London Plan. London.gov.uk, London.
39. Rimal, B. P., Van, D. P. & Maier, M. (2016). Mobile-edge computing vs. centralized cloud computing in fiber-wireless access networks. In *2016 IEEE Infocom 5G & Beyond Workshop*, San Francisco, USA.

Chapter 3
Toward a Cognitive Middleware for Context-Aware Interaction in Smart Homes

Obinna Anya and Hissam Tawfik

3.1 Introduction

Over the last decade, the smart home technology has gained increasing research attention as an approach for addressing the world's growing aging population and supporting active and independent living. A smart home for the elderly integrates multiple IoT-enabled home devices and computing subsystems, including ubiquitous ICT devices, aimed at providing context-aware and ambient assisted living services to the elderly, often facilitated by remote home control. From an interaction design perspective, a smart home donates a physical space within which an elderly user cocreates and shares information with intelligent home devices in a way that corresponds to, and meaningfully caters for, the needs and values of the user. An overarching principle in smart home design is the seamless integration of a wide range of technologies such as wireless sensor networks, IoT devices, big data, ambient intelligence, and cyber-physical systems, as well as increased automation of activities of daily living with the goal of providing the elderly with services for active aging and social inclusion as well as technologies for overcoming physical and cognitive limitations imposed by old age. As a result, smart homes are most often characterized in terms of their level of augmentation through ICT, sensor-based technologies, and IoT devices, as well as their degree of enriched user support through automation and artificial intelligence.

Although the value of the smart home market is projected to grow from 39.9 billion USD in 2016 to 79.9 billion USD by 2022 [a] and several prototype applications for elderly care have been proposed [13–15], the landscape of smart home solutions

---

O. Anya (✉)
IBM Research, San Jose, CA, USA
e-mail: obanya@us.ibm.com

H. Tawfik
Leeds Beckett University, Leeds, UK

M. Dastbaz et al. (eds.), *Technology for Smart Futures*,
DOI 10.1007/978-3-319-60137-3_3

for elderly care has been to date one with very little or no commercial applications. A major obstacle, among other factors, to the uptake of smart homes for the elderly is the lack of adequate models for managing interaction between users and system [1]. Several context-aware applications with the capability to detect human activity and offer different levels of personalized services have been developed. However, studies have shown that the elderly faces a uniquely different set of challenges when it comes to engaging with digital technologies. For example, owing to declining physical and cognitive abilities in old age, elderly users have a different attitude to technology than younger users [6] and are often challenged by pervasive and frequency changes in device user interfaces [3].

This paper presents a framework for the design of effective models of interaction for smart home applications for the elderly through the use of a cognitive middleware for managing the landscape of interaction between an elderly person and IoT devices in a smart home. The cognitive middleware will primarily serve to mediate and manage how an elderly person engages with a smart device – it learns about the person's lifestyle interests, habits, and preferences, as well as their contexts and situations, and consequently determines suitable interaction models for a given user context. Enabling effective and meaningful interaction between elderly persons and intelligent devices in a smart home will require approaches that are based on clearly defined models of context and system software architecture, as well as a systematic application of usability and user-centered design principles geared toward tackling such challenges as: how do we design smart home systems that accommodate and adapt to an elderly person's level of physical and cognitive capabilities? How do we design smart home technologies with the ability to intelligently coordinate actions between collections of heterogeneous networked devices in a way that primarily supports the varying needs of an elderly user? The foundation for such a middleware is arguably a user modeling system that, given historical data about user lifestyle interests, habits, and preferences, as well as a sequence of user actions and data about user environment, identifies a meaningful and intuitive interaction model that corresponds to satisfying user needs in a given context. Using a middleware allows for reusability and good architectural design by abstracting the interaction mechanism away from the core purpose of a particular smart device.

We start off the rest of this paper with a brief description, in Sect. 3.2, of an example highlighting the need for a cognitive middleware for managing context-aware interaction in smart homes. In Sect. 3.3, we discuss background work followed by a discussion of the relevance of middleware in context-aware interaction in Sect. 3.4. We describe our proposed interaction model in Sect. 3.5 followed by discussion and conclusion in Sect. 3.6.

3.2 Motivating Example

Although new technologies have emerged that help us keep track of multiple things such as our health regime, social connections with friends and family, and even our favorite programs, many elderly people have difficulty integrating these technologies into their daily life. John, aged 72, lives alone in his home. His children who have moved away to different locations have made efforts to equip John's home with numerous IoT gadgets including a diabetic monitor system for testing his blood glucose level four times a day; an automatic room temperature controller; Skype and Facebook technologies for social collaboration; a neckband alert system that John has to press for emergency services, e.g., in the event of a fall; a medication time reminder system; and a FaceTime connection to John's physician.

Like most elderly people with chronic ailments, John requires routine medication with changing dosage according to his physician's prescription. At 72, John equally enjoys his favorite meals, daily newspapers, and TV channels. However, John often forgets to take his medication on time and finds it difficult to coordinate his use of different gadgets that allow him, for example, to reach his physician when he wants to or to call his family. John also has difficulty managing his meal and often makes poor food choices.

Many of the issues in supporting elderly people to live quality life independently can be improved with the use of a cognitive middleware for managing their engagement with technology devices and gadgets in their smart home. The cognitive middleware learns about John's lifestyle interests including his food choices and his usual habits, e.g., his favorite TV times and channels, as well as his changing medication prescriptions. The middleware equally has a knowledge of the different roles of each of the devices in the homes, and is able to automatically manage their interactions in the context of specific activities that John engages in at various times, for example, to trigger a FaceTime call to John's physician. The cognitive middleware is able to leverage knowledge of John's behavior, lifestyle, and preferences, connecting to a history of events, things, and places around which he has progressively built his life, in ways that ensure that the help and services offered by the smart technology are both welcomed and genuinely useful.

3.3 Background

In this section, we discuss a number of background issues pertinent to understanding the challenges of context-aware interaction in smart homes for the elderly. We review relevant research projects on smart homes for the elderly, taxonomies of interaction in smart homes, and research in context-aware computing. Finally, we highlight the key features that distinguish the approach proposed in this chapter.

3.3.1 Smart Home Projects for the Elderly

Several smart home technologies have been proposed conveying different forms of IoT-enabled devices with increasing sophistication and have gained a lot of attention as potential solutions for the provision of enhanced quality of life for the elderly [18]. Recent reviews by Alam et al. [13] and Chan et al. [14] provide a general overview of past developments, present research, and future challenges. Other related reviews, such as [15] focusing on context-aware tools and monitoring services for smart homes, highlight the role and challenges of context-aware interaction in smart homes. While remarkable progress has arguably been made in smart home research, developing usable smart home technologies for elderly use remains a huge challenge [1]. The more sophisticated the underlying algorithms and architectures of a smart home are, the more difficult it is for an elderly person to engage with a system underlying such arrays of smart tools and interfaces. From a programming perspective, it is almost impossible for a designer to program all contingency plans for every context that an elderly person could possibly be in without the system performing in unexpected ways. Employing a machine learning approach is equally challenged because of the inherent difficulty in collecting training examples with appropriate scenarios for how an elderly person lives and behaves for all possible contextual situations. A fundamental problem then becomes: how do we design smart homes that employ sophisticated algorithms and yet are simple enough for an elderly person with varying degrees of physical and cognitive impairments to easily and comfortably interact with in a way that creates *value*?

Over the years, the task of building smart homes has integrated other approaches in healthcare for increased success. One example is telemedicine or "telehomecare" where clinical consultations and treatments are provided virtually to an elderly in a smart home via ICT technologies – a technique that has been used for disease prevention and monitoring of the elderly. Others include "the intelligent room," which combines robotics, vision and speech technology, and agent-based architectures in order to provide computation and information services for people engaged in day-to-day activities with the goal of pulling the computer out into the real world of people [40]; "integrated home systems," which provide for the elderly a single human-machine interface to household systems and gadgets [41]; "rehabilitation integrated systems" for better access and ergonomics for the elderly [43]; and the "adaptive house," which uses advanced neural network to control such things as room temperature, heating, and lighting without previous programming by the residents [44]. Recent approaches leverage advances in context awareness, e.g., the context-sensitive rule-based architecture for a smart home environment [45], and social computing, e.g., the aware community, which allows an elderly in a smart home to engage with other people via social networks as well as mobile technologies, the service-oriented paradigm, and cloud-based infrastructures [32]. In addition, research in smart home has served as an environment for scientific study of home life, particularly the relationships between space and information, e.g., through approaches that integrate learning into everyday activity in a smart home [32], in order to provide better personalized services for the elderly.

Despite the huge progress in smart home research, realizing the vision of smart home presents a number of challenges. Some of these challenges stem from the fact that at any specific time, a smart home is required to generate data about the environment such as temperature, humidity, medication level, and the status of doors, windows, and lights and about the location and behaviors of its inhabitants such as sleeping, cooking, watching TV, etc. As noted by [26], the central issue becomes how to fuse data from multiple sources in order to form a meaningful interpretation of a situation, in relation to other information sources such as electronic health records as well as information about user needs, and subsequently provide personalized care to the elderly inhabitant. A related challenge in smart homes is that the technologies often appear too intrusive and technologically too sophisticated and unusable for an elderly person [23]. Overall, existing smart homes hardly adapt to the personal lifestyles of the elderly [2].

3.3.2 Taxonomies of Interaction in Smart Homes

Previous studies have categorized interactions in smart homes. A taxonomy based on sociotechnical and cyclic interactions is proposed in [4]. The authors observe that the problem of interaction in smart homes lies at two extremes. On one end, smart homes enable users to share digital artifacts and services across a home network and, at the other end, enable collections of networked devices to coordinate their actions. The former refers to *sociotechnical interaction*, and the latter *cyclic interaction*. While relevant in enabling us to understand interactions in the smart home, the taxonomy does very little to address the question of how to enable effective and useful user-system interaction in the smart home. A taxonomy for interservice interactions proposed by Kolberg et al. [5] appears particularly relevant in understanding how IoT-enabled devices interact with one another in a smart home. The taxonomy includes *multiple trigger interaction*, which refers to two services controlling the same IoT-enabled device; *shared trigger interaction*, which refers to a situation where an event is sent to two different services that consequently perform conflicting actions; *sequential action interaction*, in which a service sends a request to an IoT device that in turn sends notifications to another services; and finally *missed trigger interaction*, in which a service prevents the operations of another service.

Dewsbury et al. [7] presented two categories of interactive devices in a smart home, namely, active devices and passive devices. Active devices are those with which the user can directly use and interact with, e.g., light switches or control panels, whereas passive devices are those that the user cannot directly interact with, but which can detect user action and enable predetermined functions, e.g., sensors. From an interaction design perspective, these two categories present opportunities for the design of two broad interaction models for managing the elderly person's *active* and *passive interactions* with smart home devices. The idea of active and passive interactions resonates with the distinction between explicit and implicit interaction [9]. The model is based on how the interaction is initiated, where explicit interaction is user-triggered through a direct input device and implicit interaction is sensor-triggered. Another relevant taxonomy of interactive smart home devices appears in [8]. The authors' categorization is based on how technology is made manifest in a home appliance and includes *information appliances*, which are stand-alone IoT smart home devices with well-defined self-contained interactive functionalities; *interactive household objects*, which merge interactive capabilities with existing home devices to offer new forms of interaction; and *augmented furniture*, which adds interactive capabilities to different furniture items in the home. In [6], Kim et al. proposed a taxonomy of interaction in a smart home based on user intention and attention. It includes *explicit interaction*, where a user issues an explicit command to a smart home computing system in order to achieve a specific purpose; *perceivable interaction*, where the smart home system activates an IoT device based on an acquired sensor input, which presents a clear output that captures user attention; and finally *calm interaction*, where an observed sequence of user actions is interpreted by the smart home system and automatically applied to perform actions (outputs) that are invisibly and seamlessly integrated with the home environment.

3.3.3 Context-Aware Computing

Context-aware computing is fundamentally concerned with the problem of context capture, representation, and adaptation in interactive systems and has been the focus of several studies in human-computer interaction, artificial intelligence, and ubiquitous computing. At the core of most of the research is an investigation of what context is and what it means for a system to be context-aware. We do not intend to carry out a complete review of context in this paper; interested readers might refer to [12, 17, 19, 20] as well as [21] for a survey of context-aware middleware architectures. Here, we analyze two threads of inquiry that have become prominent in discussions of context, which are particularly relevant in discussing middle architectures for context-aware interaction.

The first thread of inquiry, which we refer to as the bottom-up approach, views context as "any information that can be used to characterize the situation of an entity, i.e., a person, place, or object that is considered relevant to the interaction between a user and an application, including the user and applications themselves" [17]. This approach relates to the more traditional view of context as a concept for characterizing and representing physical objects in computational forms, i.e., identify objects within an environment that are relevant to given situation, and consequently encode them in a formalism appropriate for computer-based representation and interpretation. Thus, a system is context-aware if it uses context to provide relevant information and/or services to the user, where relevancy depends on the user's task [17]. However, the problem with this approach is that it delimits the scope of the problem of context-aware computing to that of "search- and location-based services" [19]. Interestingly, the dimensions of context are much larger than physical location. For instance, Kirsh [20] notes that in tracking context, we need to go beyond the superficial attributes of who and what is and where and when, to consider the highly structured amalgam of informational, physical, and conceptual resources that comprise "the state of digital resources, people's concepts and mental state, task state, social relations, and the local work culture" (p. 305). To be context-aware, systems and services in a smart home environment must go beyond the physical dimensions of context; they must understand the elderly users – what they know (e.g., their interests), who they know, who and what they trust, and what they value [18].

The second thread of investigation of context in computer science relates to the top-down view of context. Central to this thread is the concept of embodied interaction [11], which argues that context emerges out of interaction among objects in the world. In this perspective, context is viewed not as information but as a relation, and, as people interact with objects or devices in the world, actions do not occur in specific contexts, but rather context is actively produced, maintained, and enacted as interactions unfold [11]. What is regarded as context is therefore not a static feature that can be captured and encoded, but a dynamic feature that is determined by the setting, actors, and interaction. The interactional view of context is a notion drawn from social science and makes the following fundamental assumptions, namely, (1) context arises from activity, (2) context is an occasioned property, (3) contextuality is a relational property, and (4) the scope of contextual features is defined dynamically [12]. From the perspective of human-computer interaction, designing from the viewpoint of embodied interaction provides strong underpinnings for facilitating the process of designing for errors, making interactions visible and transparent, designing for reduced user cognitive load, and maintaining consistency across a range of application interfaces.

The distinction between these two threads of investigation of context reflects the essential difference between context as a representational problem and context as an interactional problem. While the view of context as a representational problem allows us to analyze context as a set of changing features within a smart home environment, relating to context as an interactional problem would enable us to adequately understand a set of user actions that correspond to a single information need. The foundation for the cognitive middleware for context-aware interaction in a smart home constitutes a modeling of the way elderly users experience their engagement with IoT-enabled devices in a smart home, such that a given set of user actions identifies coherent contexts and predicts user need. The prediction process would center around the following basic concepts:

1. *Habits:* Habits are actions that an elderly user takes on a regular basis, for example, reading books, cooking, etc. We identify two categories of actions that give rise to habits: short-term actions and long-term actions. Habits are important for determining and modeling lifestyle of an elderly person.
2. *Interests:* Interests include changing needs of an elderly user and like habits may occur on short- (i.e., may span days, weeks, or few months) or long-term basis (i.e., may span several months or years or often may not have an end).

We model an elderly person's interaction within a smart home environment using the concept of an *interaction landscape*. The idea of an interactive landscape refers to the structure cocreated as a result of the goal-oriented interactions among users and IoT devices in a smart home environment, for example, in the process of addressing a given user need. Interactive landscape draws upon related concepts, such as activity landscape [25], task environment, and enactive landscape [24]. It is meant to capture the goal- or activity-dependent nature of a sequence of interactions between users and/or autonomous devices.

3.4 Why Middleware for Context-Aware Interaction?

A middleware denotes a software framework used to abstract the heterogeneity of low-level system layers (e.g., hardware, operating systems) for the purpose of reducing the complexity of developing high-level layers (e.g., applications). For example, in distributed applications middleware is traditionally used to provide services to software applications beyond those available from the operating system primarily by enabling a common protocol for communication and object sharing. In context-aware computing, the concept of middleware has been applied as a mechanism for context management, for example, to build context-aware applications which are able to adapt their services to users' changing needs and situations. In this work, our goal is to utilize middleware services – given the peculiar challenges elderly people face in interacting with digital technologies as a result of age-related impairments – to adapt interactions between IoT-enabled smart home devices and elderly home occupants to the changing needs of the user. In what follows, we outline a number of interaction requirements and challenges for smart homes for the elderly. The list is not exhaustive, but is rather aimed to provide further motivations for a cognitive middleware for context-aware interaction in IoT-enabled smart homes.

- **Constant interface:** Today's technology industry is famous for designing devices to last just for a couple years, and nearly every iteration comes with significant changes in the user interface. We change the way interfaces look and feel on a regular basis and constantly add new functions and new ways of interacting with digital devices. For example, computer input has shifted from keyboard, mouse, and conventional computer display to gestures, speech, body movement, haptic feedback devices, ambient sound generator, and large smart displays. We have, in less than a decade, transformed the design of mobile phones from a conventional communication device to a smart device with support for an infinite number of apps and functionalities. But, as people age they would prefer devices with a constant interface [3]. For example, elderly people with short-term memory loss typically struggle to adapt to rapid changes. A cognitive middleware for context-aware interaction in smart homes will potentially provide a layer of user interface consistency for an elderly person regardless of advances in device functions and features.
- **Lifestyle-oriented and personalized support:** Despite the growing trend to develop and deliver technology solutions fast and with rapid upgrades, research has shown that there is still a considerable need to create solutions that enable users to have long-lasting relationships with technology. A key challenge from the perspective of design is how to enable personalized support through lifestyle-oriented digital interactions.
- **Augmentative interaction:** A major concern in IoT-enabled smart homes for the elderly is to automatically map available services to user needs via context-aware user interfaces, how to manage device heterogeneity at the level of device communication, for example, through a common interaction protocol, and how to predict user intentions and appropriately augment user actions. Addressing these

concerns requires that our design approaches are able to provide answers to a number of fundamental questions, including what does a user want to do in a given context? What infrastructure services within the smart home would cater for those needs? How would the user want the services delivered?

- **Context-based multimodal interaction:** As noted in Sect. 3.3, interaction with digital devices in a smart home can take different forms and guises. They can be explicit (e.g., via keyboard) and can demand attention (e.g., through alert triggers) but can also be peripheral and even invisible [6]. As such, they can also use natural user interfaces such as conversational interfaces or gesture-based interaction. Several studies have shown that developing adaptive, natural, and multimodal user friendly human-computer interfaces is a major challenge in building future homes for assisted elderly living. A middleware with the capability to determine the best user interaction mode based on user context is, therefore, highly needed in today's IoT-enabled smart homes.

3.5 Proposed Model

In [22, 23], we presented the conceptual design of the ACTiVAGE (ACTiVe AGeing sErvices) system – a context-aware lifestyle-oriented framework for supporting personalized elderly care and independent living. The system combines an understanding of user lifestyles, personal preferences and beliefs, as well as knowledge of user context in order to offer a model of ICT-enabled support for independent living (Fig. 3.1). It includes systematic data capture and rigorous analytics for uncovering the actual and complete picture of the elderly person's lifestyles for the purpose of building a formal representation of the lifestyle concept for system design.

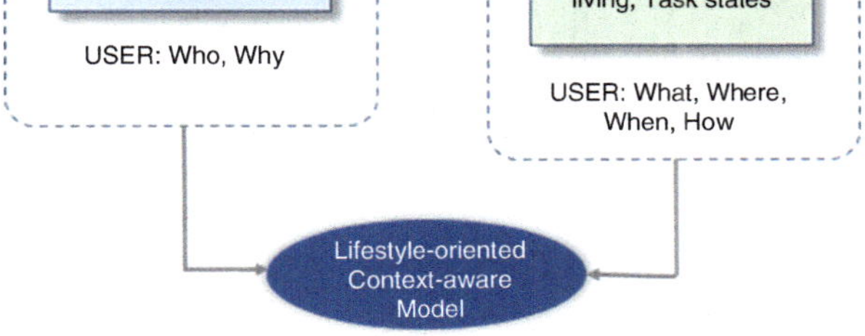

Fig. 3.1 Overview of the lifestyle-oriented context-aware model for the ACTiVAGE system

Drawing upon our experience in designing the ACTiVAGE system as well as the foregoing discussions in Sects. 3.3 and 3.4, we describe a cognitive middleware for supporting and managing interaction between users and intelligent devices in smart homes for the elderly. The goal of the middleware is to enable an elderly person to interact meaningfully with IoT-enabled devices in a smart home by providing a common protocol to ensure that interactions between elderly users and IoT-enabled smart devices are adapted to changing user contexts. As noted earlier, the foundation for the cognitive middleware proposed in this work is a user-centered situation model that, given a sequence of user actions, identifies coherent contexts.

3.5.1 Architectural Design

The architectural design for the cognitive middleware system consists of three layers, namely, the application services layer, the cognitive and context adaptation layer, and the data services layer (see Fig. 3.2). Overall, the middleware system is designed to enable the following capabilities:

1. Provide a common framework to support meaningful interaction among several IoT devices in a smart home based on knowledge of the elderly person's lifestyles, interests, and preferences.
2. Predict the most appropriate interaction mode (e.g., speech, gesture, human-in-the-loop, use of a graphic reminder or imagery, etc.) for servicing user needs in a given situation.
3. Provide explicit feedback loop to notify an elderly user about any actions before (feed-forward) and after action occurrence (feed-backward), e.g., by monitoring and displaying a set of context cues relevant to different user activities and situations.

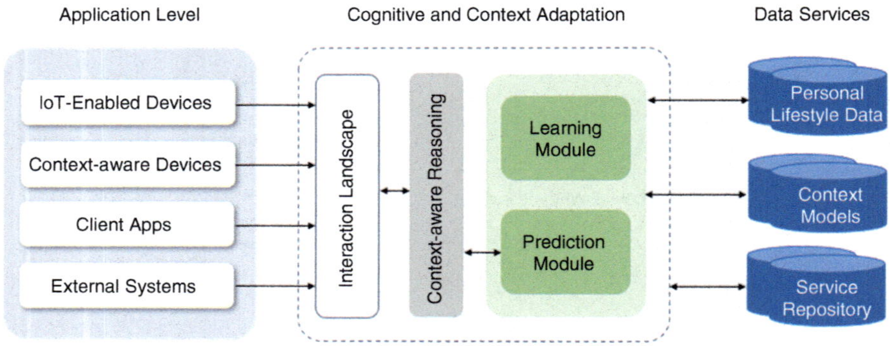

Fig. 3.2 High-level architecture of the cognitive middleware system for context-aware interaction

4. Provide alert and control capabilities to the elderly user as well as their family, friends, and caregiver, e.g., using different activity-driven messaging and collaboration services.

The cognitive and context adaptation layers include the interaction landscape, the context-aware reasoning module, and the learning and prediction modules. As noted earlier, the interaction landscape provides a structure for making sense of interactions among users and devices in an IoT-enabled smart home environment. It is responsible for gathering, processing, and interpreting the contexts implicit in interactions among various devices and/or users in a smart home, for example, gathering interaction contexts via raw sensor data (e.g., accelerometer, beacon, etc.) from IoT devices and processing by determining the category to which a specific interaction belongs (e.g., whether it is user-enacted, system-triggered, or embedded/calm interaction). In order to implement this structure in a computer system, a formalization of the interaction representation is needed. We propose the use of ontology frameworks, such as the top-level ontology framework for smart environments introduced by Ye et al. [26]. The framework allows linking of the meaning implicit in low-level information (such as a sensor detects a rise in room temperature) to higher-level information that is of interest to smart home applications. In addition, such a framework provides generic rules for facilitating generic tasks within a domain, including detecting inconsistency of information and generating new knowledge such as new activities and new relationships between devices, apps, systems, and contexts in the application level (Fig. 3.2).

The learning and predictive modules together form the cognitive module. The goal of the cognitive module is to predict the most appropriate interaction requirements and services for an elderly user based on in-depth mining of aggregated data from different sources of data that describe an individual's lifestyles, usual habits, and personal preferences based on prevailing contexts within the smart home. In the learning phase, data about users' lifestyles (stored in the personal lifestyle data store) are preprocessed into a set of feature vectors suitable for the design of active aging services [23]. The learning module discovers patterns in user's daily routine activities based on heterogeneous information about their lifestyles, interests, habits, locations, time, etc. For example, the K-means clustering algorithm could be implemented to discover such patterns. Subsequently, semantic annotation could be used to determine the best K clusters in combination with algorithmic methods, such as the elbow method and silhouette analysis.

Based on the patterns discovered during the learning phase, the prediction module determines mostly user interaction requirements and services for a given situation. This module could be implemented using artificial neural networks. Some of the suggestions of the prediction module will be automatic based on rules defined in the service repository. The context-aware reasoning phase is responsible for inferring relations between activities generated out of interactions in the interaction landscape and suggested actions by the prediction module. For example, *if the interaction landscape indicates that user has dozed off on the couch and its medication time, then call family member or caregiver.*

3.5.2 Discussion

As argued in the introduction, capturing, representing, and reasoning about the contexts in which interactions occur is a complex task. In addressing this problem, the cognitive middleware proposed in this paper seeks to enable the creation of relevant context events out of interactions among users and IoT devices in a smart home, which context-aware applications can subscribe to and be notified of for the purpose of providing adaptive services to satisfy user interaction needs. Consider once again the motivating example in Sect. 3.2. Models of John's usual habits, for example, his favorite meals, daily newspapers, and TV channels as well as the times for his favorite TV programs, will be stored in the personal lifestyle database in the data service layers. Models of John's home environments and activity contexts will be stored in the context models, whereas available services to meet John's needs will be stored in the service repository. Based on real-life data, the cognitive middleware system is able to predict if John is realizing his daily routine and what devices he needs at any point in time to achieve that effectively or if there might be apparent incidents that should be reported and who are the most likely people to contact. To achieve this requires models of human behavior and daily habits; however, big data analytics and the growing interest in collecting and analyzing data about several aspects of our life provide the means for realizing this [10].

This paper has focused on a high-level description of the proposed cognitive middleware architecture and lacks specific details about the implementation of the proposed architecture. Further work will focus on developing specific details of the framework as well as prototype and evaluate systems based on it within a real-world application context and will possibly leverage a number of off-the-shelf middleware platforms, such as SOCAM, Gaia, the Context Toolkit, and Cobra [21]. Our goal here has been to highlight a number of issues from the perspectives of analytics, context awareness, and cognitive computing that need to be addressed for the architectural design of middleware systems that combine knowledge of user lifestyle and knowledge of user activity context for context-aware interaction in smart homes for the elderly. However, there are several challenges and problems to be considered, including privacy issues and the need for a standardized communication protocols among various IoT devices.

3.6 Conclusion

Realizing many of the challenges of independent living for the elderly requires middleware architectures to be not only "context-aware" but also "smart and cognitive." The central goal of the middleware system is to introduce a cognitive mechanism – that learns the habits, lifestyles, and preferences of an elderly person – into the traditional middleware so as to provide a homogeneous interface involving interactive context management capabilities for adapting application behaviors to the changing environments and interaction requirements of an elderly person in a smart home. By leveraging the ACTiVAGE framework to offer a set of lifestyle-oriented active aging e-services, this approach will make a novel contribution to healthcare for the elderly through an approach that exploits the lifestyle concept integrated with middleware architectures for context-aware computing as a foundation for enabling new interaction design approaches for elderly people in smart homes, ultimately improving the usability of the smart technology for elderly care.

[a] http://www.marketsandmarkets.com/PressReleases/home-automation-control-systems.asp

References

1. Schneiderman, B., Plaisant, C., Cohen, M., Jacobs, S., Elmqvist, N., & Diakopoulos, N. (2016). Grand challenges for HCI researchers. *ACM Interactions, XXIII*(5), 24–25.
2. Eckl, R., & MacWilliams, A. (2009). Smart home challenges and approaches to solve them: A practical industrial perspective. Proceedings of Intelligent Interactive Assistance and Mobile Multimedia Computing: International Conference, IMC 2009, Rostock-Warnemünde, Germany, November 9–11, Springer, pp. 119–130.
3. Ebling, M. R. (2014). Pervasive change challenges elders. *IEEE Pervasive Computing, 3*(2), 2–4.
4. Callaghan, V., Zamudio, V., & Chin, J. S. Y. (2009). Understanding interactions in the smart home, *PerAda Magazine*.
5. Kolberg, M., Magill, E., Marples, D., & Tsang, S. (2002). Feature interactions in services for Internet personal appliances. *IEEE International Conference on Communications, ICC 2002, 4*, 2613–2618.
6. Kim, H. S., Kim, H. C., & Ji, Y. G. (2015). Understanding the Elders' interaction with smart home in Korea. In *Proceedings of the 33rd Annual ACM Conference Extended Abstracts on Human Factors in Computing Systems (CHI EA '15)* (pp. 2067–2072). New York: ACM.
7. Dewsbury, G., and Taylor, B., & Edge, M. (2001). *The Process of Designing Appropriate Smart Homes: Including the User in the Design*. Workshop on Ubiquitous Computing for Domestic Environments, Nottingham University.
8. Rodden, T., & Benford, S. (2003). The evolution of buildings and implications for the design of ubiquitous domestic environments. In *Proceedings of the SIGCHI Conference on Human Factors in Computing Systems (CHI '03)* (pp. 9–16). New York: ACM..
9. Moeller, S., Engelbrecht, K., Hillmann, S., & Ehrenbrink, P. (2014). New ITG guideline for the usability evaluation of smart home environments. In Speech Communication; 11. ITG Symposium; Proceedings of (2014), pp. 1–4.
10. Anya, O., & Tawfik, H. (2016). Leveraging big data analytics for personalised elderly care: Opportunities and challenges. In D. Al-Jumeily, A. Hussain, C. Mallucci, & C. Oliver (Eds.), *Applied computing in medicine and health: Emerging topics in computer science and applied computing*. Amsterdam: Elsevier.
11. Dourish, P. (2001). *Where the action is: The foundations of embodied interaction*. Cambridge: MIT Press.
12. Dourish, P. (2004). What we talk about when we talk about context. *Journal Personal and Ubiquitous Computing, 8*(1), 19–30.
13. Alam, M. R., Reaz, M. B. I., & Ali, M. A. M. (2012). A review of smart homes—Past, present, and future. *IEEE Transactions on Systems, Man, and Cybernetics, Part C (Applications and Reviews), 42*(6), 1190–1203.
14. Chan, M., Estève, D., Escriba, C., & Campo, E. (2008). A review of smart homes: Present state and future challenges. *Computer Methods and Programs in Biomedicine, 91*(1), 55–81.

15. Robles, R. J., & Kim, T. (2010). Review: Context aware tools for smart home development. *International Journal of Smart Home, 4*(1), 1–11.
16. Ni, Q., Hernando, A. B. G., & de la Cruz, I. P. (2015). The Elderly's independent living in smart homes: A characterization of activities and sensing infrastructure survey to facilitate services development. *Sensors, 15*, 11312–11362.
17. Abowd, G. D., Dey, A. K., Brown, Peter J., Davies, N., Smith, M., & Steggles, P. (1999). Towards a better understanding of context and context-awareness. In Hans-Werner Gellersen (Ed.), *Proceedings of the 1st international symposium on Handheld and Ubiquitous Computing (HUC '99)* (pp. 304–307). Springer-Verlag: Heidelberg.
18. Stajano, F., Rehman, K., & Coulouris, G. (2007). An architecture for interactive context-aware applications. *IEEE Pervasive Computing, 6*(undefined), 73–80.
19. Mehra, P. (2012). Context-aware computing: Beyond search and location-based services. *IEEE Internet Computing, 16*(2), 12–16.
20. Kirsh, D. (2001). The context of work. *Human Computer Interaction, 16*(2–4), 305–322.
21. Li, X., Eckert, M., Martinez, J., & Rubio, G. (2015). Context-aware middleware architectures: Survey and challenges. *Sensors, 15*, 20570–20607.
22. Tawfik, H. & Anya, O. (2013). Towards lifestyle-oriented and personalised e-services for elderly care. In: *Proceedings of the 6th conference on Developments in eSystems Engineering (DeSE)*. UAE: IEEE; 2013. p. 357e60. http://dx.doi.org/10.1109/DeSE.2013.69.
23. Anya, O., Tawfik, H., & Secco, E. L. (2016). A context-aware architecture for personalized elderly care in smart environments, *1st EAI International Conference on Smart Grid Inspired Future Technologies*, Liverpool, UK, May 19–20, 2016.
24. Kirsh, D. (2013). Embodied cognition and the magical future of interaction design. *ACM Transactions on Computer-Human Interaction (TOCHI), 20*(1), 30. doi:10.1145/2442106.2442109.
25. Anya, O., Tawfik, H., Amin, S., Nagar, A., & Shaalan, K. (2010). Context-aware knowledge modelling for decision support in e-health, *The 2010 International Joint Conference on Neural Networks (IJCNN)*, Barcelona, pp. 1–7.
26. Ye, J., Stevenson, G., & Dobson, S. (2011). A top-level ontology for smart environments. *Pervasive and mobile computing, 7*(3), 359–378. doi:10.1016/j.pmcj.2011.02.002.

Chapter 4
A Path Planning Approach of an Autonomous Surface Vehicle for Water Quality Monitoring Using Evolutionary Computation

M. Arzamendia, D. Gregor, D.G. Reina, and S.L. Toral

4.1 Introduction

Ypacarai Lake is the greatest inland freshwater of Paraguay, located 25 km east from the capital city, Asuncion (Fig. 4.1). According to the latest bathymetry studies performed by the Hydroinformatics International Center (CHI) of the Itaipu Dam, the area of the lake is about 64 km^2 [1]. Its relevance is given by the strategic importance for the preservation of local ecosystems, like the wetlands, and also because it is a main reservoir of freshwater for the local population.

As a consequence of the continuous disposals of wastes without treatment from industries and cities around the lake particularly in the last decades, the lake has suffered from periodic appearance of cyanobacteria. Since 2012 it has been in an aggressive form. This problem is also known as the blue-green algae bloom and not only gives a fetid smell to the lake, but also it is considered toxic for human health.

The solution will require many actions in a long-term period, and a continuous monitoring of the conditions of the lake should accompany this process. Since 2014, manual monitoring has been performed every 2 months at six locations in the lake. Additionally, three fixed monitoring stations provide continuous monitoring [2]. However, an accurate characterization of the lake is still limited. To solve this shortcoming, this work proposes a more flexible monitoring solution with the help of an autonomous surface vehicle (ASV).

M. Arzamendia • D. Gregor
Research Department, Faculty of Engineering, National University of Asuncion, San lorenzo, Paraguay
e-mail: marzamendia@ing.una.py; dgregor@ing.una.py

D.G. Reina (✉) • S.L. Toral
Electronic Engineering Department, University of Seville, Seville, Spain
e-mail: dgutierrezreina@us.es; storal@us.es

M. Dastbaz et al. (eds.), *Technology for Smart Futures*,
DOI 10.1007/978-3-319-60137-3_4

Fig. 4.1 Location of Ypacarai Lake

ASVs have been already considered for performing environmental monitoring tasks [3]. These vehicles are similar to ships; however, they do not require a crew on it to operate it. Consequently, their size may be smaller than a conventional one. This leads to a reduction of costs not only in the construction of the vehicle but also for its operation. The ASVs have been developed in different environments like academic labs, corporation, and government; for applications like bathymetric mapping, surveillance, or scouting; and for robotic research [4]. Basically an ASV consists of a structure that floats over the water (hull) and the entire payload that goes with it, which are mainly electronic components to perform control and measurement operations. The most typical structure in the ASV is the catamaran type (Fig. 4.2).

The ASV can operate via a remote control (RC). However, it is also expected that it travels along a predefined path, which is determined by way points. As the ASV travels along its path, it will be measuring the conditions of the lake with the help of limnological sensors. The number of possible paths is extremely large, so the problem can be viewed as a combinatorial problem. Particularly, this problem can be simplified as a traveling salesman problem (TSP) [5], but instead of cities, there are a number of beacons installed equally spaced along the border of the lake.

Fig. 4.2 The ROAZ II (2007), an example of ASV [3]

The beacons are data collection points of the information retrieved by the ASV. By maximizing the distance traveled, instead of minimizing it as the classic TSP, we would obtain a good coverage of the lake for evaluating its conditions. TSP is not a trivial problem, so brute force algorithms are inadequate to find the best solution. Instead, metaheuristics have been effectively used to find a good enough solution.

Genetic algorithm (GA) [6] is one of the mature metaheuristic optimization techniques, which has been frequently and effectively applied to a variety of problems from different research disciplines. It is an evolution-inspired search method developed by John Holland in the 1960s and adopts the basic Darwinian principle of "Survival of the fittest." It can be applied to a large variety of problems as long as their possible solutions meet the basic requirements of being encoded as a combination of their building blocks and having an objective function.

The main contributions of this book chapter are:

- The path planning of an ASV to monitor the Ypacarai Lake conditions using the TSP model and an evolutionary approach
- The demonstration of the goodness of the proposed path planning through a simulation performance evaluation

This chapter continues as follows. Section 4.2 presents the backgrounds of this research. It starts describing the problem of harmful algal bloom monitoring in lakes, then an overview of autonomous surface vehicles is provided and finally this section ends with some related work of ASV mobility based on AI (Artificial Intelligence). Section 4.3 describes of the problem of Ypakarai Lake which motivates this research and the proposed approach to deal with this problem. In section 4.4 the simulation environment and the obtained results are discussed. Finally, the conclusions of this work are presented in section 4.5, as well as some considerations about future work.

4.2 Backgrounds

4.2.1 Harmful Algal Bloom Monitoring in Lakes

4.2.1.1 The Environmental Problem

Bloom-forming cyanobacteria (CyanoHABs) are harmful from environmental, ecological, and human health perspectives by outcompeting beneficial phytoplankton, creating low-oxygen conditions (hypoxia, anoxia), and producing cyanotoxins [7]. The presence of cyanobacteria is a symptom of the degradation of the lake that is in a eutrophic state, i.e., with an excess of nutrients (phosphorus and nitrogen) that facilitates the appearance of plants. Eutrophication is mainly due to discharge of raw sewage into the lake, dumping of domestic and industrial waste, and fertilizer and chemicals from farms. Global warming has also a positive influence in their appearance. There are two types of cyanobacteria that are particularly toxic, the *Microcystis* sp. and *Anabaena* sp. These produce the microcystin toxin which affects the liver. There have been several severe cases of algae blooming, for example, in Lake Victoria (Africa) [8], Lake Erie (USA) [9], Lake Taihu (China) [10], etc. In Lake Erie, one of the great lakes located at the north of USA, suffered a blooming during 2011, reaching a peak of 5000 km^2 (of the 25,662 km^2 of the lake) [8]. Levels of microcystin were found from 0.1 ug/L (micrograms per liter) to 8.7 ug/L, while the World Health Organization (WHO) recommends the exposure to 1 ug/L. In [8] it is concluded that trends in agricultural practices, increased precipitation, weak lake circulation, and quiescent conditions yielded to the record-breaking blooming. In Lake Taihu (2,338 km^2), the main water repository for 10 million people, the levels of microcystin reached up to 44 ug/L in the water [9]. A prediction system was developed with the help of 18 fixed monitoring systems and periodic manual sampling. The fixed stations contain sensors for water temperature, turbidity, PH, chlorophyll a, phycocyanin, conductivity, dissolved oxygen, and other meteorological measurements. From all these, the phycocyanin sensor is the main indicator of the presence of cyanobacteria in the lake. The predicted model included algal biomass (chlorophyll a concentration), wind velocity, and precipitation. The effectiveness of the predicted model was compared with satellite image data, resulting in an accuracy of 82.2 ± 11.7%.

4.2.1.2 Hydrography Characteristics in Paraguay

Rivers are very valuable resources in a landlocked country like Paraguay. Two large rivers traversed the country, the Paraguay and the Parana Rivers. Many tributaries are derived from them that benefit one of the main economic activities which is the agriculture. They are not only for the agriculture but also because historically they have been the main transport mean for international trading and also a natural border with the neighbor countries. These reasons motivate the interest for developing an ASV to be applied in the following ways: first, for monitoring the quality of water of the river, not only because of its importance for the agriculture but also as a scarce nature resource that is indispensable for human life (only 3% of all the water in the planet is freshwater); second, as a bathymetry tool to help the barge convoys during the drought season of the year when the level of the river is low, minimizing the risk of strand in shallow water; and finally, as a surveillance vehicle at the borders, to prevent illegal activities like smuggling. For example, the reservoir of the hydroelectric Itaipu Dam, located at the border between Paraguay and Brazil, is surrounded by two national parks of 43,044 hectares which is a region prone to smuggling, drug trafficking, and poaching.

4.2.2 Autonomous Surface Vehicles

4.2.2.1 Types of ASV and Their Applications

Autonomous surface vehicles (ASVs), autonomous surface craft (ASC), and unmanned surface vehicles (USVs) are terms used to refer to the same type of vehicle, that is, vehicle capable of traveling over bodies of water (sea, rivers, or streams) without onboard crew. Throughout this work the term ASV will be adopted. The ASVs are intended to be used in a variety of applications in the aquatic environment. Typical applications include exploration, surveillance, environmental monitoring, search and rescue (SAR), and bathymetry.

ASVs are similar to other autonomous vehicles, like unmanned aerial vehicles (UAVs), unmanned ground vehicles (UGVs), or unmanned underwater vehicles (UUVs); however, they are adapted to aquatic environments. Even though it does not have the popularity of UAVs, they are less complex because they move in a 2D plane and not in a 3D plane like UAVs or UUVs. UUVs operate also in aquatic environments; however, they have limitations with respect to their communication range during their operation because electromagnetic waves are highly attenuated. Acoustic communication is used instead. In some cases, ASV can provide support to the UUV by acting as a communication relay with the remote control center at the coast.

The structure of an ASV is composed of a hull, which can be a kayak, catamaran (twin hull), and trimaran (triple hull). Then, there are the propulsion and power system provided by a propeller or water jet and a rudder for the direction of the vehicle. The guiding, navigation, and control (GNC) system is the main component as it consists of the hardware and software for managing the ASV. The data collection equipment is composed of the sensors that are related to the GNC, such as the inertial motion units (IMUs) and the global positioning system (GPS) in addition to cameras, radar, and sonar, and also the sensor related to the application that the ASV is used for. For example, sensors such as PH, temperature, dissolved oxygen, turbidity, etc. are used to evaluate the quality of water [11].

One of the first recorded ASVs in the literature is the Deep Ocean Logging Platform with Hydrographic Instrumentation and Navigation (DOLPHIN) developed by International Submarine Engineering Ltd., in 1983 in Canada. Since the 1990s ASVs acquire more relevance with the development of several prototypes, like the ARTEMIS or SCOUT developed at MIT. During the first decade of the new millennium, other countries also developed their own ASVs at their universities, for example, the ROAZ I and II (Portugal), SESAMO (Italy), or SPRINGER (UK). Since the DOLPHIN, 56 prototypes of ASVs have been surveyed in Liu et al. [11].

Most of these ASVs are stand-alone vehicles with a length of few meters. But instead of using a single vehicle for achieving the tasks, a concept of using a team of vehicles was introduced in this decade. These teams of robots could be heterogeneous or homogeneous teams. Heterogeneous teams are like the one presented in Pinto et al. [12] where a team of robots is composed of a UAV and an ASV for riverine environmental applications. This ASV is a catamaran of 4.5 m that piggybacks a small-sized UAV. The idea behind this association is that the UAV extends the perception capability of the ASV, i.e., the field of vision (FOV). From the information of both vehicles, a cost map is built for navigation. In Machado et al. [13] two types of ASV are tested for SAR operations under the EU-ICARUS project. It involves the ROAZ, a catamaran-type ASV, and the SWIFT ASV, a water jet-based ASV. The ROAZ with a greater size and autonomy will carry the SWIFT near the disaster point. When this point is reached, the SWIFT gets off the ROAZ and will be directed to the disaster point with high speed, carrying on a life raft for four persons. A third example of heterogeneous network is seen in Busquets et al. [14] where they propose the operation of an ASV near and AUV to improve the communication capabilities of the AUV with the command ground center at the coast. Homogeneous teams present a different approach. The teams consist of similar ASVs which form a swarm. But differently from the first ASVs built, the ASV members are smaller with less computation capabilities. In this way the cost of building a swarm is balanced with the cost of deploying one single ASV. The advantage of this strategy is that the tasks assigned to the swarm can be finished faster. Valada et al. [15] have introduced a Cooperative Research Program (CRP) intended to work in environmental monitoring of lakes and rivers. They performed tests with teams of up to five ASVs in the Taal Lake (Batangas, Philippines), Gowanus Canal (New York, USA), and Allegheny River (Pennsylvania, USA). Another swarm of ASVs is presented in Duarte et al. [16]. They propose neural network-based controllers for robotic tasks in a swarm of ten ASVs. They achieve similar performance of the simulated controllers. Field tests in a small water body connected to the Tagus River (Lisbon, Portugal).

4.2.2.2 ASV for Environmental Monitoring

As the focus of this work is environmental monitoring, then related works of this type of application are reviewed. Sukhatme et al. [17] has been one of the first works to implement a monitoring system in a lake of the microbial communities in an aquatic ecosystem. The system consists of a mix of ten stationary buoys and one ASV. While the buoys provide high-resolution temporal sampling and low-resolution temporal sampling, the mobile boat provides high-resolution spatial sampling. The devices are equipped with temperature and chlorophyll sensors. Dunbabin et al. [18] also propose a heterogeneous system of an autonomous catamaran integrated to a 50-node floating sensor network for water quality monitoring in a lake. The catamaran is built with an arm that can take samples from down to 20 m in the water. Also, the arm has a multipurpose probe that can measure environmental variables such as temperature, conductivity, chlorophyll, turbidity, dissolved oxygen, and incident radiation. The wireless sensor network provides non-line-of-sight communication with the remote operators. Regarding the problem of algae bloom in lakes, Pradalier et al. [19] present a novel ASV designed to take measurements within a range of depths with its custom-made winch. The paper describes the conception and also shows field results on the way-point navigation mode. Also, the first results of a sampling campaign for monitoring algae blooms in Lake Zurich are presented. Combining surface mobility with depth measurements within a single robot allows a fast deployment in remote location, which is a cost-efficient solution for lake sampling.

4.2.3 *Mobility of ASV Based on AI*

4.2.3.1 Path Planning

Robot path planning is one of the main research topics in robotics. It can be defined as the problem of finding a route between two positions in a mobile space, considering that the route should be collision-free (feasible) and satisfy certain optimization criteria [20]. The criteria can be the minimization of distance, time, or energy consumption; however, Euclidean distance is normally the first condition because it is related to other aspects such as time and energy (minimum path length). In most cases, reducing distance also represents reducing time and energy.

The planning can be labeled global or local. Global is considered when there is complete knowledge of the space in which the vehicle will move, i.e., all the feasible and unfeasible regions in the area are known. If there is only partial knowledge of the environment or if it is a dynamic environment, then local planning is performed. The global and local planning are also referred as offline and online planning, because in the first case, all the planning is calculated before the robots start moving with all the information available. As a difference, the online planning is calculated as the vehicles travel in the environment, which means that online planning needs a faster response than offline planning. In other words, online planning has a reactive behavior. Usually a combination of both is required; an initial path will be defined by the global planning, but an online planning is also carried when unexpected situations occur, like an unforeseen obstacle.

The final path solution consists of a sequence of intermediate way points. These way points can be defined by dividing the environment in a uniform grid or lattice. The way point could be represented as the center of each of the squares that forms the grid. The simplest solution consists of the way points that are closest to the straight line between the source and destination; however, this is not the normal case because the environment presents constrains in the form of obstacles or unfeasible regions. There are many approaches to solve this problem. For example, graph search algorithms like Dijkstra's algorithm or its heuristic derived A* (A-star) algorithm are used. The disadvantage of these algorithms is that they perform an exhaust search, and the execution time increases exponentially with the size of the problem

[21]. An alternative is to use metaheuristic algorithms which balance the execution time and the optimality of the solution. In this group are included Tabu Search (TS), ant colony optimization (ACO), genetic algorithms (GAs), and particle swarm optimization (PSO). These algorithms might find a solution of a path with sharp edges. If a smoother path is desired, then Bezier curves or B-spline curves can be used at the expense of introducing greater complexity [22].

Another very popular technique is the use of artificial potential fields, in which the movements of the robot (represented as a particle) are governed by a field, which is usually composed of two components, an attractive potential drawing the robot toward the goal and a repulsive potential pushing the robot away from obstacles. The main drawback with potential field techniques is their susceptibility to be trapped at local minima [23].

The metaheuristic approach is chosen in this work due to its simplicity and efficiency and particularly the genetic algorithms, which belong to the group of evolutionary algorithms (EAs).

4.2.3.2 Path Planning as the TSP

In the previous section, it was said that path planning consists of finding a sequence of way points while minimizing some characteristic like the distance. The traveling salesman problem (TSP), a famous mathematical problem, resembles very much this idea. The aim of the TSP is given a list of m cities, where the cost of traveling from city i to city j is c_{ij}, finding a route that visits all m cities once while minimizing the total cost of the route [24]. Despite TSP being very easy to describe, it is a challenging problem because of the complexity to find an optimal solution, due to the large number of possible solutions, expressed as $(m\text{-}1)!/2$. For instance, for a set of 15, cities have more than 43×10^9 solutions. This problem is known as non-polynomial (NP) complete problem because to date there is not a known polynomial time algorithm that can solve it [25]. Then brute force algorithms are impractical for use in this type of problems. Instead heuristics approach has been used giving satisfactory results.

TSP has been studied thoroughly, and it has become a benchmark where new combinatorial optimization methods are often applied. Currently, one of the main effective works on solving the TSP is suggested by Helsgaun [26], which is based on the Lin-Kernighan heuristic, and achieving optimal solutions for 13,509 city problems. There are different ways of mathematically representing the problem, but looking at it as a permutation problem, and considering P_n as the collection of all permutations of the set $\{1, 2, \ldots, n\}$, the TSP is the searching of $\pi = (\pi(1), \pi(2), \ldots, \pi(n))$ in P_n such that

$$c_{\pi(n)\pi(1)} + \sum_{i=1}^{n-1} c_{\pi(i)\pi(i+1)} \tag{4.1}$$

is minimized, where c is the cost or distance between two cities. This is known as the tour length [25].

There have been different works that have considered TSP as the base for the path planning problem. In [27], an AUV is used to collect the data from an underwater

WSN. They admit that the task of the AUV visiting the nodes is also a TSP and that there are many optimization techniques for solving it. However, due to significant error and uncertainty of the UUV while traveling underwater, they opt to consider the next node to visit the closest to the current one, not guaranteeing that the total path is minimized. Ergezer and Leblebicioglu [28] have considered as TSP one of the steps for its 3D path planning for UAVs for maximum information collection [3D path planning for UAVs for maximum information collection]. In the first step of their algorithm after selecting the desired regions (DRs), the visiting sequence is determined by pattern search, which they consider sufficient if a static environment is considered. Savuran and Karakaya [29] have applied the modeling of a constrained version of the TSP, which is the vehicle routing problem (VRP) to be used with a UAV and a mobile depot. The UAV should take off and land again in the same mobile depot after visiting a maximum number of fixed targets. Alnasser and Bennaceur [21] have evaluated the performance of the GA in a path planning using TSP and compared it with Tabu Search, showing the efficiency of GA in terms of solution quality and execution time.

4.2.3.3 Genetic Algorithms for Solving Path Planning Problems

Genetic algorithm is the most widespread of a group of optimization techniques known as evolutionary algorithm (EAs) or evolutionary computations (ECs) whose operation is based on biological processes [30]. There are techniques that have similar principles with GA and that were even proposed earlier, like evolutionary programming (EP) or evolutionary strategies (ES). Since then, these have inspired the development of many new ones.

As an optimization technique, GA has been applied for optimizing path planning, as in Alvarez et al. [31], where they study the path planning of an autonomous underwater vehicle (AUV) in dynamic environments. They introduce two new operators in the GA algorithm. First, the initial population is formed from the best individuals obtained from running the GA a few generations. The second operator is the incorporation of random immigrants into the new population at each generation. The path is found such that the energy of the AUV is minimized. Tuncer et al. [32] apply GA to a path planning problem of a mobile robot in dynamic environment, i.e., an environment with obstacles. In this work they also propose a mutation operator that increases the diversity of population and avoids a premature convergence. In the already mentioned work of Ergezer and Leblebicioglu [28], after solving the TSP for the DR, the GA is used to maximize a variable called collected information (CI) in the DR and at the same time avoid flying over forbidden regions (FR). Roberge et al. [33] compare the performance of the GA versus the PSO for the path planning of a UAV in 3D environment. Furthermore, they reduce the execution time by using the "single-program, multiple-data" parallel programming paradigm. After simulating in 3D real-based environments, it is observed that the GA performance is better in most cases. Ramirez et al. [34] have used a multi-objective GA for mission planning of a fleet of UAVs and a set of ground control stations (GCS). The algorithm has been tested optimizing different variables of the mission, such as the makespan, fuel consumption, and distance. It is seen that GA has been applied to solve the path planning in different types of autonomous vehicles like an AUV, UAV, and mobile robots. One of the objectives of the work is to expand the application of this technique in the environment of the ASVs.

4.3 Ypacarai Lake: A Case Study

During a monitoring campaign in 2005/2006, the presence of *Microcystis aeruginosa* was observed with levels above the 20,000 cel/ml recommended by the WHO. However, the levels of the microcystin did not surpass the safe levels (1 ug/L) [35]. Again, between October 2012 and April 2014, a new monitoring campaign was carried out. The results show elevated levels of nitrogen and phosphorus and the presence of *Microcystis aeruginosa*, especially during summer.

In 2014, with the objective of stopping the degradation, the government of Paraguay through the Itaipu Hydroelectric Dam provided support to several activities that will help in the recovery of the Ypacarai Lake. Among others, the activities were a new bathymetry study of the lake, the construction of a permanent laboratory of water in the city of San Bernardino, a consulting for the socioeconomic viability of a disposal treatment plant for the Yukyry stream, and a new monitoring campaign that will last for 2 years between 2014 and 2016. This campaign was assigned to the Multidisciplinary Technological Research Center (CEMIT) of the National University of Asuncion (UNA). It consists of performing 12 analyses operations every 2 months of the conditions of the Ypacarai Lake at six points and some streams that reached the lake (eight points). They are more detailed studies than the measurements of the fixed monitoring stations that provide physicochemical analysis of the water. In addition, the periodic monitoring provides biological analysis. The latest available report is from the tenth analysis in June of 2016. This report shows that the levels of phosphorus of all analysis (this one and the previous) have reached levels for considering the lake is in eutrophic state according to a definition of Vollenweider. However the quantity of cyanobacteria has varied at different times of the year [36].

4.3.1 Covered Problem

Ypacarai Lake is located approximately between latitudes 25° 15′ and 25° 22′ (14 km) south and between longitudes 57° 17′ and 57° 23′ west (10 km). The largest distance between two points at the shore is about 14 km. The area that should be covered is around 90 km^2. Its average depth is about 1.31 m with a maximum value of 3 m [1]. This simplifies the sampling mechanism when comparing with [17] and [18], because there is no need for submerging a probe for several meters under water.

For achieving the coverage task with a single drone, a wireless network of n beacons around the lake will serve as way points that the drone should visit at least once. For this particular case $n = 60$, as shown in Fig. 4.3. The distance between two

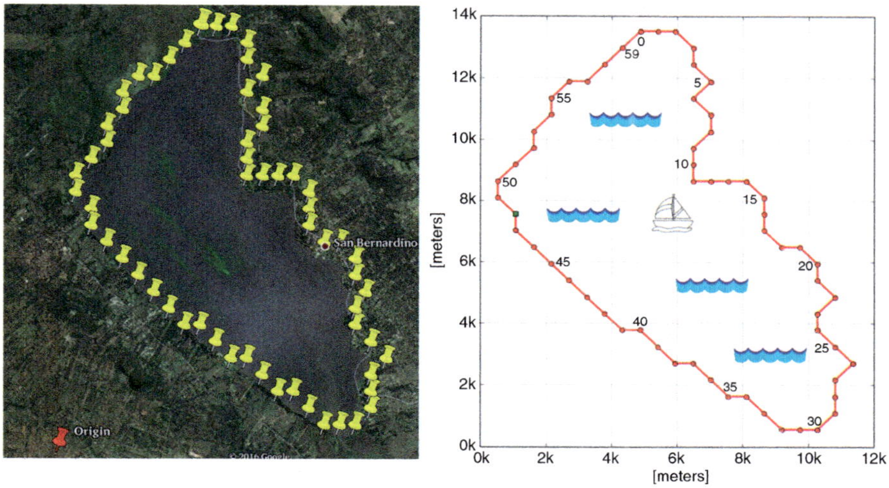

Fig. 4.3 (**a**) Beacon distribution and (**b**) numbering convention

consecutive beacons is less than 1 km. This distance is the expected wireless transmission range of the transceiver installed at the beacons.

In our problem, the beacons are not for reference only but also for receiving the collected data which is transmitted from the drone. Eventually the beacon will transmit updated information for the drone. Once the data is received by a beacon, it will be forwarded through a multi-hop route to the gateway. In Fig. 4.3b it is indicated as a green square.

In order to identify each beacon, they are assigned a number between 0 and 59. Fig. 4.3b shows a numbering convention that will be used throughout the paper. In this graph, the coordinates are translated from degrees, minutes, and seconds to meters, considering 1 s = 30 m. One unit in both axes of the graph is equivalent to 1 m. The origin of this graph is taken at point 25° 22′ 21″ S and 57° 22′ 57″ W, as shown as a red pin in Fig. 4.3. A candidate for the base station is beacon *b48* (W 57 22′ 21″ and S 25 18′ 9″), which is located at Aregua city. Another alternative is beacon *b18* (W 57 17′ 51″ and S 25 18′45″), which is located at San Bernardino city.

4.3.2 Proposed Approach

The GA technique is used to find the solution offline before the drone starts its travel. Another assumption is that there are no obstacles at the lake.

GA was coined by John Holland in his book *Adaptation in Natural and Artificial Systems* [1975]. One of Holland's students, Ken De Jong, in his doctoral thesis in the same year provided a thorough description of the capabilities of the GA. In 1985 another PhD student of Holland, David Fogel acquired notoriety with his thesis and the application of GA in gas pipeline optimization and later in 1989 with the book *Genetic Algorithms in Search, Optimization, and Machine Learning* [37].

When applying GA to an optimization problem, all candidates are encoded into finite-length strings of alphabets [38]. GA adopts many terms from biology to describe its operation. For example, the candidate solutions are called chromosomes, the alphabets genes and the values of the alphabet alleles. For the case of TSP, a chromosome represents a route and a gene a city. The optimization mechanism is given through several generations, in which at each generation new individuals (candidate solutions) are created from previous generations' individuals. Survival of the fittest describes the natural selection of the Darwinian evolution theory. The biological concept of fitness is defined meaning reproduction success, meaning that an individual with higher fitness has more probabilities of passing its genetic characteristics to future generations. Then a fitness function that describes the quality of the individual should be defined. If the optimization problem has only one parameter, then it is a single-objective problem (SOP), while if it has different parameters, it is a multi-objective problem (MOP). Normally in MOP the optimization parameters are conflicting, for example, in a two-parameter optimization problem, one parameter cannot be optimized without affecting the performance of a second parameter. For example, a multi-objective problem could be the maximization of the distance traveled by an autonomous vehicle while minimizing its energy consumed. A trade-off is necessary, and this is represented by the Pareto front that will allow the user to select the solution from a set of possible combinations (non-dominated points) [39].

There are two genetic operators that are applied to the existing population for the creation of new individuals, the crossover operator and the mutation operator. The crossover operator combines the genetic information of current individuals, while the mutation operator introduces new information through small modifications. In GA, crossover operator is considered the main operator. Another important concept is the population, which is the set of individuals to whom the operators are applied so that new individuals are incorporated to at each new generation.

Algorithm 1 presents the pseudocode of the GA used:

```
Make initial population
Compute the evaluation function of each individual.
WHILE NOT stop DO
        FOR Population_Size/2 DO
                Select parents from the population
                Produce children from the selected parents
                Mutate the individuals.
                Compute the evaluation function of each
                child
                Extend the population adding the children
                to it.
    Reduce the extended population.
        IF Stop Criteria THEN
                Stop = True
```

Fig. 4.4 Path representation (not necessarily in order)

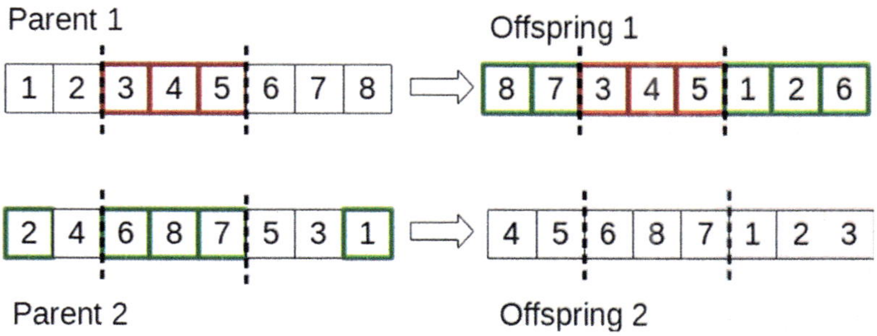

Fig. 4.5 Crossover operation

First, the population is created and the quality of each individual is evaluated. Then, the evolution is performed by creating new generations until a given stop condition is satisfied. This condition may be a fixed number of generations or not observing any improvements in the quality of individuals after a certain number of generations.

Each individual will be represented by the path representation, which is the most natural way to express the solution in a TSP. However, there are other types of representation, like the binary or matrix representations. The path representation indicates that in a list of n cities, if the city i is the *j-th* element of the list, then the city i is the *j-th* city to be visited. An example of an individual would be as shown in Fig. 4.4, where b_i is one of the 60 beacons of the network.

This representation is adequate if we would like to solve the classic TSP. However our problem includes further constrains; for example, a direct path from one beacon to any other is not always possible because the path might pass through land, as in the case of the path between beacon 30 and beacon 5 *(b30-b5)*. This restriction is implemented in the GA by penalizing it and assigning a fitness function equals of -1 to all the routes that are in the same situation.

The new generation starts with the selection of pairs of individuals to which a crossover operator will be applied. As a result, two new individuals are created with information of each parent. This type of crossover is used when the chromosomes are represented by an ordered list, and a direct swap may produce invalid individuals.

A simple example of the ordered crossover technique is next presented with two individuals represented by a list of eight unique elements (Fig. 4.5): First, two crossing points are selected, for example, between the third and fifth elements of the list for both parents. This is indicated in Parent 1 in red squares. This part will be copied exactly and at the same position in Offspring 1. The rest of the missing elements will be added starting from the second crossing point in the same order as they appear in the other parent (Parent 2). In Fig. 4.5, the missing elements are indicated

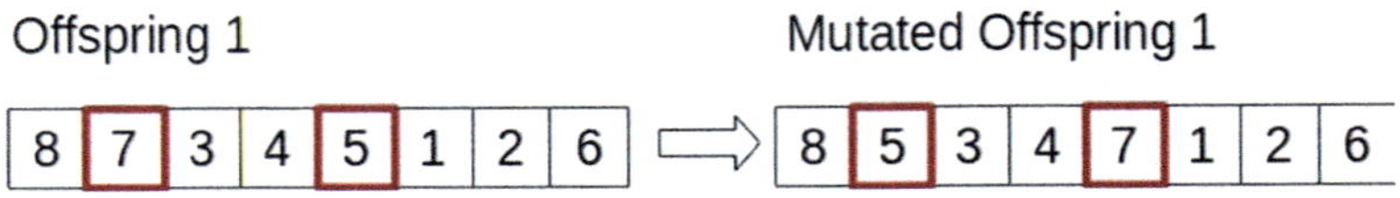

Fig. 4.6 Mutation operation

in green squares, and they are added to Offspring 1 following the order *1*, *2*, *6*, *8* and *7*. Offspring 2 is created in the same way, but it is based on Parent 2 and adds the missing elements as they appeared in Parent 1.

Finally, a mutation operator is applied. The operation selects two positions randomly and inter-exchanges their values between them. Figure 4.6 shows the application of the operator to the Offspring 1 obtained in Fig. 4.5.

Both operators, crossover and mutation, have certain probability to occur, Pc and Pm, respectively. While the crossover produces individuals with the information of their parents, the mutation is used to explore the vicinity of potential solutions obtained from the crossover. Next the fitness function is calculated of each new individual.

The fitness function is given by the total path length, which is the sum of distances d between a beacon and the next beacon to visit in the sequence defined by the individual. Deriving from Eq. (4.1), the fitness function is represented as Eq. (4.2):

$$fitness = d_{b_{n-1}b_0} + \sum_{n-2}^{i=0} d_{b_i b_{i+1}}, if\ route = valid - 1, otherwise \tag{4.2}$$

4.4 Simulation Results

This section includes the simulation results obtained. The section has been divided into two parts. First, we describe the simulation environment used to evaluate the proposed approach. Second, we show the simulation results obtained.

4.4.1 Simulation Environment

The problem was simulated in a Debian 7.9 server with the following specifications, an Intel Xeon v3 E5-2603 1.6 GHz CPU and 64 GB RAM memory. The model was implemented in Python version 2.7.6 [40], where the GA was implemented with the module Distributed Evolutionary Algorithm for Python [41] version 1.0.2.

The list of simulation parameters used is shown in Table 4.1.

Table 4.1 Simulation parameters

Parameter	Value
Number of simulations	100
Population size	100
Number of generations	5,000
Crossover	Ordered (OX1)
Selection	Tournament (size 3)
Mutation	Shuffle index (indpb = 0.05)
Crossover probability P_c	0.8
Mutation probability P_m	0.2
Elitism rate	0.2

The ordered crossover exploits a property of the path representation, which is that the order of cities (not their positions) is important [25]. As it was mentioned in Sect. 3, it constructs a sub-tour of one parent while preserving the relative order of cities of the other parent. The tournament selection with size 3 means that three individuals are selected and entered into a tournament against each other. The fittest individual in the group wins the tournament and is selected as the parent [38]. The shuffle index operator shuffles the attributes (genes) of an individual. Each element is exchanged with another element randomly according to an independent probability *inpb* (0.05) [42]. The elitism rate is the amount of the best population that will be preserved for the new generation. This mechanism is used to guarantee that the best individuals are not lost during the evolution procedure due to the probabilistic selection. The 20% of best individuals passes directly to the next generation. The population size selected is big enough to ensure a high initial exploration. The number of generations used is sufficient for the convergence of the genetic algorithm as it will be shown in the next subsection.

4.4.2 *Simulation Results*

The simulations of the GA applied to the classic TSP (no constrains) are compared with the GA applied to the TSP with some constrains included (land crossing constrains). Figure 4.8a presents the results of applying the GA to the TSP for the distributed beacons. The initial beacon is chosen randomly, and it is indicated as a red square, while the other beacons are blue dots. In this case, beacon *b21* is selected as starting point. It is observed that the lake is better covered around its center, because more routes pass through this area. This is expected because the best next beacon to visit is the one at the opposite side of the lake that has not been visited yet. After simulations, the fitness of the best individual, i.e., the distance traveled is 576,106 m. The evolution of the solution of the best individual is shown in Fig. 4.8a. Here it is seen that the solution converges after 1000 generations.

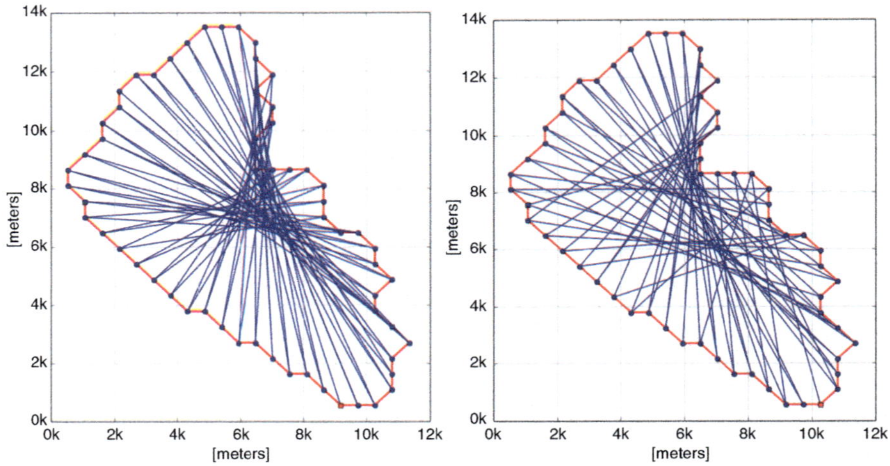

Fig. 4.7 GA applied to our problem, (**a**) no constrains and (**b**) land-crossing constrains

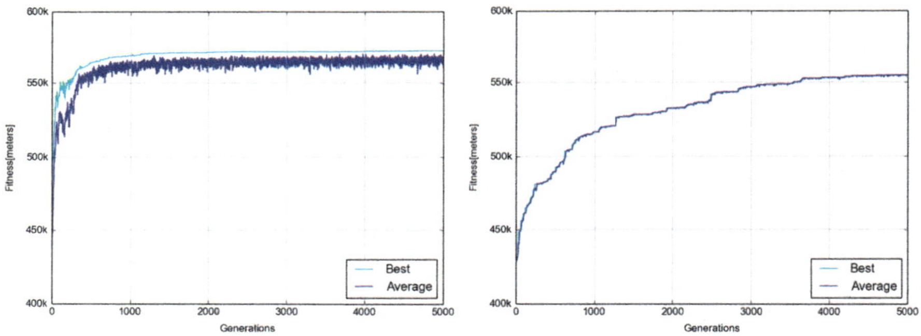

Fig. 4.8 GA applied to our problem, (**a**) no constrains and (**b**) land-crossing constrains

This solution includes some invalid routes, i.e., routes that cross the land. This restriction was included in the algorithm and simulated again, obtaining a solution shown in Fig. 4.7b. The initial beacon is beacon *b5*, and it is seen that all the routes remain inside the lake. However, the number of generations to achieve this solution is much higher than in the previous case, needing more than 3000 generations (Fig. 4.8b). The average solution is calculated considering only the valid individuals, and it is seen that this number is very close to best solution, almost overlapping it. Table 4.2 presents the statistics calculated from the multiple simulations. It is seen that due to the constrains of the lake, the traveled distance was reduced from 574,335 m down to 554,754 m, about a 5% reduction. It is important to notice that constrains are related to the shape of the lake.

A summary of these simulations are presented in Table 4.2. This table presents the following scenarios:

Table 4.2 Simulations comparison

Scenario	Generations for convergence[a]	Best solution (m)	Average (m)	Standard dev. (m)
1	1,000	576,106	573,740	1,152
2	3,000	554,754	539,655	5,354

[a]Approximately

- Scenario 1: GA on TSP without constrains
- Scenario 2: GA on TSP with land-crossing constrains

The statistical data is related to the simulation that achieved the best result of the 100 simulations set.

It is seen that in the unconstrained scenario (Scenario 1), there is a faster convergence and a better solution (longest path). However, introducing the constrains (Scenario 2) reduces the best solution, but still the algorithm is capable of finding a solution, even though it takes more generations to find it. The obtained results also show that the solution found is quite robust since the standard deviation is significantly low.

4.5 Conclusions

An overview has been given about the application of ASV as a tool for environmental monitoring in inland water rivers, especially as a tool that helps in diagnosing the quality of water to prevent the appearance of harmful algal blooms (HABs). Ypacarai Lake is an important lake that is exposed to the problem, and even though actions are already carried out about the monitoring of its conditions, still it requires further work, and it will benefit from the use of a new type of technologies like the ASV.

The path planning of an ASV is presented as a TSP, and GA is applied to solve this NP problem. A ring of beacons was proposed to be used as the cities of the TSP. However constrains were introduced because not all the routes between beacons (cities) are feasible. The simulations have shown that an optimal solution is found, even though constrains were introduced due to the shape of the lake. It was found that this was only represented in a reduction of 5% when compared to the solution with no constrains, and the number of generations for convergence was increased from 1000 to 3000. As future work, this problem will be translated from a single-objective problem (SOP) to a multi-objective problem (MOP). Other aspects like the energy consumption and the energy source available will be introduced in the problem. Also, other metaheuristics techniques like the particle swarm optimization or the ant colony optimization will be compared to evaluate the performance of the GA.

References

1. Hydroinformatics International Center (CHI), & Itaipu Hydroelectric. (2014). August. Technical Report – Bathymetry Studies of Ypacarai Lake.

2. Hydroinformatics International Center (CHI), & Itaipu Hydroelectric. (2016). Sistema de Monitoreo, Control y Estudios de la Cuenca del Lago Ypacarai [online]. Available at: http://hidroinformatica.itaipu.gov.py/gestiondecuenca/py/ypacarai/. Accessed 1 Nov 2016.
3. Yuh, J., Marani, G., & Blidberg, D. R. (2011). Applications of marine robotic vehicles. *Intelligent Service Robotics, 4*(4), 221–231.
4. Manley, J. E. (2008). Unmanned surface vehicles, 15 years of development. In *OCEANS 2008* (pp. 1–4). IEEE Press, Piscataway, NJ, USA.
5. Punnen, A. P. (2007). The traveling salesman problem: Applications, formulations and variations. In *The traveling salesman problem and its variations* (pp. 1–28). New York: Springer.
6. Chatterjee, S., Carrera, C., & Lynch, L. A. (1996). Genetic algorithms and traveling salesman problems. *European Journal of Operational Research, 93*(3), 490–510.
7. Paerl, H. W. (2014). Mitigating harmful cyanobacterial blooms in a human-and climatically-impacted world. *Life, 4*(4), 988–1012.
8. Ochumba, P. B., & Kibaara, D. I. (1989). Observations on blue-green algal blooms in the open waters of Lake Victoria, Kenya. *African Journal of Ecology, 27*(1), 23–34.
9. Michalak, A. M., Anderson, E. J., Beletsky, D., Boland, S., Bosch, N. S., Bridgeman, T. B., Chaffin, J. D., Cho, K., Confesor, R., Daloğlu, I., & DePinto, J. V. (2013). Record-setting algal bloom in Lake Erie caused by agricultural and meteorological trends consistent with expected future conditions. *Proceedings of the National Academy of Sciences, 110*(16), 6448–6452.
10. Qin, B., Li, W., Zhu, G., Zhang, Y., Wu, T., & Gao, G. (2015). Cyanobacterial bloom management through integrated monitoring and forecasting in large shallow eutrophic Lake Taihu (China). *Journal of Hazardous Materials, 287*, 356–363.
11. Liu, Z., Zhang, Y., Yu, X., & Yuan, C. (2016). Unmanned surface vehicles: An overview of developments and challenges. *Annual Reviews in Control, 41*, 71–93.
12. Pinto, E., Marques, F., Mendonça, R., Lourenço, A., Santana, P., & Barata, J. (2014). An autonomous surface-aerial marsupial robotic team for riverine environmental monitoring: Benefiting from coordinated aerial, underwater, and surface level perception. In *Robotics and Biomimetics (ROBIO), 2014 IEEE international conference on* (pp. 443–450). IEEE Press, Piscataway, NJ, USA.
13. Machado, D., Martins, A., Almeida, J. M., Ferreira, H., Amaral, G., Ferreira, B., Matos, A., & Silva, E. (2014). Water jet based autonomous surface vehicle for coastal waters operations. In *2014 Oceans-St. John's* (pp. 1–8). IEEE Press, Piscataway, NJ, USA.
14. Busquets, J., Zilic, F., Aron, C., & Manzoliz, R. (2013). AUV and ASV in twinned navigation for long term multipurpose survey applications. In *OCEANS-Bergen, 2013 MTS/IEEE* (pp. 1–10). IEEE Press, Piscataway, NJ, USA.
15. Valada, A., Velagapudi, P., Kannan, B., Tomaszewski, C., Kantor, G., & Scerri, P. (2014). Development of a low cost multi-robot autonomous marine surface platform. In *Field and service robotics* (pp. 643–658). Berlin/Heidelberg: Springer.
16. Duarte, M., Costa, V., Gomes, J., Rodrigues, T., Silva, F., Oliveira, S. M., & Christensen, A. L. (2016). Evolution of collective behaviors for a real swarm of aquatic surface robots. *PloS One, 11*(3), e0151834.
17. Sukhatme, G. S., Dhariwal, A., Zhang, B., Oberg, C., Stauffer, B., & Caron, D. A. (2007). Design and development of a wireless robotic networked aquatic microbial observing system. *Environmental Engineering Science, 24*(2), 205–215.
18. Dunbabin, M., Grinham, A., & Udy, J. (2009). An autonomous surface vehicle for water quality monitoring. In *Australasian Conference on Robotics and Automation (ACRA)* (pp. 2–4). Australian Robotics and Automation Association. Sidney, Australia.
19. Pradalier, C., Posch, T., Pernthaler, J., & Siegwart, R. (2012). Design and application of a surface vessel for autonomous inland water monitoring. *IEEE Robotics & Automation Magazine*, vol. 19, no.1, pp. 1–9.
20. Sedighi, K. H., Ashenayi, K., Manikas, T. W., Wainwright, R. L., & Tai, H. M. (2004). Autonomous local path planning for a mobile robot using a genetic algorithm. In *Evolutionary Computation, 2004. CEC2004. Congress* on (Vol. 2, pp. 1338–1345). IEEE Press, Piscataway, NJ, USA.
21. Alnasser, S., & Bennaceur, H. (2016). An efficient Genetic Algorithm for the global robot path planning problem. In *Digital Information and Communication Technology and its Applications*

(DICTAP), 2016 sixth international conference on (pp. 97–102). IEEE Press, Piscataway, NJ, USA.
22. Song, B., Song, B., Wang, Z., Wang, Z., Sheng, L., & Sheng, L. (2016). A new genetic algorithm approach to smooth path planning for mobile robots. *Assembly Automation, 36*(2), 138–145.
23. Pêtrès, C., Romero-Ramirez, M. A., & Plumet, F. (2012). A potential field approach for reactive navigation of autonomous sailboats. *Robotics and Autonomous Systems, 60*(12), 1520–1527.
24. Hoffman, K. L., Padberg, M., & Rinaldi, G. (2013). Traveling salesman problem. In *Encyclopedia of Operations Research and Management Science* (pp. 1573–1578). Springer. New York. USA.
25. Larrañaga, P., Kuijpers, C. M. H., Murga, R. H., Inza, I., & Dizdarevic, S. (1999). Genetic algorithms for the travelling salesman problem: A review of representations and operators. *Artificial Intelligence Review, 13*(2), 129–170.
26. Helsgaun, K. (2000). An effective implementation of the Lin–Kernighan traveling salesman heuristic. *European Journal of Operational Research, 126*(1), 106–130.
27. Dunbabin, M., Corke, P., Vasilescu, I., & Rus, D. (2006). Data muling over underwater wireless sensor networks using an autonomous underwater vehicle. In *Proceedings 2006 IEEE International Conference on Robotics and Automation, 2006. ICRA 2006.* (pp. 2091–2098). IEEE Press, Piscataway, NJ, USA.
28. Ergezer, H., & Leblebicioğlu, M. K. (2013). 3D path planning for UAVs for maximum information collection. In *Unmanned Aircraft Systems (ICUAS), 2013 International Conference on* (pp. 79–88). IEEE Press, Piscataway, NJ, USA.
29. Savuran, H., & Karakaya, M. (2015). Efficient route planning for an unmanned air vehicle deployed on a moving carrier. *Soft Computing,* vol. 20, no. 7, pp 2905–2920.
30. Alba, E., & Cotta, C. (2006). Evolutionary algorithms. In *Handbook of bioinspired algorithms and applications* (pp. 3–19). CRC Press. Boca Raton, USA.
31. Alvarez, A., Caiti, A., & Onken, R. (2004). Evolutionary path planning for autonomous underwater vehicles in a variable ocean. *IEEE Journal of Oceanic Engineering, 29*(2), 418–429.
32. Tuncer, A., & Yildirim, M. (2012). Dynamic path planning of mobile robots with improved genetic algorithm. *Computers and Electrical Engineering, 38*(6), 1564–1572.
33. Roberge, V., Tarbouchi, M., & Labonté, G. (2013). Comparison of parallel genetic algorithm and particle swarm optimization for real-time UAV path planning. *IEEE Transactions on Industrial Informatics, 9*(1), 132–141.
34. Ramirez-Atencia, C., Bello-Orgaz, G., R-Moreno, M. D., & Camacho, D. (2016). Solving complex multi-UAV mission planning problems using multi-objective genetic algorithms. *Soft Computing,* pp. 1–18.
35. Secretariat of Environment (SEAM), & Presidency of the Republic of Paraguay. (2014). Technical Report – Informaciones sobre las actividades humanas que condicionan la calidad de agua del rio Paraguay, lago Ypacarai y otros afluentes.
36. Multidisciplinary Technological Research Center (CEMIT), & National University of Asuncion (UNA). (2016). June. Technical Report – 10th Sampling Campaign from Ypacarai Lake.
37. Reeves, C. (2003). Genetic algorithms. In *Handbook of metaheuristics* (pp. 55–82). Springer. New York. USA.
38. Sastry, K., Goldberg, D. E., & Kendall, G. (2014). Genetic algorithms. In *Search methodologies* (pp. 93–117). Springer. New York. USA.
39. Reina, D. G., León-Coca, J. M., Toral, S. L., Asimakopoulou, E., Barrero, F., Norrington, P., & Bessis, N. (2014). Multi-objective performance optimization of a probabilistic similarity/dissimilarity-based broadcasting scheme for mobile ad hoc networks in disaster response scenarios. *Soft Computing, 18*(9), 1745–1756.
40. Van Rossum, G., & Drake, F. L. (2011). *The python language reference manual.* Network Theory Ltd. UK.
41. Fortin, F. A., Rainville, F. M. D., Gardner, M. A., Parizeau, M., & Gagné, C. (2012). DEAP: Evolutionary algorithms made easy. *Journal of Machine Learning Research, 13*(Jul), 2171–2175.
42. DEAP documentation. [Online]. http://deap.readthedocs.org/en/1.0.x/

Part II
"Smart Living" Case Studies

Chapter 5
Big Data and Data Science Applications for Independent and Healthy Living

Robert Keight, Dhiya Al-Jumeily, Abir Hussain, Paul Fergus, and Jamila Mustafina

5.1 Introduction

Paralleling the state of human progress, developments in healthcare reflect a deeply entrained drive to improve the parameters governing our own existence, including both those which threaten to disrupt our biological functions and followed by those which limit our ability to improve the effectiveness of the former [1, 2]. The technology of the past has allowed us to improve the conditions of our environment and to undertake limited medical interventions in the absence of a direct understanding of disease-causing mechanisms [2]. It is the arrival of the modern era that has opened unprecedented understanding of biological systems and disease mechanisms [3–9], yet such depth of knowledge has also brought a wider realisation of the full complexity and scale of the systems responsible for the biological processes underpinning our existence [10–13]. It is clear that in order to rise to the unprecedented challenges presented by such novel domains, the methods at our disposal must be advanced accordingly to support the changing nature of our task frameworks. The idea that representable forms of information processing may underpin familiar (and novel) forms of intelligence, such as the human brain, raises the possibility that intelligence itself may be practically simulated in alternative settings, for example, via computation, providing a capacity to sustainably address problems of arbitrary complexity. The field of intelligent systems, a research direction within the wider field of artificial intelligence (AI), is concerned with enabling the computational resources of today for the construction of systems that may respond to problems through intelligent abstractions, whose parameters differ from human

R. Keight • D. Al-Jumeily (✉) • A. Hussain • P. Fergus
Faculty of Engineering and Technology, Liverpool John Moores University, Liverpool, UK
e-mail: D.Aljumeily@ljmu.ac.uk

J. Mustafina
Kazan Federal University, Kazan, Russia

M. Dastbaz et al. (eds.), *Technology for Smart Futures*,
DOI 10.1007/978-3-319-60137-3_5

cognition. Through the combination of computational infrastructure and the patterns of intelligence in this way, it is conceivable that the frontiers of healthcare and medicine may be sustainably advanced to address the broad challenges that underpin the current era.

Within the healthcare setting, both information requirements and data generation volumes are growing at an ever-increasing rate, owing in particular to advances in low-cost sensor technologies and rapid progress in medically relevant areas of science, for instance, the completion of the Human Genome Project [14–23] and other ongoing initiatives [24–27]. The hallmarks of such a transition include the growth of emerging disciplines, in particular those formed at the intersection of medicine, science and computational techniques. Typifying examples include nanomedicine [28, 29], pharmacogenomics [30, 31], proteomics [32, 33], translational bioinformatics [34–36], computational neuroscience [37], network medicine [38], regenerative medicine [39] and others [40, 41]. Within this rapidly advancing setting, computation represents a key enabling factor within both mainstream healthcare operations and within the space of medically related scientific disciplines, mediating an increasing depth of integration between human participants and machine processes within task frameworks. Consequently, such reliance on tight integration and advanced forms of information handling necessitates a wider space of computational operations that must be supported autonomously if feasibility is to be maintained, given the deepening remit. In particular, the fidelity and breadth of biomedical data resources are growing exponentially, while the cognitive capability of human agency remains relatively constant. Such a rapid expansion in data acquisition capacity, however, represents an unprecedented opportunity within the medical sciences and mainstream healthcare operations, promising to realise aspirational scenarios such as those ascribed by personalised medicine [36, 42–50] and related paradigm P4 medicine (predictive, preventative, personalised, participatory) [51–54]. Significantly, the growing momentum of such ambition is accompanied by a wider shift in thinking, typified by the departure from reductionist views of medicine in favour of a systems perspective [55]. The emerging research of today therefore must embrace biocomplexity at all levels [33, 56–64], employing high-throughput and automated processing to address what may now be centrally framed as an information problem [65–73]. The advancement of intelligent systems, manifested through computation, therefore plays an essential role in supporting the continued and sustainable progression of modern medical research areas, in addition to healthcare operations as a whole [74].

5.2 A Perspective on Intelligence

The phenomenon of intelligence represents an essential and pervasive aspect of our daily lives, equipping the human species with an unprecedented and distinguishing advantage over all other species on the planet [75]. It is clear from observation that the capacity for intelligence profoundly increases the potential for achievement and actualisation of living systems, implying trivially that the arbitrary creation of such a capacity would trigger a transformative epoch in human history [76–79]. Despite such a profound and recognised presence throughout history and across fields and disciplines, no satisfactory definition has as yet been established to completely characterise the basis for the phenomenon [80, 81]. Numerous definitions of varying generality have been proposed in the literature, both formal and informal, by researchers and practitioners grounded in various disciplines. Legg and Hutter [81] undertook a survey of informal definitions of intelligence in 2007, considering 71 separate definitions drawn from a diverse cross section of the available literature, including results grounded in psychology, artificial intelligence and other sources such as encyclopedias. The authors analyse points of consensus, reporting that among the range of definitions suggested, intelligence is predominantly perceived in terms of three primary dimensions: (1) the property held by an agent as it interacts with its environment; (2) the relation of such a property, as described in the former, to the success of the agent in respect to some objective; and followed by (3) the capability of such an agent to adapt to novel situations and environments. The authors move to formulate an overarching definition, proposing that "Intelligence measures an agents ability to achieve goals in a wide range of environments".

Further researchers propose definitions to address the space of machine-based intelligence [82–86]. One such definition, based on text compression, is proposed by Mahoney in [84], while Legg and Hutter put forward a universal definition for machine intelligence in [82]. The verification of human level intelligence represents a further significant area of inquiry. The widely known "Turing test", proposed by Alan Turing in 1950 in his seminal paper entitled "Computing Machinery and Intelligence" [87], was developed as a means of testing for the presence of human level intelligence. The Turing test has prompted numerous subsequent works, resulting in both debate [88–90] and further proposals. For example, Bringsjord et al. introduce what they refer to as the "Lovelace test" [91], a proposal that stands among several suggested successors to the Turing test. Goertzel suggests that to ground such frameworks in terms of human intelligence introduces unnecessary bounds on the notion of intelligence, proposing that there is no a priori reason to suspect that the space of all possible minds is human centric [77]. The idea that the human mind is a realisation of a more fundamental phenomenon permits a broader view of intelligence that encompasses both biological and artificial systems. Within such a broadened view, we may consider a range of alternative processes that operate at differing scales and within varying domains, all of which appear to possess forms of intelligence. Indicative examples include swarm and collective intelligences [92–97], intelligent processes in primitive organisms such as slime moulds [98, 99], plant intelligence [100, 101] and so on. It remains clear that the human brain, the most complex object yet identified [102], gives rise to our current leading example of intelligence. However, the output of such intelligence is cognitively bounded and therefore fundamentally limited in its operation [103, 104]. The continued advancement of our technology depends therefore on a wider interpretation of intelligence, permitting the introduction of intelligent agency that may both expand upon and differ in capability from our own.

5.3 Computational Approaches to Intelligence in Healthcare

The general setting of artificial intelligence and machine learning encompasses a broad body of ideas, each ultimately directed towards the fulfilment of generalised intelligence within an artificial framework. Meanwhile, solutions relevant in the case of applied domains such as healthcare serve as a focus for many researchers, who seek to translate the ideas of artificial intelligence from a space of potential approaches towards realised intelligent capacity within the working components of both specialist and pervasive systems. The broad space of ideas in intelligence presents therefore an essential grounding for any discussion of the applied setting, where active components embody and inherit the developments and advancements formed principally within the artificial intelligence setting. In order to examine the healthcare-specific considerations and solutions, we therefore begin our discussion with consideration of the space of artificial intelligence, from principal methodologies, related theories and algorithmic representations. Such an exposition provides a suitable starting point from which the specific venue of healthcare, both a consumer and driver of advancements in intelligent systems, can be elaborated.

Directions in artificial intelligence may be traced to a number of prominent ideas upon which specific realisations rest, including the principal areas of symbolism, connectionism, evolutionism, analogy and the Bayesian paradigm [105]. Such ideas represent fundamental yet separate directions, whose axioms are not mutually compatible with one another. Importantly, it is thought that given sufficient data, each of the former approaches may in the limit learn any possible problem. Symbolist approaches rely upon high-level representations, taking the view that symbol manipulation may serve as an essential operational primitive. Grounded in logic and philosophy, works of influence within the symbolist paradigm include Newel and Simon [106], who hypothesise that "A physical symbol system has the necessary and sufficient means for general intelligent action". Expert systems represent a well-known symbolic approach. In contrast to the symbolic approaches, the idea of connectionism defines a space of operations grounded in sub-symbolic representations, where concepts are formed as an emergent property of interconnected simple units. The connectionist paradigm is grounded principally in neuroscience, where the neurons of the biological brain serve as a source of inspiration for information processing techniques. McCulloch and Pitts first introduced the idea of the artificial neuron in their seminal paper in 1943, since which time a succession of developments have led to the widespread use of artificial neural network (ANN) models, including the more recent extension of deep learning.

Evolutionism, also referred to as evolutionary computation, is another important direction in artificial intelligence. Similarly to connectionism, evolutionism is grounded in fundamental biological function, within the sense of a natural process.

In contrast with the connectionism paradigm, evolutionism represents a principle of operation derived from Darwinian evolution, which may be viewed as a form of global stochastic search. Techniques within the evolutionary paradigm include genetic algorithms (GA), genetic programming (GP) and evolution strategies (ES), in addition to evolutionary programming (EP) [107]. A further paradigm in artificial intelligence is given by the analogy approach, which is grounded in the inference of similarities. The principal algorithm used within this direction is the support vector machine (SVM), the present form of which was introduced by Cortes and Vapnik in 1995 [108], as built upon results from statistical learning theory [109]. The original SVM algorithm, originally a binary classifier, was extended during the same year by Vapnik to accommodate the case of regression, forming the support vector regression (SVR) algorithm [110]. The Bayesian family of approaches represent another principal approach within artificial intelligence. The Bayesian methods are grounded in Bayesian probability theory, which is attributed to Thomas Bayes in work published in 1763 [111], in which what is now known as Bayes' theorem was reported. Popular methods within such paradigm include Bayesian networks (BN), probabilistic graphical models (PGMs) and latent Dirichlet allocation (LDA) [112].

The application of computerised intelligence within the healthcare setting represents a specialist realisation of the principles found within the wider field of artificial intelligence. In order to provide a solution of relevance and utility within the healthcare setting, generalised components sufficient for intelligent utility must be suitably integrated and framed in terms of the concerns of healthcare platforms [113]. Important dimensions of healthcare fall within ethical, security and legal modalities, which collectively ensure that the unique concerns of human health remain integral to the foundation of adopted solutions. Such dimensions govern, for instance, the parameters of use for human medical data, in addition to the protocols and procedures surrounding medical decisions that must be carried out in a variety of circumstances. Intelligent solutions must therefore operate within such task frameworks, upholding roles once occupied solely by human actors. As a result, many approaches from the general domain of machine intelligence must be extended to provide features such as interpretable knowledge representation; the use of black box models alone, for example, may be insufficient in many applications concerning healthcare.

To compliment the analysis delivered in previous paragraphs, we provide a brief cross section of research exemplars, each of which serves to highlight an approach to intelligence, moreover within the context of a particular healthcare outlet or task domain. Such a series of examples may also be viewed as a limited exposition of the intelligence approaches at work within healthcare as a whole. ANNs are seen to feature across the literature [114], with variants including fuzzy ANNs [115], extreme learning machines (ELM) [116], echo state networks (ESNs) [117] and the competitive Hopfield network (CHN) [118], among others. An extension of ANN, deep learning (DL), can also be found across healthcare research domains, including convolutional neural networks (CNN) [119] and the deep belief network (DBN) [120]. Further approaches include the SVM [121], of which variations are explored by researchers, for example, in [122, 123]. Furthermore, decision trees (DTs), for example [124], represent another type of approach, also incorporating a space of extensions and enhancements such as error cost sensitivity [125] and the integration of fuzzy logic [126]. A further generalisation of the DT is the random forest (RF), as exemplified in [127], where an ensemble is constructed to enhance the performance of DTs through means of multiplicity. The Bayesian family of approaches represents an additional modality within the literature, featuring models such as naive Bayes [128] and Bayesian networks (BN) [129]. Another popular model is given by the K-nearest neighbour algorithm, a form of lazy instance learning [130] that has attracted various extension including the use of fuzzy logic [131]. In other literature, the use of alternative approaches is given, such as case-based reasoning (CBR) [132], hidden Markov models (HMM) [133], Gaussian mixture models (GMM) [134] and cellular automata (CA) [135], in addition to nature-inspired algorithms such as artificial immune systems (AIS) [136] and quantum-inspired computational intelligence (QCI) [137]. Finally, we draw the readers' attention to meta-learning and hyper-heuristic techniques, which may be used to modulate all of the former [138, 139].

5.4 Case Studies

5.4.1 Case Study: Multivariate Association Mapping for Biological Markers in Schizophrenia

Schizophrenia is the most debilitating mental health disorder. This research aims to look into the genetic aetiology of the disorder using the multiple feature association analysis. The current work undertaken in a preliminary GWA (genome-wide association) study outlines the potential for genetic risk in the disorder, therefore providing a basis for the continuation and further study of the genetic aetiology. We introduce the methodology proposed for our study, inclusive of six procedures and the resulting aims of the study.

Schizophrenia (SZ) is a psychiatric disorder which affects the cognitive functions of a person, a lax definition, the way in which the person thinks. Schizophrenia stands out as one of the most complex mental health disorders to determine and treat, considered to be a neurodevelopment disorder [140]. Schizophrenia prevalence agreeably stands at 0.5–1%, fluctuating across the world, and although not common, it presents an economic issue as the resulting service and total costs combine to conclude an expense of 4 billion pa [141]. Mental health accounted for 7.4% of disease burden worldwide in 2010 [142]; SZ (acute/active) holds the highest disability weight comparatively against all other mental health disorders [142]. It is difficult to provide a definitive description of schizophrenia given its subjective experiences in each patient; however, there are common occurrences in the symptoms experienced across the spectrum of those affected which provides the basis for diagnosis. The period length and required knowledge of the patient behaviour and thoughts necessary for diagnosis are comprehensive [143, 144]; this is to ensure the validity of the diagnosis to approach the treatment of the patient with the most appropriate course of action. Although schizophrenia has held a robust position in the psychiatric disorder family, it is becoming more apparent that we require a defining feature which uses biological evidence as its representation for diagnosis [145]. The treatment causes difficult side effects for the patient, but given the alternative option of no medication, it is a decisively better option [146]. Development in the area of schizophrenia must move forward with the clinical implications and aim to aid in the discovery of treatment and diagnostic criteria which can potentially improve the patient's experience and quality of life. This case study highlights a novel methodology which can be applied to genetic data in order to discover the underlying risk variants which can be applied in the clinical setting for the purposes of improved pharmacological solutions, diagnostic criteria and susceptibility and finally the potential to identify predisposition to the disorder.

5.4.1.1 Research Progress

Currently the work that has been undertaken remains in its early stages. The following outlines the procedure undertaken to produce our preliminary GWAS (genome-wide association study) results. Genetic data access was granted by GAIN (Genetic Association Information Network) which provided 2599 individuals from European ancestry along with 729,454 markers. We firstly performed quality control using the provided data, extracting markers and individuals that did not meet the required thresholds applied based on allele frequency, individual missing data, genotypic missing data and further assessment of data confirmation. After quality control checks, the study sample included 2383 European ancestry (EA) individuals, 1099 cases and 1284 controls with 658,834 markers remained for the association analysis phase. Logistic regression association analysis was performed using PLINK V 7.0, adjusted based on genomic control and visualised using R package qqman. The results provided insight into the associations that were present within the sample cohort for schizophrenia. As can be seen from Fig. 5.1, there are numerous SNPs

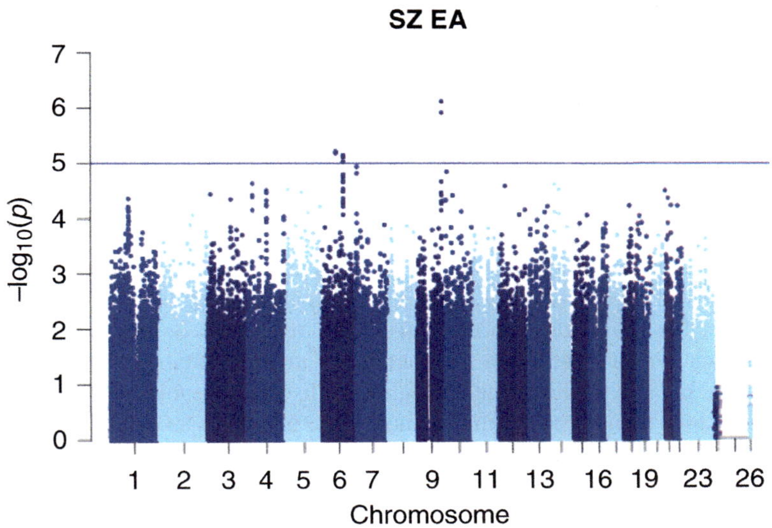

Fig. 5.1 Manhattan plot outlining associative SNPs for schizophrenia

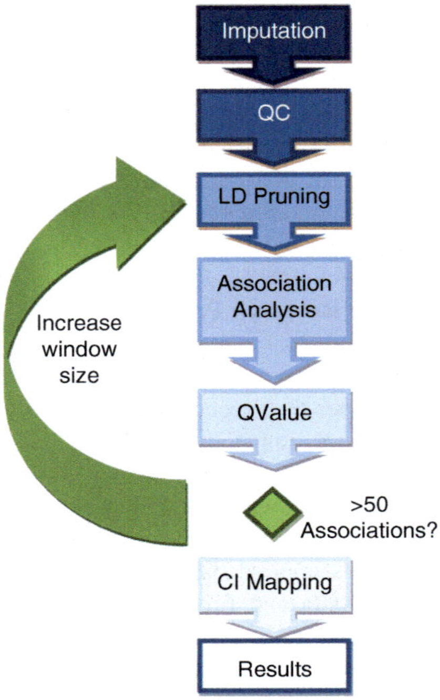

Fig. 5.2 Proposed methodology

which surpass the genome-wide significance threshold, present here at 1×10^{-5}, which is currently the standard threshold.

A framework to identify latent features within the data which could be indicative of risk factors for SZ in Fig. 5.2 that consists of six phases in which the data will be manipulated in order to uncover associative information. The first phase is imputation which will enrich our dataset to include genetic data which has not been genotyped within our provided dataset. This is becoming common practice, reducing the cost of genotyping raw data by relying on the inclusion of imputed data which is readily available and improving in density of both the markers available and the populations. Quality control is the next phase which is a necessary step in any genetic analysis approach in order to ensure the validity of the data and to reduce the potential of false discovery; a summarised description is available in the previous section within the preliminary GWAS. Our association analysis phase will consist of numerous techniques which will be adapted based on the information required for analysis; the most common and favourable technique is the additive model which includes linear regression for continuous and quantitative variables, while logistic regression is applied in the presence of discrete variables. Our next phase, Q values, aims to incorporate more modern techniques developed in order to improve the rate of false discoveries which can appear when performing association analysis as false reports for significant associative SNPs. The previous techniques applied a conservative threshold, while the Q value presents a more representative figure. This improves the validity of the results which will be obtained throughout the methodology. CI (conditional independence) mapping applies a multivariate analysis, incorporating genetic, phenotypic and environmental features to produce association maps for the discovery of highly correlated features which can provide indication of risk factors in SZ. The phenotypic information will take advantage of previous research [147] which effectively dissected the symptoms of schizophrenia into three dimensions, namely, reality distortion, psychomotor poverty and disorganisation which hold both significantly correlated associations and commonly presented as combined symptoms. These clusters incorporate the most common symptoms displayed by SZ patients and as such the predominant manifestations of the disorders and how it affects the patients.

5.4.1.2 Summary of Case Study 1

The provided information outlines the intended methodology for the analysis of genetic data with identification of potential biological markers and risk variants as an outlined aim. The method described in this book chapter combines some of the most successful techniques available in genomics currently, taking into consideration new, reliable and effective approaches. The adoption of these techniques as a method should provide insight into the relationships that are present between phenotype, genetic markers and environmental factors. Conditional independence mapping used to combine genetic, environmental and phenotypic trait data could prove capable of highlighting latent features which otherwise may not have been considered. In review, the method stands to expand the possible markers for testing purposes, eliminate the markers least likely to produce true associations, consider and decide the threshold for false discovery to produce a subset of markers and map these markers combined with phenotype and environmental information to produce an outcome which concludes the correlation found between these elements.

5.4.2 Case Study: Foetal Intrapartum Hypoxia Monitoring

Intrauterine hypoxia is the leading cause of cardiopulmonary arrest, brain damage and in severe cases death. A critical review is presented based on subjective prediction using the visual inspection of cardiotocography monitors. This is followed by a discussion on an open dataset used in the work that comprises 506 normal and 46 pathological (hypoxic) cases. Using this dataset a robust data science methodology is presented that describes the data collection and preprocessing stages including a discussion on data oversampling.

It is estimated that, worldwide, over 130 million babies are born every year – 3.6 million will die due to perinatal complication. One million of these will be intrapartum stillbirths [148]. In the UK, in 2013, there were approximately 671,255 births with 1 in every 200 being stillbirth and 300 that died in the first 4 weeks of life [149]. According to the Royal College of Obstetricians and

Gynecologists (RCOG), 500 full-term babies each year that are alive at the onset of labour are either stillborn, die within a week or suffer from brain injuries, as a result of childbirth complications . It is estimated that between 1 and 7 foetuses in every 1000 experience hypoxia, caused by delivery-induced stress [150] that leads to cardiopulmonary arrest, brain damage and in severe cases death [151]. Consequently, the early detection of severe intrauterine hypoxia episodes is needed to help avoid adverse pathological outcomes and determine when caesarean section is required.

Clinical decisions are made using the visual inspection of CTG. However, the main weakness with CTG is poor human interpretation which has led to high inter- and intra-observer variability [152]. This has led to significant works in automated CTG trace interpretation techniques that have helped to reduce perinatal mortality rates compared with human visual interpretation when used routinely and in women with increased risk of pregnancy complications [149]. In one approach the foetal heart rate and uterine contraction pairs were modelled to estimate dynamic relations using an impulse response function [153] and CTG and ST waveform analysis in the PQRST complex [154]. The results demonstrate timely intervention for caesarean and vaginal delivery and a reduction in foetal morbidity and mortality.

5.4.2.1 Materials and Methods

In this section a discussion is presented on the open dataset used and the steps taken to extract the feature vectors from the raw CTG data signals.

5.4.2.2 CTG Data Collection

Chudacek et al. conducted a comprehensive study that captured intrapartum recordings between April 2010 and August 2012 [155]. The recordings were collected from the University Hospital in Brno (UHB), Czech Republic, by obstetricians with the support of the Czech Technical University (CTU) in Prague. These records are publicly available, via the CTG-UHB database, in PhysioNet.

The CTG-UHB database contains 552 CTG recordings for singleton pregnancies with a gestational age less than 36 weeks that were selected from 9164 recordings. The STAN S21/S31 and Avalon FM 40/50 foetal monitors were used to acquire the CTG records. The dataset contains no prior known development factors; the duration of stage two labour is less than or equal to 30 min; foetal heart rate signal quality is greater than 50% in each 30 min window; and the pH umbilical arterial blood sample is available. In the dataset 46 caesarean section deliveries are included (determined by pH < 7) and the rest are vaginal deliveries. Each recording begins no more than 90 min before delivery. Each CTG record contains the FHR time series (measured in beats per minute) and uterine contraction (UC) signal (measured in mmHg) each sampled at 4 Hz. The FHR was obtained from an ultrasound transducer attached to the abdominal wall.

Preprocessing

Each of the 552 FHR signal recordings was filtered using a sixth-order low-pass Butterworth filter with fc = 4 Hz and a cut-off frequency of 0.034 Hz. To correct the phase distortion introduced by a one-pass filter, a two-pass filter (forward and reverse) was used to filter each of the signals. Noise and low-quality artefacts were removed using cubic Hermite spline interpolation [156].

Feature Extraction

Through initial empirical studies, a number of features were evaluated and ranked using recursive feature elimination [157] and the top seven selected for the classification tasks considered in this paper. The remainder of this section discusses these features in more detail.

Long-term time correlations or self-affinity measures of FHR signals have proved in previous studies to be a useful coefficient for separating normal and pathological cases (hypoxic cases) [158]. In this work detrended fluctuation analysis (DFA) is used to achieve this where the returned exponent value indicates the presence or absence of fractal properties, i.e. self-similarity.

Root mean square (RMS) is not typically used in FHR analysis; however, our empirical studies show that it is possible to estimate short-term variability between accelerations and deceleration [159].

Poincare plots are used to quantify heart rate variability [160]. In FHR the difference between two beats is given by NN and the two basic descriptors of the plot are SD1 and SD2. Fundamentally, SD1 and SD2 are directly associated with the standard deviation of NN and the standard deviation of the successive difference of the NN interval.

Sample entropy is used to quantify the nonlinear dynamics of the FHR and quantify the loss of complexity in the FHR. It is a useful indicator that the foetus is deprived of oxygen or not as the case may be [161].

Short-term variability (STV) and long-term variability (LTV) are further indicators used by obstetricians. The presence of both suggests an intact neuromodulation of the FHR and normal cardiac function and is one of the most reassuring measures in neonatal care [162].

Synthetic Minority Oversampling Technique

In a two-class balanced dataset, the prior probabilities will be equal for each. This is not the case for the CTU-UHB dataset given that there are 506 true negatives (majority class) and 46 true positive values (minority class). Classifiers are more sensitive to detecting the majority class and less sensitive to the minority class and this leads to biased classification [163]. Therefore, given a random sample taken from the dataset, the probability of a classifier classifying a foetus as normal will be much higher (91.6% 506/552) than the probability of it classifying a foetus as pathological (8.3%, 46/552). This imposes a higher cost for misclassifying the minority (predicting that a foetus is normal and the outcome being pathological) than the majority class (predicting a foetus is pathological and the outcome being normal).

In order to address this problem, it is necessary to resample the dataset. Various resampling techniques are available, and these include undersampling and oversampling. Undersampling reduces the number of records from the majority class to make it equal to the minority class; in this instance it would mean removing 460 records leaving us with a small dataset. In contrast, data in the minority class can be increased using oversampling. In this study, the synthetic minority oversampling technique (SMOTE) is used rather than reducing the dataset further [164]. In this work SMOTE is used to oversample the minority class and undersample the majority class.

5.4.2.3 Summary of Case Study 2

In this study, a robust data science methodology is demonstrated that combines FHR features, extracted from the signals recorded from 552 subjects contained in the CTU-UHB database. The aim of this work is to use neural networks for the classification and prediction of hypoxia.

5.4.3 Case Study: Preterm Birth Prediction

Globally, the rate of preterm births is increasing, thus resulting in significant health, development and economic problems. Current methods for the early detection of such births are inadequate. Nevertheless, there has been some evidence that the analysis of uterine electrical signals, collected from the abdominal surface, could provide an independent and easier way to diagnose true labour and detect when preterm delivery is about to occur. Using advanced machine learning algorithms, in conjunction with electrohysterography signal processing, numerous studies have focused on detecting true labour several days prior to the event. However, in this paper, the EHG signals have been used to detect preterm births. This has been achieved using an open dataset, which contains 262 records for women who delivered at term and 38 who delivered prematurely.

The World Health Organization (WHO) defines preterm birth as the delivery of any baby that has been born alive before 37 weeks of gestation. In other words, preterm births occur before 259 days of pregnancy, while term births occur between 259 and 294 days [165]. Preterm births have a significant adverse impact on the newborn, including an increased risk of death and other health defects. In particular, infant (less than 24 weeks) death rates are increasing. In 2009, preterm births accounted for approximately 7% of live births, in England and Wales [166].

During pregnancy, the monitoring of uterine contractions is vital in order to differentiate between those that are normal and those that may lead to premature birth.

The early onset of such contractions can be caused by a number of conditions, including abnormalities in the cervix and uterus, recurrent antepartum haemorrhage and infection [167]. Preterm births are a global problem. In the USA, the cost of treatment is reportedly $25.6 billion, while in England and Wales, it is estimated at 2.95 billion, annually [166]. Consequently, in the last 20 years, a great deal of research has been undertaken to detect and prevent the threat of preterm birth. This has been achieved using different monitoring techniques, which detect uterine contractility. Many approaches have focused on the use of external tocography and intrauterine pressure catheters. However, they have proven ineffective in the detection of preterm births.

5.4.3.1 Methodology

The TPEHG dataset, used in this book chapter, contains the raw EHG signals [168]. This data has been preprocessed using data segmentation, feature extraction and classification. The study in [169] illustrates how such EHG signals can be preprocessed through various frequency-related parameters. The study uses several linear and non-linear signal preprocessing techniques, via three different channels, to discern term and preterm deliveries. The preprocessing technique, used in [169], passed the EHG signal through a Butterworth filter configured to filter 0.8–4 Hz, 0.3–4 Hz and 0.3–3 Hz frequencies. However, [170] found that uterine electrical activity occurred within 1 Hz and that the maternal heart rate was always higher than 1 Hz. Furthermore, 95% of the patients measured had respiration rates of 0.33 Hz or less. In this work, the raw TPEHG signals have been passed through the same Butterworth filter to focus on data between 0.34 and 1 Hz.

Raw Data Collection

The raw EHG signals, obtained from the PhysioNet database [168], have been recorded using four bipolar electrodes. These have been adhered to the abdominal surface and spaced at a horizontal and vertical distance of 2.5 cm to 7 cm apart. The total number of records in the EHG dataset is 300 (38 preterm records and 262 term records). Each of the signals was either recorded early, <26 weeks (at around 23 weeks of gestation), or later, $\geq$ 26 weeks (at around 31 weeks). Within the dataset, three signals have been obtained simultaneously, per record. This has been achieved by recording through three different channels.

Feature Extraction

Given that the uterus is a muscle, this work investigates whether techniques used to capture EMG activity can also work as well on EHG activity. Several feature extraction techniques have been utilised from [171, 172] to extract features from the records on Channel 3. Thirteen features have been extracted and used in the experiment,

Table 5.1 SMOTE TPEHG signal (262 term and 262 preterm)

Classifiers	Sensitivity (%)	Specificity (%)	AUC (%)
BPXNC	79	58	72
LMNC	82	69	82
PERLC	46	67	63
RBNC	85	80	90
RNNC	86	72	83
VPC	98	2	50
DRBMC	59	55	56

which are described in Table 5.1. In this list, power spectrum is calculated using the fast discrete Fourier transform, while N denotes the length of the EHG signal. In the sample entropy, the parameters of m from (2, 3, 4) and r from (0.1, 0.125, 0.15, 0.175, 0.2). The main difference between [172, 173] and our work is in the analysis of the electrical activity in the uterus, rather than other muscle activities.

Feature Selection

After calculating the features, feature vectors have then been generated. This has been achieved using several measures, including statistical significance, linear discriminant analysis using independent search (LDAi), linear discriminant analysis using forward search (LDAf), linear discriminant analysis using backward search (LDAb) and Gram-Schmidt (GS) analysis. The results were promising, with the best four features being sample entropy, waveform length, log detector and variance.

Oversampling of EHG Signals

The TPEHG dataset is unbalanced and contains 262 term and 38 preterm records. This has a significant impact on machine learning algorithms, as classifiers are more prone to detect the majority class. Therefore, given that there are more term records, the probability of detecting a preterm record is low. To address this issue, the minority class (preterm) is oversampled using the synthetic minority oversampling technique (SMOTE). The technique is effective in solving class skew problems [174]. Using the 38 preterm records that are already available, SMOTE has been utilised to generate 262 preterm records. The oversampled results will then be compared with the original feature set that has been extracted from the database (262 term and 38 preterm).

Experiment Design

Seven advanced artificial neural network classifiers have been evaluated in this study. This includes the back-propagation trained feed-forward neural network classifier (BPXNC), Levenberg-Marquardt trained feed-forward neural network classifier (LMNC), voted perceptron classifier (VPC), radial basis function neural network classifier (RBNC), random neural network classifier (RNNC), perceptron linear classifier (PERLC) and discriminative restricted Boltzmann machine classifier (DRBMC).

Results

This section presents the classification results for term and preterm delivery records.

This has been achieved using the extracted feature set from the 0.34–1 Hz filter on Channel 3. Using the 80% holdout technique and k-fold cross-validation, the initial validation results have been presented. This provides a baseline for comparison against all subsequent evaluations that have been performed, using the oversampled dataset.

The preterm records have been oversampled using SMOTE. This algorithm allows the dataset to become balanced by oversampling the minority class (38 preterm records) to 262, which equals the 262 term samples.

Summary of Case Study 3

In conclusion, our experiment shows that using this new feature set has improved the results of the RNNC, LMNC and RBNC classifiers, which all generated accuracies above 80%. In particular, with an accuracy of 90%, the RBNC network generated the best results. This is due to the properties of the network being an effective multidimensional structure, which provides an alternative to polynomial values. Furthermore, the MLP network trained by the Levenberg-Marquardt (LM) classifier showed an improved accuracy of 82%. This training algorithm approximates Newton's method of least-squares optimization and is an efficient learning algorithm. Algorithms like DRBMCs didn't perform well because they are usually used for feature extraction and initialization procedures for other neural network architectures rather than standalone classifiers. These results are encouraging and suggest that the approach posited in our experiments shows a line of enquiry worth pursuing.

5.4.4 Case Study: Applied Machine Learning Approaches for the Clinical Data Analysis of Sickle Cell Disease

Sickle cell disease (SCD) is a genetic disease, which can negatively impact life expectancy due to a red blood cell (RBC) abnormality. The disease is caused by a group of ancestral disorders that have affected a protein inside the RBC called haemoglobin. SCD can be easily inherited by children through the genes responsible for sickle haemoglobin (Hb S), either from both parents or from one of them measured with abnormal haemoglobin [175]. The World Health Organization (WHO) reported that seven million babies born each year around the world suffer either from an inherited disease or from a congenital anomaly [176]. Even more worrying is that 5% of the population are affected by haemoglobin disorders, primarily, SCD and thalassemia [177]. Recent research has confirmed the assured effects of a drug called hydroxyurea/hydroxycarbamide in association with modifying the disease phenotype [178]. The medical practice of managing SCD modifying therapy is a source of considerable difficulty and time consumption for medical staff. In order to curtail the significant medical variability presented by such difficult process, clinicians need to advance adherence to therapy, which is regularly poor and subsequently results in fewer benefits and elevated risks to patients. Up to the present, the new trend of machine learning classifiers is essential for the analysis of data within the healthcare domain. Machine learning offers a large number of services for prognostic and diagnostic problems in clinical societies. ML is used for analysis of the importance of medical parameters in terms of their integrations for prognosis, overall patient management and therapy and support provision and most importantly to predict the disease progression [179].

The datasets utilised in our study for SCD were collected within a 10-year period. Each sample contains 12 features deemed important factors for predicting the SCD. These features have significant effects on the blood test. In order to collect a large amount of data, the local hospital in the city of Liverpool has supported this research with a number of patient records for obtaining better accuracy and services. The resulting dataset comprised 1168 sample points, with a single target variable describing the hydroxyurea medication dosage in milligrams. To facilitate our classification study, the target dosage was discretised into six bins, denoted classes 1 through 6. It formed through dividing the output range (in milligrams) into membership intervals of equal size: class 1 [$140 \leq Y < 423$ mg], class 2 [$423 \leq Y < 657$ mg], class 3 [$657 \leq Y < 932$ mg], class 4 [$932 \leq Y < 1208$ mg], class 5 [$1208 \leq Y < 1431$ mg] and class 6 [$1431 \leq Y \leq 1700$ mg].

The empirical setup includes the design of the test environment utilised in our studies, the configuration of each models. The performance evaluation is used to measure the results of the machine learning models. This study is composed of trained models using five types of integrated machine learning approaches: the Levenberg-Marquardt neural network algorithm, back-propagation trained feed-forward neural net classifier, random forest classifier, support vector classifier and trainable decision tree classifier. These five models are suitable to act as comparators for the purpose of acquiring high performance. The linear model used comprises a linear transformation with a single-layer neural network at each class output unit. To obtain performance estimates for the respective models, we ran each simulation 50 times and calculated the mean of the responses. Finally, the random oracle model (ROM) is utilised to find random case performance by the assignment of random responses for each class.

Our classifier evaluation consists of both out-of-sample (testing) diagnostics and in-sample (training), involving sensitivity, specificity, precision, the F1 score, Youden's J statistic and overall classification accuracy. Additionally, the classifiers were characterised using the area under the curve (AUC) and receiver operating characteristic (ROC) plots and where the classification ability across all operating points was ascertained.

5.4.4.1 Results

We now analyse the results from a number of experiments that have been implemented in this study. We make use of ROC and AUC for drawing comparisons between the models tested. When comparing the results of the classifiers, we found that the LEVNN outperformed the other models tested, illustrating capability both in generalising and in fitting the training data to the ground truth examples. The calculated means of AUCs for the LEVNN model obtained for six classes during training yielded an area of 1, which considered such an optimal outcomes in comparison to 0.99 over the test sample. Classes 1 to 6 were found to illustrate optimal consistent generalisation and performance from the training to the test sets for this classifier. It was discovered that the BPXNC model was able to yield an average AUC of 0.993, outperforming the RF classifier and showing an overall rank of second place. The RFC produced an average AUC of 0.973, ranking third overall, outperforming the lower ranking models by a reasonable margin. The TREEC yield an average AUC of 0.95. Eventually, SVC produced the lowest rate in comparison with other classifiers with 0.666. All four of the top performing models, the LEVNN, BPXNC, RFC and TREEC, obtained nearly ideal AUCs and represent viable candidates for future use. These models produced exceptional results in terms of both training and generalisation. The average test of AUCs for the LNN ranged 0.85, which is seen to demonstrate performance significantly below that of the other models. Moreover, as expected, the average test AUCs for the ROM is 0.524, which matches our expectation for uninformed guessing.

5.4.4.2 Summary of Case Study 4

We have reported on an empirical investigation into the use of various machine learning algorithms to classify the level of dosage for SCD medication. In this study, various model architectures are used for analysing the medical datasets obtained for SCD patients. The main purpose of this research is to examine the effectiveness of these models in terms of training and testing, investigating if such architectures could enhance classification results. It was found through experimental investigation, including the usage of SCD datasets and approaches such as LEVNN, BPXNC, RFC and TREEC that the analysis of medical datasets is viable and produces precise results. The results gained from a range of models during our experiments have demonstrated that the proposed Levenberg-Marquardt neural network classifier produced the best results with an AUC of 0.99, in comparison with other models.

5.4.5 Case Study: Data Quality Control to Genetic Data for Type 2 Diabetes

The global growth in incidence of type 2 diabetes (T2D) has been a pandemic and has recently emerged as a major international health concern. As such, understanding the aetiology of type 2 diabetes is vital. Researchers pointed out that T2D results from the convergence of genetics, environment, diet and lifestyle risk factors; however, genetic susceptibility has been established as a key component of risk. This has led researchers to investigate the genetic variants (single nucleotide polymorphisms (SNPs)) associated with an increased susceptibility to type 2 diabetes and related traits using a number of data science approaches. However, in order for the study and the results to be reliable, quality control/filtering steps need to be conducted prior to further association analysis to eliminate bias into the study.

Genome-wide association studies (GWAS) are usually used for investigating the genetic architecture of human disease. More specifically, to identify common single nucleotide polymorphisms (SNPs), which is a single base pair change in the genetic code and it is the main cause of human genetic variability [180], that influence human traits [181]. With the increase of genotypic technologies, information related to human genome is growing very fast, although studies with very large sample sizes have more power but are also more likely to be subjected to experiment errors. Such errors include low-quality DNA samples, differences in DNA quality that cause differences in the frequency of missing genotype call rate, errors in sample identification (sex identification problem) and poorly performing SNP assays [182]. Consequently, these errors can generate systematic bias into the study, leading to increase the number of false-positive and false-negative associations [183]. Researchers in [184] have demonstrated that loci with low minor allele frequency are more likely to result in false findings. Thus many studies related to GWAS have excluded SNPs with MAF <10% [185]. Therefore, to decrease the effect of systematic bias, it is particularly important to perform quality control measures / filters to detect and remove markers and individuals for whom the genotypic quality is problematic [186]. These critical steps are especially important prior to conducting any statistical analysis.

5.4.5.1 Background

Researchers have defined quality control (QC) as steps taken to monitor and control the quality of a GWAS data as it is being produced [187]. The need for careful QC of genotypic data is paramount to produce the subset of reliable markers and sample to serve as a rigorous ground for the subsequent association analysis. Several publications perform and address various aspects and steps into type 2 diabetes case-control studies.

In [188] a systematic meta-analysis was performed on a case-control study to investigate the role of potassium inwardly rectifying channel, subfamily J, member

11 (KCNJ11) variation particularly E23K polymorphism (rs5219) in susceptibility to type 2 diabetes (T2D). In this meta-analysis, 56,349 T2D cases, 81,800 controls and 483 family trios were collected from 48 published studies. The statistical methods used within the approach included the standard Q-statistic test; subgroup analyses (ethnicity, sample size, BMI, age and sex) were performed to explore whether the variation in these studies was due to heterogeneity. Furthermore, for family-based association studies, the transmission disequilibrium test (TDT) was employed to analyse effect size of the polymorphism. Moreover, the Z-test was used to determine the significance of overall odds ratio (OR). The study also conducted risk allele frequency (RAF) and population attributable risk (PAR) for a comprehensive observation of the effect of the E23K variant on T2D at population level. This study suggested that a modest but statistically effect of the 23 K allele of rs5219 polymorphism in susceptibility to T2D, particularly in East Asians and Caucasians. However, the contribution of this genetic variation to T2D in other ethnic populations (e.g. Indian, African, American, Jews and Arabian) appears to be relatively low.

Seven novel type 2 diabetes susceptibility loci were identified in [189]. The researchers suggested that in order to further understand the genetic basis of T2D susceptibility, they aggregated several published meta-analysis of GWAS. These studies contain 26,488 cases and 83,964 controls of East Asian, South Asian, European, Mexican and Mexican American ancestries. By combining GWAS across ancestry groups using trans-ethnic meta-analysis, researchers in this study found that under such a model, they observed significant improvement in detecting novel complex trait loci for the disease. Furthermore, with this approach, there was an enhancement in fine-mapping resolution of causal variants by leveraging differences in local linkage disequilibrium structure between ethnic groups.

In [190] researchers performed a case-control study of 400 type 2 diabetes cases and controls of South Indian population to analyse and outline the association of potassium inwardly rectifying channel, subfamily J, member 11 (KCNJ11) gene on risk of T2D. The study also conducted a systematic review and meta-analysis for KCNJ11 (rs5219) polymorphism in 3831 cases and 3543 controls aggregated from five published reports from South Asian and East Asian population. In this study, odds ratio (OR) was employed as the measure of association of KCNJ11 polymorphisms (rs5219, rs5215, rs41282930, rs1800467) and T2D with its corresponding 95% confidence interval (CI). Moreover, Cochran's Q, I2 statistics were utilised to assess for heterogeneity within and between the eligible studies. The resulting evidence therefore showed that KCNJ11 rs5215, C-G-C-C haplotype and two loci analysis (rs5219 vs rs1800467) have a significant association with T2D; however, copy number variation (CNV) analysis did not show significant variation between T2D cases and control subjects. Furthermore, the meta-analysis of the study suggested that KCNJ11 (rs5219) polymorphism is associated with risk of T2D in East Asian and global population; however, this outcome is inapplicable to South Asian population.

While in [191], the aims of the case-control study were focused on the investigation of the differences in association of peroxisome proliferator-activated receptor gamma coactivator 1-alpha (PPARGC1A) gene with T2D risk among population with African origins. The study includes adults aged >30 years old from African Americans (cases = 124, controls = 122) and Haitian Americans (cases = 110, controls = 116). The statistical methods used within this study included standard summary statistics such as chi-square goodness of fit test that was employed to check genotype counts for each SNP for Hardy-Weinberg equilibrium. Furthermore, student t-test was used to compare between cases and controls considering demographic (age, sex, BMI, smoking status) and clinical information. Logistic regression approach was also used to calculate adjusted and unadjusted odd ratios (OR) with 95% confidence interval CI. The result suggested that adjusted logistics regression pointed out that SNP rs7656250 and rs4235308 showed protective association with T2D in Haitian Americans. While in African Americans, SNP rs4235308 showed significant risk association with T2D.

5.4.5.2 Data Quality Control and Results

In this paper, we conducted quality control processes by using PLINK toolset and standard statistical software R. As a starting point, we identified individuals with discordant sex information in our dataset.

We conducted missing genotype rate per individual in the dataset. Generally low genotyping call rate refers to low DNA quality or low sample concentration. We generated a plot as shown in Fig. 5.1, using the proportion of missing SNPs per individuals that is plotted on the x-axis and the observed heterozygosity rate per individual which is plotted on y-axis (Fig. 5.3).

Based on the plot, we observed most samples without missing genotype and a few with missing genotype rate; therefore, we decided to set thresholds at which to exclude samples. Thirty-seven samples with a genotype failure rate > 0.005 and heterozygosity rate were omitted from our dataset.

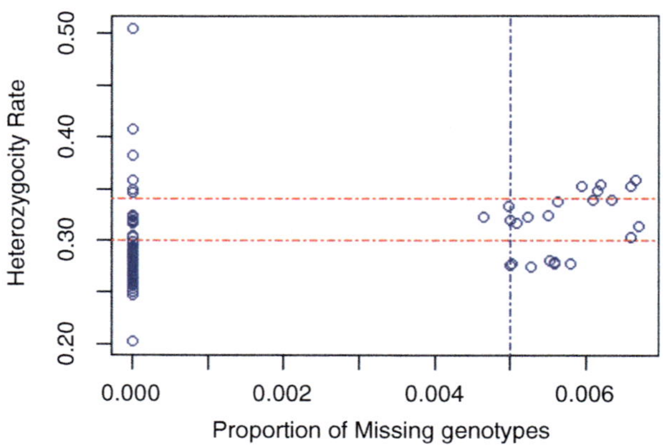

Fig. 5.3 Genotype failure rate vs. heterozygosity rate. Dashed lines indicate quality control thresholds and the dot represents the observed samples

Removing SNPs from the study is substantially critical as each SNP may correlate with disease risk. Therefore, considerable caution needs to be taken to decide at which thresholds these SNP must to remove from study. In our dataset, we tested the dataset against the identification of high missing data rate for each marker. Based on the results from F MISS column which represents the proposition of individual missing for this marker, we found that a few SNPs need to be removed, while the rest remained for further analysis.

We also performed a statistical summary to check SNPs with low minor allele frequency (MAF). According to the results from frequency test, we chose to remove MAF threshold of <0.01. We further conducted the Hardy-Weinberg equilibrium (HWE) test for all SNPs in the study for the controls. Based on the results, we removed those markers with P-value <0.001.

Following those statistical summary test steps, 3471 individuals and 430,336 markers remained in the study from the original sample of 3503 and 500,568, respectively.

5.4.5.3 Summary of Case Study 5

The interpretation of GWAS of case-control association requires the proper use of summary statistic tests for quality assessment and control to reduce bias in the results. This study used an existing data from the Wellcome Trust Case Control Consortium (WTCCC) type 2 diabetes to perform quality assessment control steps. The dataset is in PLINK format and the size of the files is very large to be handled; thus the dataset was analysed using rigorous data science techniques using PLINK toolset and R packages. This first led us to store large genome data in a more efficient structure and thus improved our methodology to handle this data. Second, the inquiry enhanced our understanding of the data structure and features and helped us to modify the data in a more efficient way. A series of quality control steps were performed to remove data items that introduce bias to the study.

Although this study has shown that the steps conducted for filtering the dataset are promising, a number of outstanding inquiries remain for this type of data cleaning. Specifically, even after considering comprehensive quality control filtering, it is impossible to identify and eliminate all inherent genotyping errors in the dataset. As the genotyping errors of some SNPs and samples are not significantly poor to be marked as a red indicator in the quality control filtering step, such data elements will remain in the dataset under the current method.

5.4.6 Case Study: Obesity Genetics and Classification

In the past decades, the improvement and evolution of high-throughput genomics has moved biologists into the big data domain due to the exponential growth of the amount of biological data they handle. Life scientists are dealing with massive datasets, encountering challenges with handling, processing and moving information that were once the domain of other areas such as astronomy. However, this raises another issue, the extraction of useful information from these data. Such an issue is one of the most important challenges in bioinformatics. In this study we investigate the use of complex high-volume data for the analysis of obesity in humans, with a view to gaining insight into the genetic factors associated with the condition.

The global prevalence of obesity has reached epidemic proportions [192–195]. According to the World Health Organization (WHO), approximately 2.8 million people die each year as a consequence of being overweight or obese [196, 197]. Obesity is a major risk for other chronic diseases which include diabetes, cardiovascular diseases and cancer [198–200]. Consequently, it is high on the political agenda of many countries.

Advances in human genomics have provided significant opportunities, and research suggests that it might be possible to quantify an individual's susceptibility to obesity from an early age and manage risk as individuals progress through life [201]. Given that we can sequence the human genome and new genotyping and sequencing technology is available, it is possible to analyse whole genetic sequences and detect diseases and associated traits [202]. Therefore, combining personalised medicine with genomic information and integrating it into medical care and individualised risk assessments will allow us to mitigate the long-term effects of obesity and its associated comorbidities [203]. This is being made possible through advances in bioinformatics [204, 205], data science [206, 207] and advanced machine learning algorithms, such as deep learning [208, 209].

The information gathered from genome-wide association studies (GWAS) – which permit the analysis of a larger number of genetic variants for association with traits of interest – has served to improve knowledge and understanding of complex diseases. Thousands of SNPs have been associated with diseases and other complex traits [210]. However, the effect of these genetic variants alone is not sufficient for clinical utility, where epistatic interactions – interactions between genetic variants – are also needed [211]. Inter- and intragene interaction is considered important in the development of many complex diseases [212]. Therefore, approaches that take into account the complexity of genotype-phenotype relationships that are characterised by multiple gene-gene and gene-environment interactions are needed [213]. The complexity of these interactions supports the utilisation of machine learning and data mining techniques [214].

Our case study explores these ideas further and proposes a robust methodology to combine state-of-the-art bioinformatics and data science to investigate genetic profiling and risk factor assessment for obesity and type 2 diabetes (T2D). The motivation for this research is to identify strong genetic markers for obesity and type 2 diabetes for use in decision support systems. This study utilises data science to automatically build a dataset using publicly available demographic and genetic information provided by individuals that focuses on obesity and T2D and related genetic variants.

5.4.6.1 Proposed Methodology

Our current study presents a method for the collection, manipulation and identification of obesity and T2D-related genetic variants using publicly available data sources provided by direct-to-consumer genetic testing (DTCGT) users and the National Human Genome Research Institute (NHGRI) Catalog. This study extends our previous work [215] for the identification of obesity-related genetic variants in extremely obese subjects.

We focused on generating our own database using web scraping techniques. We collected information about 164 participants from the Personal Genome Project (PGP) which included demographic information and full genome profiles. Participants were aged between 28 and 79 years of age (average age 45.88) and were all from the USA. The average height, body mass and BMI of all participants were 1.747 m, 77.7 kg and 25.43, respectively. 73.17% of the population were male and 26.3% female. Information about whether the participants had T2D or not was also collected. Additional features were generated using information from existing columns. These include body mass index (BMI), constructed from the weight and height variables calculated using the metric formula, BMI = weight (kg)2/height (m). A status feature was also generated from the BMI result. Following the WHO classification for BMI, five standard weight status categories associated with BMI ranges for adults were derived: underweight, normal range, overweight, obese and extremely obese. Extremely obese is commonly divided into obese I, obese II and obese III, but we grouped them into one category for convenience (Fig. 5.4).

First, we used a manually curated and publicly available database with results from GWAS – the NHGRI Catalog – for the identification of obesity and T2D-related genetic variants or SNPs. Single nucleotide polymorphisms (SNPs) are now recognised as the main cause of human genetic variability and are already a valuable resource for mapping complex genetic traits. Metadata from the NHGRI Catalog was accessed using the R/Bioconductor package gwascat and the hg19 genome

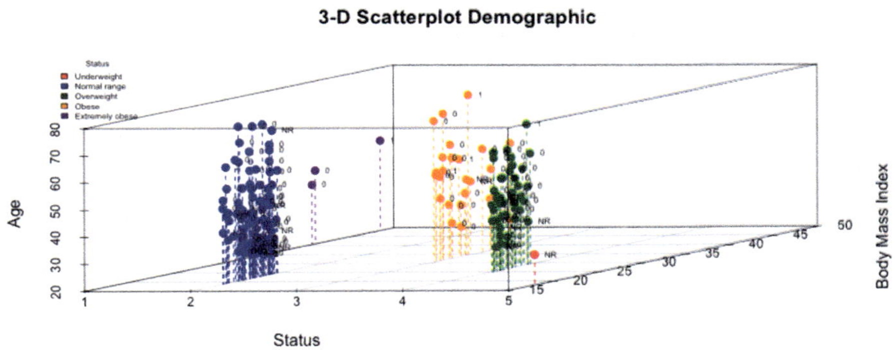

Fig. 5.4 Distribution of participants based on the BMI status, BMI and age. Each point has the label T2D 0,1 and NR if the participant did not have T2D or had T2D or the information was not reported

build ebicat37 data. Subsequently, we identified the frequency of SNPs or variants of interest in the 164 examined profiles, and these were indexed in the NHGRI Catalog. This mapping allows us to link SNP traits in our samples with risk alleles indexed in the NHGRI Catalog. This process resulted in a set of traits that best represent phenotypes related to obesity and T2D.

The information from the NHGRI Catalog and the 164 samples were analysed using rigorous data science techniques, which led to a significant reduction in the initial number of observations and the identification of SNPs that could help to explain the development of obesity and T2D. Missing values and no relational data items were removed from the study. The variables rsid and genotype as well as SNPS and STRONGEST.SNP.RISK.ALLELE were the key for the identification of obesity-related SNPs in our samples. Refer to [215] for more details.

The dataset we created was built from research-grade data (i.e. not for clinical use), and the conductors of the PGP stated that many types of errors are possible.

Some of these include errors in the data failure to report or discover significant genetic issues and ambiguous or false-positive findings. Our study in polygenic obesity is SNP based which was located either within or near the genes reported in the NHGRI Catalog. The method of analysis and SNP identification is built upon the assumption that all the SNPs identified in the genetic profiles of the 164 participants were statistically significant according to the NHGRI Catalog. However, it is highly probable that there are additional as yet to be identified genetic variants that will emerge from other GWAS and will be included in the NHGRI Catalog.

Based on the results obtained and after a closer look at the genetic variants present in the participants with different BMI classification status, we identified a set of SNPs that we considered relevant. These SNPs were identified predominantly in participants classified as overweight, obese or extremely obese. We eventually termed risk category.

5.4.6.2 Summary of Case Study 6

Understanding obesity as a complex disease is an arduous task. In an attempt to better understand how obesity is developed, we sought to identify the main obesity and T2D-related variants present in the genetic profiles of the participants collected from the PGP. This was achieved using information contained in the NHGRI Catalog as a reference for the identification of risk variants in our samples. The methodology proposed in our study investigated a solid foundation for identifying SNPs and their associated risks and in this sense provides a foundation for much more complex genetic analytics.

The potential for variants identified in GWAS to predict the risk of complex diseases is problematic. At the moment, currently known variants do not fully explain the risk of disease occurrence to be of clinical use in predictive systems. However, the presence of specific obesity-related SNPs could help us to gain a better understanding about what SNPs we need to look for when predicting obesity.

SNPs in this study have been studied in isolation. We did not investigate whether they appear collectively or disjointedly to cause obesity or T2D. While the results show specific genetic variants that could serve as good discriminators in the investigation of classification studies, more analysis with a higher representation of samples must be carry out. We propose a set of eight SNPs to be used in future studies as features for the prediction of obesity and other comorbidities such as T2D. Further application of data science techniques and machine learning approaches will help provide a better understanding of how these genetic variants perform in predictive systems as well as how they interact with other SNPs in the development complex traits. This will be the focus of our future work.

5.5 Conclusion

In this chapter we have focused on the background, use and potential of intelligent systems and data-intensive workflows within the modern healthcare setting. In particular, we have delivered for the reader the background to intelligent systems within the modern healthcare setting, deepened through a perspective on intelligent agency itself. We subsequently provided a compact review of both established and state-of-the-art techniques that fall at the intersection of applied machine intelligence and healthcare solutions. We have then culminated our discussion in the presentation of six active case studies, whose focuses progress through the task domains of genetics, physiological signals and medication control. Our report highlights the heightening importance of artificial intelligence within the changing operational frameworks of modern healthcare, where an increasing reliance on data-intensive systems approaches is driven by the need to accomplish a shift from reactive to proactive medical outlooks, such as those codified by the P4 definition. We are in an age of unprecedented opportunity for achieving such lofty goals, where the formation of new frontiers at the intersection of computation, traditional and entirely novel disciplines has driven the rise of new fields, including the systems biomedicine, bioinformatics and genomics, which in aggregate promise to provide revolutionary insights into living systems. The presence of intelligent agency within computerised processes is now not only important but is increasingly viewed as an essential enabling factor within contemporary modes of inquiry, whose critical pathways demand information processing at a level that exceeds the bounds of the human actor. The continued growth and acceptance of intelligent agency within the computerised setting therefore promise to lead towards unprecedented advancement and expansion in healthcare solutions, transforming the wellness of populations.

Acknowledgements We would like to thank the following PhD candidates, currently studying at Liverpool John Moores University, who kindly contributed the six case studies used in this chapter: Case Study 1, Jade Hind; Case Study 2, Rounaq Abbas; Case Study 3, Ibrahim Idowu; Case Study 4, Mohammed Khalaf; Case Study 5, Basma Abdulaimma; and Case Study 6, Casimiro Aday Curbelo Montañez.

References

1. Rosen, G., & Imperato, P. J. (2015). *A history of public health*. Baltimore: JHU Press.
2. Porter, D. (1999). *Health, civilization, and the state: A history of public health from ancient to modern times*. London/New York: Psychology Press.
3. Lederberg, J. (2000). Infectious history. *Science, 288*(5464), 287–293.
4. Tomes, N. (1990). The private side of public health: Sanitary science, domestic hygiene, and the germ theory, 1870–1900. *Bulletin of the History of Medicine, 64*(4), 509.
5. Lawrence, C., & Dixey, R. (1992). Practising on principle: Joseph Lister and the germ theories of disease. In *Medical theory, surgical practice: Studies in the history of surgery* (pp. 153–215). London: Routledge.
6. Pray, L. (2008). Discovery of DNA structure and function: Watson and Crick. *Nature Education, 1*(1), 100.
7. Wells, P. (2001). Physics and engineering: Milestones in medicine. *Medical Engineering & Physics, 23*(3), 147–153.
8. Hockstein, N., Gourin, C., Faust, R., & Terris, D. J. (2007). A history of robots: From science fiction to surgical robotics. *Journal of Robotic Surgery, 1*(2), 113–118.
9. Yates, D. R., Vaessen, C., & Roupret, M. (2011). From Leonardo to da Vinci: The history of robot-assisted surgery in urology. *BJU International, 108*(11), 1708–1713.
10. Bruggeman, F. J., Westerhoff, H. V., & Boogerd, F. C. (2002). BioComplexity: A pluralist research strategy is necessary for a mechanistic explanation of the "live" state. *Philosophical Psychology, 15*(4), 411–440.
11. Covich, A. (2000). Biocomplexity and the future: The need to unite disciplines. *Bioscience, 50*(12), 1035.
12. Michener, W. K., Baerwald, T. J., Firth, P., Palmer, M. A., Rosenberger, J. L., Sandlin, E. A., & Zimmerman, H. (2001). Defining and unraveling biocomplexity. *Bioscience, 51*(12), 1018–1023.
13. Nicholson, J. K., Holmes, E., Lindon, J. C., & Wilson, I. D. (2004). The challenges of modeling mammalian biocomplexity. *Nature Biotechnology, 22*(10), 1268–1274.
14. Collins, F., Galas, D., et al. (1993). A new five-year plan for the US human genome project. *SCIENCE-NEW YORK THEN WASHINGTON, 262*, 43–43.
15. Sawicki, M. P., Samara, G., Hurwitz, M., & Passaro, E. (1993). Human genome project. *The American Journal of Surgery, 165*(2), 258–264.
16. Collins, F. S., & McKusick, V. A. (2001). Implications of the human genome project for medical science. *JAMA, 285*(5), 540–544.
17. Collins, F. S. (1999). Medical and societal consequences of the human genome project. *New England Journal of Medicine, 341*(1), 28–37.
18. Collins, F. S., Patrinos, A., Jordan, E., Chakravarti, A., Gesteland, R., Walters, L., et al. (1998). New goals for the US human genome project: 1998-2003. *Science, 282*(5389), 682–689.
19. Kevles, D. J., & Hood, L. E. (1993). *The code of codes: Scientific and social issues in the human genome project*. Cambridge, MA: Harvard University Press.
20. Collins, F. S., Morgan, M., & Patrinos, A. (2003). The human genome project: Lessons from large-scale biology. *Science, 300*(5617), 286–290.
21. Watson, J. D. (1990). The human genome project: Past, present, and future. *Science, 248*(4951), 44–49.
22. Bentley, D. R. (2000). The human genome project: An overview. *Medicinal Research Reviews, 20*(3), 189–196.
23. Venter, J. C., Adams, M. D., Myers, E. W., Li, P. W., Mural, R. J., Sutton, G. G., Smith, H. O., Yandell, M., Evans, C. A., Holt, R. A., et al. (2001). The sequence of the human genome. *Science, 291*(5507), 1304–1351.
24. Markram, H., Meier, K., Lippert, T., Grillner, S., Frackowiak, R., Dehaene, S., Knoll, A., Sompolinsky, H., Verstreken, K., DeFelipe, J., et al. (2011). Introducing the human brain project. *Procedia Computer Science, 7*, 39–42.

25. Alivisatos, A. P., Chun, M., Church, G. M., Deisseroth, K., Donoghue, J. P., Greenspan, R. J., McEuen, P. L., Roukes, M. L., Sejnowski, T. J., Weiss, P. S., et al. (2013). The brain activity map. *Science, 339*(6125), 1284–1285.
26. Insel, T. R., Landis, S. C., & Collins, F. S. (2013). The NIH brain initiative. *Science, 340*(6133), 687–688.
27. Collins, F. S., & Varmus, H. (2015). A new initiative on precision medicine. *New England Journal of Medicine, 372*(9), 793–795.
28. Wagner, V., Dullaart, A., Bock, A.-K., & Zweck, A. (2006). The emerging nanomedicine landscape. *Nature Biotechnology, 24*(10), 1211–1217.
29. Roco, M. C. (2003). Nanotechnology: Convergence with modern biology and medicine. *Current Opinion in Biotechnology, 14*(3), 337–346.
30. Evans, W. E., & Relling, M. V. (2004). Moving towards individualized medicine with pharmacogenomics. *Nature, 429*(6990), 464–468.
31. McLeod, H. L., & Evans, W. E. (2001). Pharmacogenomics: Unlocking the human genome for better drug therapy. *Annual Review of Pharmacology and Toxicology, 41*(1), 101–121.
32. Vidal, M., Chan, D. W., Gerstein, M., Mann, M., Omenn, G. S., Tagle, D., & Sechi, S. (2012). The human proteome–A scientific opportunity for transforming diagnostics, therapeutics, and healthcare. *Clinical Proteomics, 9*(1), 1.
33. Weston, A. D., & Hood, L. (2004). Systems biology, proteomics, and the future of health care: Toward predictive, preventative, and personalized medicine. *Journal of Proteome Research, 3*(2), 179–196.
34. Butte, A. J. (2008). Translational bioinformatics: Coming of age. *Journal of the American Medical Informatics Association, 15*(6), 709–714.
35. Shah, N. H., & Tenenbaum, J. D. (2012). The coming age of data-driven medicine: Translational bioinformatics' next frontier. *Journal of the American Medical Informatics Association, 19*(e1), e2–e4.
36. Yan, Q. (2010). Translational bioinformatics and systems biology approaches for personalized medicine. In *Systems biology in drug discovery and development: Methods and protocols* (pp. 167–178). New York: Humana Press.
37. Feng, J. (2003). *Computational neuroscience: A comprehensive approach*. Boca Raton: CRC press.
38. Baraba´si, A.-L., Gulbahce, N., & Loscalzo, J. (2011). Network medicine: A network-based approach to human disease. *Nature Reviews Genetics, 12*(1), 56–68.
39. Chen, C., Dubin, R., & Kim, M. C. (2014). Emerging trends and new developments in regenerative medicine: A scientometric update (2000–2014). *Expert Opinion on Biological Therapy, 14*(9), 1295–1317.
40. Tanaka, H. (2010). Omics-based medicine and systems pathology. *Methods of Information in Medicine, 49*(2), 173–185.
41. Garay, J. P., & Gray, J. W. (2012). Omics and therapy–A basis for precision medicine. *Molecular Oncology, 6*(2), 128–139.
42. Hamburg, M. A., & Collins, F. S. (2010). The path to personalized medicine. *New England Journal of Medicine, 363*(4), 301–304.
43. Cuijpers, P., et al. (2009). Prevention: An achievable goal in personalized medicine. *Dialogues Clin Neurosci, 11*(4), 447–454.
44. Harvey, A., Brand, A., Holgate, S. T., Kristiansen, L. V., Lehrach, H., Palotie, A., & Prainsack, B. (2012). The future of technologies for personalised medicine. *New Biotechnology, 29*(6), 625–633.
45. Dietel, M., & Sers, C. (2006). Personalized medicine and development of targeted therapies: The upcoming challenge for diagnostic molecular pathology. A review. *Virchows Archiv, 448*(6), 744–755.
46. Jain, K. K. (2005). Personalised medicine for cancer: From drug development into clinical practice. *Expert Opinion on Pharmacotherapy, 6*, 1463–1476.
47. Jain, K. (2002). Personalized medicine. *Current Opinion in Molecular Therapeutics, 4*(6), 548–558.

48. Kalow, W. (2002). Pharmacogenetics and personalised medicine. *Fundamental & Clinical Pharmacology, 16*(5), 337–342.
49. Mirnezami, R., Nicholson, J., & Darzi, A. (2012). Preparing for precision medicine. *New England Journal of Medicine, 366*(6), 489–491.
50. Ziegler, A., Koch, A., Krockenberger, K., & Großhennig, A. (2012). Personalized medicine using DNA biomarkers: A review. *Human Genetics, 131*(10), 1627–1638.
51. Flores, M., Glusman, G., Brogaard, K., Price, N. D., & Hood, L. (2013). P4 medicine: How systems medicine will transform the healthcare sector and society. *Personalized Medicine, 10*(6), 565–576.
52. Hood, L., & Galas, D. J. (2009). Systems biology and emerging technologies will catalyze the transition from reactive medicine to predictive, personalized, preventive and participatory (P4) medicine. *Interdisciplinary Bio Central, 1*(1), 6.
53. Hood, L., & Flores, M. (2012). A personal view on systems medicine and the emergence of proactive P4 medicine: Predictive, preventive, personalized and participatory. *New Biotechnology, 29*(6), 613–624.
54. Hood, L., & Friend, S. H. (2011). Predictive, personalized, preventive, participatory (P4) cancer medicine. *Nature Reviews Clinical Oncology, 8*(3), 184–187.
55. Ahn, A. C., Tewari, M., Poon, C.-S., & Phillips, R. S. (2006). The limits of reductionism in medicine: Could systems biology offer an alternative? *PLoS Medicine, 3*(6), e208.
56. Auffray, C., Chen, Z., & Hood, L. (2009). Systems medicine: The future of medical genomics and healthcare. *Genome Medicine, 1*(1), 2.
57. Martin-Sanchez, F., Iakovidis, I., Nørager, S., Maojo, V., de Groen, P., Van der Lei, J., Jones, T., Abraham-Fuchs, K., Apweiler, R., Babic, A., et al. (2004). Synergy between medical informatics and bioinformatics: Facilitating genomic medicine for future health care. *Journal of Biomedical Informatics, 37*(1), 30–42.
58. Miller, P. L. (2000). Opportunities at the intersection of bioinformatics and health informatics. *Journal of the American Medical Informatics Association, 7*(5), 431–438.
59. Hood, L., Balling, R., & Auffray, C. (2012). Revolutionizing medicine in the 21st century through systems approaches. *Biotechnology Journal, 7*(8), 992–1001.
60. Hood, L., Heath, J. R., Phelps, M. E., & Lin, B. (2004). Systems biology and new technologies enable predictive and preventative medicine. *Science, 306*(5696), 640–643.
61. Loscalzo, J., & Barabasi, A.-L. (2011). Systems biology and the future of medicine. *Wiley Interdisciplinary Reviews: Systems Biology and Medicine, 3*(6), 619–627.
62. Vidal, M., Cusick, M. E., & Barabasi, A.-L. (2011). Interactome networks and human disease. *Cell, 144*(6), 986–998.
63. Noble, D. (2002). The rise of computational biology. *Nature Reviews Molecular Cell Biology, 3*(6), 459–463.
64. Kholodenko, B., Yaffe, M. B., & Kolch, W. (2012). Computational approaches for analyzing information flow in biological networks. *Science Signaling, 5*, 2002961.
65. Issa, N. T., Byers, S. W., & Dakshanamurthy, S. (2014). Big data: The next frontier for innovation in therapeutics and healthcare. *Expert Review of Clinical Pharmacology, 7*(3), 293–298.
66. Ohno-Machado, L. (2012). Big science, big data, and a big role for biomedical informatics. *Journal of the American Medical Informatics Association: JAMIA, 19*(e1), e1.
67. Martin-Sanchez, F., Verspoor, K., et al. (2014). Big data in medicine is driving big changes. *Yearbook of Medical Informatics, 9*(1), 14–20.
68. Cattell, J., Chilukuri, S., & Levy, M. (2013). *How big data can revolutionize pharmaceutical R&D*. New York: McKinsey & Company.
69. Sejnowski, T. J., Churchland, P. S., & Movshon, J. A. (2014). Putting big data to good use in neuroscience. *Nature Neuroscience, 17*(11), 1440–1441.
70. Andreu-Perez, J., Poon, C. C., Merrifield, R. D., Wong, S. T., & Yang, G.-Z. (2015). Big data for health. *IEEE Journal of Biomedical and Health Informatics, 19*(4), 1193–1208.
71. Costa, F. F. (2014). Big data in biomedicine. *Drug Discovery Today, 19*(4), 433–440.

72. Chaussabel, D., & Pulendran, B. (2015). A vision and a prescription for big data-enabled medicine. *Nature Immunology, 16*(5), 435–439.
73. Slaymaker, M., Power, D., Russell, D., Wilson, G., & Simpson, A. (2008). Accessing and aggregating legacy data sources for healthcare research, delivery and training. In *Proceedings of the 2008 ACM Symposium on Applied Computing* (pp. 1317–1324). ACM.
74. Patel, V. L., Shortliffe, E. H., Stefanelli, M., Szolovits, P., Berthold, M. R., Bellazzi, R., & Abu-Hanna, A. (2009). The coming of age of artificial intelligence in medicine. *Artificial Intelligence in Medicine, 46*(1), 5–17.
75. Vitousek, P. M., Mooney, H. A., Lubchenco, J., & Melillo, J. M. (1997). Human domination of Earth's ecosystems. *Science, 277*(5325), 494–499.
76. Good, I. J. (1965). Speculations concerning the first ultraintelligent machine. *Advances in Computers, 6*(99), 31–83.
77. Goertzel, B. (2007). Human-level artificial general intelligence and the possibility of a technological singularity: A reaction to Ray Kurzweil's The Singularity Is Near, and McDermott's critique of Kurzweil. *Artificial Intelligence, 171*(18), 1161–1173.
78. Bostrom, N. (2014). *Superintelligence: Paths, dangers, strategies*. Oxford: Oxford University Press.
79. Kurzweil, R. (2000). *The age of spiritual machines: When computers exceed human intelligence*. New York: Penguin.
80. Steels, L. (1996). "The origins of intelligence".
81. Legg, S., Hutter, M., et al. (2007). A collection of definitions of intelligence. *Frontiers in Artificial Intelligence and applications, 157*, 17.
82. Legg, S., & Hutter, M. (2007). Universal intelligence: A definition of machine intelligence. *Minds and Machines, 17*(4), 391–444.
83. Legg, S., & Hutter, M. (2006). A formal measure of machine intelligence. arXiv preprint cs/0605024.
84. Mahoney, M. V. (1999). Text compression as a test for artificial intelligence. In *AAAI/IAAI* (p. 970), AAAI Press.
85. Hernandez-Orallo, J. (2000). Beyond the Turing test. *Journal of Logic, Language and Information, 9*(4), 447–466.
86. Yonck, R. (2012). Toward a standard metric of machine intelligence. *World Future Review, 4*(2), 61–70.
87. Turing, A. M. (1950). Computing machinery and intelligence. *Mind, 59*, 433–460.
88. Whitby, B. (1996). Why the turing test is ai's biggest blind alley. In *Machines and thoughts* (pp. 53–62). Oxford: Clarendon Press.
89. Hayes, P., & Ford, K. (1995). Turing test considered harmful. *In IJCAI, 1*, 972–977.
90. Searle, J. R. (1980). Minds, brains, and programs. *Behavioral and Brain Sciences, 3*(03), 417–424.
91. Bringsjord, S., Bello, P., & Ferrucci, D. (2003). Creativity, the Turing test, and the (better) Lovelace test. In *The Turing test* (pp. 215–239). Dordrecht: Springer.
92. Krause, J., Ruxton, G. D., & Krause, S. (2010). Swarm intelligence in animals and humans. *Trends in Ecology & Evolution, 25*(1), 28–34.
93. Park, H.-J., Kim, B. K., & Lim, K. Y. (2001). Measuring the machine intelligence quotient (MIQ) of human-machine cooperative systems. *IEEE Transactions on Systems, Man, and Cybernetics-Part A: Systems and humans, 31*(2), 89–96.
94. Parpinelli, R. S., & Lopes, H. S. (2011). New inspirations in swarm intelligence: A survey. *International Journal of Bio-Inspired Computation, 3*(1), 1–16.
95. Franks, N. R., Pratt, S. C., Mallon, E. B., Britton, N. F., & Sumpter, D. J. (2002). Information flow, opinion polling and collective intelligence in house– Hunting social insects. *Philosophical Transactions of the Royal Society of London B: Biological Sciences, 357*(1427), 1567–1583.
96. Tereshko, V., & Loengarov, A. (2005). Collective decision making in honeybee foraging dynamics. *Computing and Information Systems, 9*(3), 1.

97. Bonabeau, E., Dorigo, M., & Theraulaz, G. (1999). *Swarm intelligence*. Oxford: Oxford University Press.
98. Nakagaki, T., Yamada, H., & T´oth, A. (2000). Intelligence: Maze-solving´ by an amoeboid organism. *Nature, 407*(6803), 470–470.
99. Adamatzky, A., Armstrong, R., Jones, J., & Gunji, Y.-P. (2013). On creativity of slime mould. *International Journal of General Systems, 42*(5), 441–457.
100. Trewavas, A. (2005). Plant intelligence. *Naturwissenschaften, 92*(9), 401–413.
101. Trewavas, A. (2005). Green plants as intelligent organisms. *Trends in Plant Science, 10*(9), 413–419.
102. Bishop, C. M. (1994). Neural networks and their applications. *Review of Scientific Instruments, 65*(6), 1803–1832.
103. Simon, H. A. (1957). *Models of man; social and rational*. New York: Wiley.
104. Klingberg, T. (2000). Limitations in information processing in the human brain: Neuroimaging of dual task performance and working memory tasks. *Progress in Brain Research, 126*, 95–102.
105. Domingos, P. (2015). *The master algorithm: How the quest for the ultimate learning machine will remake our world*. New York: Basic Books.
106. Newell, A., & Simon, H. A. (1976). Computer science as empirical inquiry: Symbols and search. *Communications of the ACM, 19*(3), 113–126.
107. Back, T., Hammel, U., & Schwefel, H.-P. (1997). Evolutionary computation: Comments on the history and current state. *IEEE Transactions on Evolutionary Computation, 1*(1), 3–17.
108. Cortes, C., & Vapnik, V. (1995). Support-vector networks. *Machine Learning, 20*(3), 273–297.
109. Vapnik, V. N. (1999). An overview of statistical learning theory. *Neural Networks, IEEE Transactions on, 10*(5), 988–999.
110. Vapnik, V. (1995). The nature of statistical learning theory. *Data Mining and Knowledge Discovery, 6*, 1–47.
111. Bayes, M., & Price, M. (1763). An essay towards solving a problem in the doctrine of chances. by the late rev. mr. bayes, frs communicated by mr. price, in a letter to john canton, amfrs. *Philosophical Transactions (1683–1775), 53*, 370–418.
112. Barber, D. (2012). *Bayesian reasoning and machine learning*. Cambridge University Press.
113. Cios, K. J., & Moore, G. W. (2002). Uniqueness of medical data mining. *Artificial Intelligence in Medicine, 26*(1), 1–24.
114. Papik, K., Molnar, B., Schaefer, R., Dombovari, Z., Tulassay, Z., & Feher, J. (1998). Application of neural networks in medicine – A review. *Medical Science Monitor, 4*(3), MT538–MT546.
115. Subasi, A. (2006). Automatic detection of epileptic seizure using dynamic fuzzy neural networks. *Expert Systems with Applications, 31*(2), 320–328.
116. Kim, J., Shin, H. S., Shin, K., & Lee, M. (2009). Robust algorithm for arrhythmia classification in ECG using extreme learning machine. *Biomedical Engineering Online, 8*(1), 1.
117. Wajid, S. K., Hussain, A., & Luo, B. (2014). An efficient computer aided decision support system for breast cancer diagnosis using echo state network classifier. In *Computational Intelligence in Healthcare and e-health (CICARE), 2014 IEEE Symposium on* (pp. 17–24). IEEE.
118. Cheng, K.-S., Lin, J.-S., & Mao, C.-W. (1996). The application of competitive Hopfield neural network to medical image segmentation. *IEEE Transactions on Medical Imaging, 15*(4), 560–567.
119. Roth, H. R., Yao, J., Lu, L., Stieger, J., Burns, J. E., & Summers, R. M. (2015). Detection of sclerotic spine metastases via random aggregation of deep convolutional neural network classifications. In *Recent advances in computational methods and clinical applications for spine imaging* (pp. 3–12). Cham: Springer.
120. Hammerla, N. Y., Fisher, J., Andras, P., Rochester, L., Walker, R., & Plo¨tz, T. (2015). PD disease state assessment in naturalistic environments using deep learning. In *AAAI* (pp. 1742–1748).

121. Akay, M. F. (2009). Support vector machines combined with feature selection for breast cancer diagnosis. *Expert Systems with Applications, 36*(2), 3240–3247.
122. Razzaghi, T., Roderick, O., Safro, I., & Marko, N. (2016). Multilevel weighted support vector machine for classification on healthcare data with missing values. *PloS One, 11*(5), e0155119.
123. Keller, B. M., Nathan, D. L., Wang, Y., Zheng, Y., Gee, J. C., Conant, E. F., & Kontos, D. (2012). Estimation of breast percent density in raw and processed full field digital mammography images via adaptive fuzzy c-means clustering and support vector machine segmentation. *Medical Physics, 39*(8), 4903–4917.
124. Karaolis, M. A., Moutiris, J. A., Hadjipanayi, D., & Pattichis, C. S. (2010). Assessment of the risk factors of coronary heart events based on data mining with decision trees. *IEEE Transactions on Information Technology in Biomedicine, 14*(3), 559–566.
125. Freitas, A., Costa-Pereira, A., & Brazdil, P. (2007). Cost-sensitive decision trees applied to medical data. In *International Conference on Data Warehousing and Knowledge Discovery* (pp. 303–312). Springer.
126. Khan, M. U., Choi, J. P., Shin, H., & Kim, M. (2008). Predicting breast cancer survivability using fuzzy decision trees for personalized healthcare. In *2008 30th Annual International Conference of the IEEE Engineering in Medicine and Biology Society* (pp. 5148–5151). IEEE.
127. Khalilia, M., Chakraborty, S., & Popescu, M. (2011). Predicting disease risks from highly imbalanced data using random forest. *BMC Medical Informatics and Decision Making, 11*(1), 51.
128. Wei, W., Visweswaran, S., & Cooper, G. F. (2011). The application of naive Bayes model averaging to predict Alzheimer's disease from genome-wide data. *Journal of the American Medical Informatics Association, 18*(4), 370–375.
129. Gevaert, O., De Smet, F., Timmerman, D., Moreau, Y., & De Moor, B. (2006). Predicting the prognosis of breast cancer by integrating clinical and microarray data with Bayesian networks. *Bioinformatics, 22*(14), e184–e190.
130. Shouman, M., Turner, T., & Stocker, R. (2012). Applying k-nearest neighbour in diagnosing heart disease patients. *International Journal of Information and Education Technology, 2*(3), 220.
131. Hayat, M., & Khan, A. (2012). Discriminating outer membrane proteins with fuzzy K-nearest neighbor algorithms based on the general form of Chou's PseAAC. *Protein and Peptide Letters, 19*(4), 411–421.
132. Holt, A., Bichindaritz, I., Schmidt, R., & Perner, P. (2005). Medical applications in case-based reasoning. *The Knowledge Engineering Review, 20*(03), 289–292.
133. Cooper, B., & Lipsitch, M. (2004). The analysis of hospital infection data using hidden Markov models. *Biostatistics, 5*(2), 223–237.
134. Thomas, E., Temko, A., Lightbody, G., Marnane, W., & Boylan, G. (2010). Gaussian mixture models for classification of neonatal seizures using EEG. *Physiological Measurement, 31*(7), 1047.
135. McKinney, B. A., Reif, D. M., Ritchie, M. D., & Moore, J. H. (2006). Machine learning for detecting gene-gene interactions. *Applied Bioinformatics, 5*(2), 77–88.
136. Polat, K., S¸ahan, S., & Gu¨ne¸s, S. (2006). A new method to medical diagnosis: Artificial immune recognition system (AIRS) with fuzzy weighted pre-processing and application to ECG arrhythmia. *Expert Systems with Applications, 31*(2), 264–269.
137. Manju, A., & Nigam, M. J. (2014). Applications of quantum inspired computational intelligence: A survey. *Artificial Intelligence Review, 42*(1), 79–156.
138. Vilalta, R., Giraud-Carrier, C., & Brazdil, P. (2010). Meta-learning-concepts and techniques. In *Data mining and knowledge discovery handbook* (pp. 717–731). New York: Springer.
139. Burke, E. K., Hyde, M., Kendall, G., Ochoa, G., Ozcan, E., & Qu, R. (2009). *A survey of hyper-heuristics* (Computer science technical report no. NOTTCS-TR-SUB-0906241418-2747). School of Computer Science and Information Technology, University of Nottingham.
140. Gupta, S., Kulhara, P., et al. (2010). What is schizophrenia: A neurodevelopmental or neurodegenerative disorder or a combination of both? A critical analysis. *Indian Journal of Psychiatry, 52*(1), 21.

141. McCrone, P. R., Dhanasiri, S., Patel, A., Knapp, M., & Lawton-Smith, S. (2008). *Paying the price: The cost of mental health care in England to 2026*. London: King's Fund.
142. Whiteford, H. A., Degenhardt, L., Rehm, J., Baxter, A. J., Ferrari, A. J., Erskine, H. E., Charlson, F. J., Norman, R. E., Flaxman, A. D., Johns, N., et al. (2013). Global burden of disease attributable to mental and substance use disorders: Findings from the global burden of disease study 2010. *The Lancet, 382*(9904), 1575–1586.
143. Tandon, R., Gaebel, W., Barch, D. M., Bustillo, J., Gur, R. E., Heckers, S., Malaspina, D., Owen, M. J., Schultz, S., Tsuang, M., et al. (2013). Definition and description of schizophrenia in the DSM-5. *Schizophrenia Research, 150*(1), 3–10.
144. Gallagher, S. (2004). Neurocognitive models of schizophrenia: A neurophenomenological critique. *Psychopathology, 37*(1), 8–19.
145. Craddock, N., & Owen, M. J. (2010). The Kraepelinian dichotomy–going, going... But still not gone. *The British Journal of Psychiatry, 196*(2), 92–95.
146. Elkis, H. (2016). Treatment-resistant schizophrenia. *Psychiatric Clinics of North America, 39*(2), 239–265.
147. Bilder, R. M., Mukherjee, S., Rieder, R. O., & Pandurangi, A. K. (1985). Symptomatic and neuropsychological components of defect states. *Schizophrenia Bulletin, 11*(3), 409.
148. Warren, J. B., Lambert, W. E., Fu, R., Anderson, J. M., & Edelman, A. B. (2012). Global neonatal and perinatal mortality: A review and case study for the Loreto Province of Peru. *Research and Reports in Neonatology. Dovepress, 2*, 103–113.
149. Brown, R., Wijekoon, J. H., Fernando, A., Johnstone, E. D., & Heazell, A. E. (2014). Continuous objective recording of fetal heart rate and fetal movements could reliably identify fetal compromise, which could reduce stillbirth rates by facilitating timely management. *Medical Hypotheses, 83*(3), 410–417.
150. Rees, S., & Inder, T. (2005). Fetal and neonatal origins of altered brain development. *Early Human Development, 81*(9), 753–761.
151. Rees, S., Harding, R., & Walker, D. (2008). An adverse intrauterine environment: Implications for injury and altered development of the brain. *International Journal of Developmental Neuroscience, 26*(1), 3–11.
152. Ugwumadu, A. (2014). Are we (mis) guided by current guidelines on intrapartum fetal heart rate monitoring? Case for a more physiological approach to interpretation. *BJOG: An International Journal of Obstetrics & Gynaecology, 121*(9), 1063–1070.
153. Pinto, P., Bernardes, J., Costa-Santos, C., Amorim-Costa, C., Silva, M., & Ayres-de Campos, D. (2014). Development and evaluation of an algorithm for computer analysis of maternal heart rate during labor. *Computers in Biology and Medicine, 49*, 30–35.
154. Kessler, J., Moster, D., & Albrechtsen, S. (2014). Delay in intervention increases neonatal morbidity in births monitored with cardiotocography and ST-waveform analysis. *Acta Obstetricia et Gynecologica Scandinavica, 93*(2), 175–181.
155. Chud´aˇcek, V., Spilka, J., Burˇsa, M., Jank, P., Hruban, L., Huptych, M., & Lhotska´, L. (2014). Open access intrapartum CTG database. *BMC Pregnancy and Childbirth, 14*(1), 1.
156. Kreyszig, E. (2005). *Advanced engineering mathematics* (9th ed.p. 816). New York: wiley.
157. Gysels, E., Renevey, P., & Celka, P. (2005). SVM-based recursive feature elimination to compare phase synchronization computed from broadband and narrowband EEG signals in brain–computer interfaces. *Signal Processing, 85*(11), 2178–2189.
158. Abry, P., Roux, S. G., Chud´aˇcek, V., Borgnat, P., Gon,calves, P., & Doret, M. (2013). Hurst exponent and intrapartum fetal heart rate: Impact of decelerations. In *Proceedings of the 26th IEEE International Symposium on Computer-Based Medical Systems* (pp. 131–136). IEEE.
159. Schiermeier, S., van Leeuwen, P., Lange, S., Geue, D., Daumer, M., Reinhard, J., Gro¨nemeyer, D., & Hatzmann, W. (2007). Fetal heart rate variation in magnetocardiography and cardiotocography–a direct comparison of the two methods. *Zeitschrift fur Geburtshilfe und Neonatologie, 211*(5), 179–184.
160. Spilka, J., Chud´aˇcek, V., Koucky`, M., Lhotsk´a, L., Huptych, M., Jank, P., Georgoulas, G., & Stylios, C. (2012). Using nonlinear features for fetal heart rate classification. *Biomedical Signal Processing and Control, 7*(4), 350–357.

161. Signorini, M. G., Fanelli, A., & Magenes, G. (2014). Monitoring fetal heart rate during pregnancy: Contributions from advanced signal processing and wearable technology. *Computational and Mathematical Methods in Medicine, 2014*, 1–10.
162. of the European Society of Cardiology, T. F, et al. (1996). Heart rate variability standards of measurement, physiological interpretation, and clinical use. *European Heart Journal, 17*, 354–381.
163. Blagus, R., & Lusa, L. (2002). SMOTE for high-dimensional Class Imbalanced data. *BMC Bioinformatics, 16*, 321–357.
164. Chawla, N. V., Bowyer, K. W., Hall, L. O., & Kegelmeyer, W. P. (2002). SMOTE: Synthetic minority over-sampling technique. *Journal of Artificial Intelligence Research, 16* 321–357.
165. Organization, W. H, et al. (2012). *Born too soon: The global action report on preterm birth*. Geneva: World Health Organization.
166. Pooley, J. (2013). *Gestation-specific infant mortality in England and Wales, 2011*. London, UK: UK National Statstics.
167. Lucovnik, M., Maner, W. L., Chambliss, L. R., Blumrick, R., Balducci, J., Novak-Antolic, Z., & Garfield, R. E. (2011). Noninvasive uterine electromyography for prediction of preterm delivery. *American Journal of Obstetrics and Gynecology, 204*(3), 228–2e1.
168. PhysioNet. (2012). *The Term-Preterm EHG Database (TPEHG-DB)*. physionet.org.
169. Leman, H., Marque, C., & Gondry, J. (1999). Use of the electrohysterogram signal for characterization of contractions during pregnancy. *IEEE Transactions on Biomedical Engineering, 46*(10), 1222–1229.
170. Maner, W. L., Garfield, R. E., Maul, H., Olson, G., & Saade, G. (2003). Predicting term and preterm delivery with transabdominal uterine electromyography. *Obstetrics & Gynecology, 101*(6), 1254–1260.
171. Fele-Zorˇz, G., Kavˇsek, G., Novak-Antoliˇ c, ˇ. Z., & Jager, F. (2008). A comparison of various linear and non-linear signal processing techniques to separate uterine EMG records of term and pre-term delivery groups. *Medical & Biological Engineering & Computing, 46*(9), 911–922.
172. Phinyomark, A., Limsakul, C., & Phukpattaranont, P. (2009). A novel feature extraction for robust EMG pattern recognition. arXiv preprint arXiv:0912.3973.
173. Phinyomark, A., Nuidod, A., Phukpattaranont, P., & Limsakul, C. (2012). Feature extraction and reduction of wavelet transform coefficients for EMG pattern classification. *Elektronika ir Elektrotechnika, 122*(6), 27–32.
174. Richman, J. S., & Moorman, J. R. (2000). Physiological time-series analysis using approximate entropy and sample entropy. *American Journal of Physiology-Heart and Circulatory Physiology, 278*(6), H2039–H2049.
175. Weatherall, D. J. (2013). The role of the inherited disorders of hemoglobin, the first molecular diseases, in the future of human genetics. *Annual Review of Genomics and Human Genetics, 14*, 1–24.
176. Weatherall, D. J. (2010). The importance of micromapping the gene frequencies for the common inherited disorders of haemoglobin. *British Journal of Haematology, 149*(5), 635–637.
177. Weatherall, D. J. (2010). The inherited diseases of hemoglobin are an emerging global health burden. *Blood, 115*(22), 4331–4336.
178. Kosaryan, M., Karami, H., Zafari, M., & Yaghobi, N. (2014). Report on patients with non transfusion-dependent β-thalassemia major being treated with hydroxyurea attending the Thalassemia Research Center, Sari, Mazandaran Province, Islamic Republic of Iran in 2013. *Hemoglobin, 38*(2), 115–118.
179. Magoulas, G. D., & Prentza, A. (2001). Machine learning in medical applications. In *Machine Learning and its applications* (pp. 300–307). Berlin: Springer.
180. Consortium, G. P., et al. (2010). A map of human genome variation from population-scale sequencing. *Nature, 467*(7319), 1061–1073.
181. Bush, W. S., & Moore, J. H. (2012). Genome-wide association studies. *PLoS Computational Biology, 8*(12), e1002822.

182. Burton, P. R., Clayton, D. G., Cardon, L. R., Craddock, N., Deloukas, P., Duncanson, A., Kwiatkowski, D. P., McCarthy, M. I., Ouwehand, W. H., Samani, N. J., et al. (2007). Genome-wide association study of 14,000 cases of seven common diseases and 3,000 shared controls. *Nature, 447*(7145), 661–678.
183. Tabangin, M. E., Woo, J. G., & Martin, L. J. (2009). The effect of minor allele frequency on the likelihood of obtaining false positives. In *BMC proceedings* (Vol. 3, p. S41). BioMed Central Ltd.
184. Lam, A. C., Schouten, M., Aulchenko, Y. S., Haley, C. S., & de Koning, D.-J. (2007). Rapid and robust association mapping of expression quantitative trait loci. In *BMC Proceedings* (Vol. 1, p. S144). BioMed Central Ltd.
185. Florez, J. C., Manning, A. K., Dupuis, J., McAteer, J., Irenze, K., Gianniny, L., Mirel, D. B., Fox, C. S., Cupples, L. A., & Meigs, J. B. (2007). A 100K genome-wide association scan for diabetes and related traits in the Framingham Heart Study: Replication and integration with other genome-wide datasets. *Diabetes, 56*, 3063–3074.
186. Anderson, C. A., Pettersson, F. H., Clarke, G. M., Cardon, L. R., Morris, A. P., & Zondervan, K. T. (2010). Data quality control in genetic case-control association studies. *Nature Protocols, 5*(9), 1564–1573.
187. Laurie, C. C., Doheny, K. F., Mirel, D. B., Pugh, E. W., Bierut, L. J., Bhangale, T., Boehm, F., Caporaso, N. E., Cornelis, M. C., Edenberg, H. J., et al. (2010). Quality control and quality assurance in genotypic data for genome-wide association studies. *Genetic Epidemiology, 34*(6), 591–602.
188. Qiu, L., Na, R., Xu, R., Wang, S., Sheng, H., Wu, W., & Qu, Y. (2014). Quantitative assessment of the effect of KCNJ11 gene polymorphism on the risk of type 2 diabetes. *PloS One, 9*(4), e93961.
189. Consortium, D. S., Consortium, D. M., Mahajan, A., Go, M. J., Zhang, W., Below, J. E., Gaulton, K. J., Ferreira, T., Horikoshi, M., Johnson, A. D., et al. (2014). Genome-wide trans-ancestry meta-analysis provides insight into the genetic architecture of type 2 diabetes susceptibility. *Nature Genetics, 46*(3), 234–244.
190. Phani, N. M., Guddattu, V., Bellampalli, R., Seenappa, V., Adhikari, P., Nagri, S. K., Sydney, C., Mundyat, G. P., Satyamoorthy, K., Rai, P. S., et al. (2014). Population specific impact of genetic variants in KCNJ11 gene to type 2 diabetes: A case-control and meta-analysis study. *PloS One, 9*(9), e107021.
191. Cheema, A. K., Li, T., Liuzzi, J. P., Zarini, G. G., Dorak, M. T., & Huffman, F. G. (2015). Genetic associations of PPARGC1A with type 2 diabetes: Differences among populations with African origins. *Journal of Diabetes Research, 2015*, 921274.
192. James, W. (2008). WHO recognition of the global obesity epidemic. *International Journal of Obesity, 32*, S120–S126.
193. Speliotes, E. K., Willer, C. J., Berndt, S. I., Monda, K. L., Thorleifsson, G., Jackson, A. U., Allen, H. L., Lindgren, C. M., Luan, J., Ma¨gi, R., et al. (2010). Association analyses of 249,796 individuals reveal 18 new loci associated with body mass index. *Nature Genetics, 42*(11), 937–948.
194. Harbron, J., Van der Merwe, L., Zaahl, M. G., Kotze, M. J., & Senekal, M. (2014). Fat mass and obesity-associated (FTO) gene polymorphisms are associated with physical activity, food intake, eating behaviors, psychological health, and modeled change in body mass index in overweight/obese Caucasian adults. *Nutrients, 6*(8), 3130–3152.
195. Freemark, M. S. (2010). *Pediatric obesity*. New York: Springer.
196. Poloz, Y., & Stambolic, V. (2015). Obesity and cancer, a case for insulin signaling. *Cell Death & Disease, 6*(12), e2037.
197. Bullen, V., & Feenie, V. (2015). The human cost of failing to address obesity. *British Journal of Obesity, 1*, 19–24.
198. Rao, K. R., Lal, N., & Giridharan, N. (2014). Genetic & epigenetic approach to human obesity. *The Indian Journal of Medical Research, 140*(5), 589.
199. Hung, C.-F., Breen, G., Czamara, D., Corre, T., Wolf, C., Kloiber, S., Bergmann, S., Craddock, N., Gill, M., Holsboer, F., et al. (2015). A genetic risk score combining 32 SNPs is associ-

ated with body mass index and improves obesity prediction in people with major depressive disorder. *BMC Medicine, 13*(1), 1.

200. Van Dijk, S., Molloy, P., Varinli, H., Morrison, J., Muhlhausler, B., Buckley, M., Clark, S., McMillen, I., Noakes, M., Samaras, K., et al. (2015). Epigenetics and human obesity. *International Journal of Obesity, 39*(1), 85–97.
201. Loos, R. J. (2012). Genetic determinants of common obesity and their value in prediction. *Best Practice & Research Clinical Endocrinology & Metabolism, 26*(2), 211–226.
202. Pirmohamed, M. (2011). Pharmacogenetics: Past, present and future. *Drug Discovery Today, 16*(19), 852–861.
203. Mardis, E. R. (2008). The impact of next-generation sequencing technology on genetics. *Trends in Genetics, 24*(3), 133–141.
204. Sung, W.-K. (2012). Bioinformatics applications in genomics. *Computer, 45*(6), 57–63.
205. Samish, I., Bourne, P. E., & Najmanovich, R. J. (2015). Achievements and challenges in structural bioinformatics and computational biophysics. *Bioinformatics, 31*(1), 146–150.
206. Higdon, R., Haynes, W., Stanberry, L., Stewart, E., Yandl, G., Howard, C., Broomall, W., Kolker, N., & Kolker, E. (2013). Unraveling the complexities of life sciences data. *Big Data, 1*(1), 42–50.
207. Rudin, C., Dunson, D., Irizarry, R., Ji, H., Laber, E., Leek, J., McCormick, T., Rose, S., Schafer, C., Van der Laan, M., et al. (2014). *Discovery with data: Leveraging statistics with computer science to transform science and society*. ASA (American Statistical Association).
208. Tanwani, A. K., Afridi, J., Shafiq, M. Z., & Farooq, M. (2009). Guidelines to select machine learning scheme for classification of biomedical datasets. In *Evolutionary computation, machine learning and data mining in bioinformatics* (pp. 128–139). Berlin/Heidelberg: Springer.
209. Sheehan, S., & Song, Y. S. (2016). Deep learning for population genetic inference. *PLoS Computational Biology, 12*(3), e1004845.
210. Rabbani, B., Tekin, M., & Mahdieh, N. (2014). The promise of whole-exome sequencing in medical genetics. *Journal of Human Genetics, 59*(1), 5–15.
211. Upstill-Goddard, R., Eccles, D., Fliege, J., & Collins, A. (2012). Machine learning approaches for the discovery of gene–gene interactions in disease data. *Briefings in Bioinformatics, 14*, bbs024.
212. Manolio, T. A. (2010). Genomewide association studies and assessment of the risk of disease. *New England Journal of Medicine, 363*(2), 166–176.
213. Moore, J. H., Asselbergs, F. W., & Williams, S. M. (2010). Bioinformatics challenges for genome-wide association studies. *Bioinformatics, 26*(4), 445–455.
214. Niel, C., Sinoquet, C., Dina, C., & Rocheleau, G. (2015). A survey about methods dedicated to epistasis detection. *Frontiers in Genetics, 6*, 285.
215. Montan˜ ez, C. A. C., Fergus, P., Hussain, A., Al-Jumeily, D., Abdulaimma, B., & Al-Askar, H. (2016). A genetic analytics approach for risk variant identification to support intervention strategies for people susceptible to polygenic obesity and overweight. In *International Conference on Intelligent Computing* (pp. 808–819). Springer.

Chapter 6
A Comprehensive Framework for Elderly Healthcare Monitoring in Smart Environment

M. Al Hemairy, M. Serhani, S. Amin, and M. Alahmad

6.1 Introduction

Population aging is being experienced in nearly all the countries across the globe. The two main contributors are enhanced life expectancy and declining fertility. According to a report published by the United Nations, the global share of older people (aged 60 years or over) increased from 9.2% in 1990 to 11.7% in 2013 and will continue to grow upwards reaching 21.1% of the world population by 2050. The number of older persons (aged 60 years or over) is expected to be more than double, from 841 million people in 2013 to more than 2 billion in 2050 [49].

In most of the developed countries, the population is aged which results in low old-age support ratio, that is to say, the number of working-age adults per older person in the population. In this respect, the developing countries are also going into that direction within the next few decades [49]. While presently the old-age support ratio is favorable for the developing countries, the absolute numbers are not. At present, over two-thirds of the older persons live in developing countries. The growth rate of the older population in the less developed regions is growing faster than in the more developed regions and will continue to outpace that in the more developed regions; therefore it is believed that by 2050, nearly eight in ten of the world's older population will live in the less developed regions [62].

M. Al Hemairy (✉) • S. Amin
Faculty of Engineering and Computing, Coventry University, Coventry, UK
e-mail: SalihM@Coventry.ac.uk; S.Amin@coventry.ac.uk

M. Serhani
College of Information Technology, UAE University, Al Ain, Abu Dhabi, UAE
e-mail: serhanim@uaeu.ac.ae

M. Alahmad
College of Engineering, UAE University, Al Ain, Abu Dhabi, UAE
e-mail: M.Alahmad@uaeu.ac.ae

M. Dastbaz et al. (eds.), *Technology for Smart Futures*,
DOI 10.1007/978-3-319-60137-3_6

The modernity and rapid social change has affected the social structure. Nowadays, the atomic family in UAE, for example, has become the dominant family structure, and women are joining the workforce in ever-larger numbers. Such social change has, in many cases, weakened the bonds used to be the icon of UAE society. Caring for an aging family member can be a very stressful responsibility and, sometimes, entails a big financial burden as well.

It is well understood that the prevalence of noncommunicable diseases and disabilities increases as the population ages. Undoubtedly, an aging population leads to an increasing burden on the healthcare infrastructure, and rendering effective and timely healthcare in an efficient manner has emerged as a pressing challenge across the globe. All of the above clearly underlines a need for a multi-faceted strategy for innovation in healthcare management. This will enable healthcare services to perform in an effective and efficient manner in order to tackle the challenge of aging world population [62].

6.2 Background on Pervasive Health Monitoring

Due to current developments in ubiquitous computing, healthcare services availability and quality have noticeably increased [52, 60]. Pervasive healthcare concept is providing healthcare to anyone, at any time, and anywhere [59]. The integration of mobile technology and broadband communications and the proliferation of innovative medical devices have resulted in the development of pervasive healthcare. This development provides healthcare services to anyone at any time, beyond the restrictions of place, time, and quality. Pervasive healthcare does not evolve from the evolution of technology alone, but also from the patient's awareness and acceptance to embrace this technology [32, 50, 51].

Recently, many pervasive healthcare electronic devices have undergone experimental trials in hospitals as well as in patients' homes. Examples of these include sensors and video cameras, which use Wi-Fi Bluetooth, Infrared, or cellular GSM/3G networks to gain access to the Internet and reveal remotely their potential. Increasing instances of these devices were installed in healthcare facilities [59] as well as in patients' homes. Research on Pervasive Healthcare Technology (PHT) started in early 2000 to improve patient self-sufficiency, independent and healthcare mobility through continuous monitoring [59], which use evolving ubiquitous computing technologies and advance communication systems.

The traditional noninvasive PHT often requires patient's physical engagement in embracing medical devices at certain time and place. However, recent developments of pervasive monitoring systems focus on automated and un-obstructive PHT, which is not restricted by time and place. This is an extension of previous definition of pervasive healthcare from [59] as PHT is not only presented to anyone, at any time, and anywhere but also autonomously and unobtrusively. An earlier PHT experience used video-telephony [57] to provide live and interactive video communications through POTS (plain old telephone service) for its wide availability and relatively low costs [34]. Using video-telephony, healthcare professionals can review therapies and provide support in real time. More importantly, this approach minimizes the gap by allowing care providers to monitor the patient's emotional and mental status and not only the simple physiological information and vital signs [46].

Some other types of PHT are enabled by portable sensors, which integrate wireless technology and clinical devices. Examples of these devices include teledevices, such as tele-ECG (electrocardiogram, device used to measure heartbeat rate), and ring sensors that are generally carried by patients to benefit from PHT services. Data from ECG, pulse rate, respiration rate, and oxygen saturation levels are collected and forwarded to healthcare providers automatically [58, 59]. This continuously monitored data can provide important clinical insights for timely and accurate diagnosis regularly by recording of the vital signs such as blood pressure or heartbeat and enable end-to-end monitoring between the patient at any location with his medical caregiver and physician. Figure 6.1 [59] shows an example of comprehensive health monitoring system. Advanced pervasive devices that automatically collect multiple clinical indicators have already shown a successful deployment in body sensor network systems [44].

Pervasive health technology system equipped with multiple sensors is able to collect, process, and wirelessly transmit the received data via a secured link to a computer for further analysis. PHT devices that do not require patients to wear teledevices have also been developed in past years. For example, mattresses,

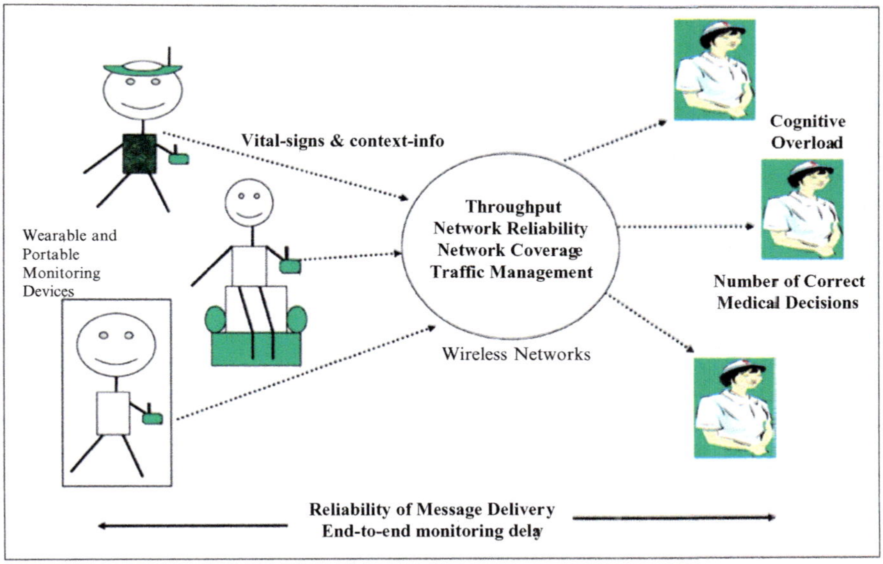

Fig. 6.1 Comprehensive health monitoring system [59]

toilets, kitchen appliances, and cloths are embedded with monitors to sense sleep pattern, body weight, body temperature, and pulse rate [9, 22]. Further experiments are done on advanced tele-sensing systems to gather scattered vital signs from the body [63].

These systems can gather multiple clinical parameters and are able to operate autonomously without disturbing the normal lives of patients. PHT is built on a widely deployed wireless network and advanced computing technologies. PHT solutions have focused mainly on risk disease management [4]. However, growing market incentives in a wide range of healthcare fields are propelling the development and consumption of PHT. Besides, this practice made profound influences on special healthcare to cater for the elderly, the disabled, the underserved, and the critically ill population of patients [61].

Healthcare benefits from the recent technology revolution are enormous. A group of researchers at "*Partners Healthcare in Boston*" has done several studies looking at using technology to conduct "*virtual visits*" between patients and primary care providers. In two studies, they showed that both patients and primary care providers felt that a "virtual visit" was indeed a useful alternative to the traditional in-person visit [34].

There is growing evidence that the benefits of pervasive healthcare are worth considering. One obvious benefit is that pervasive healthcare can often replace a physical visit, which is a tremendous convenience for the patient [47].

A survey involving 2000 physicians showed that 7% of the physicians report that they are using online conferencing systems to communicate with patients [42]. Among PHT developments, tele-monitoring is the fastest growing category that highly suits particularly aged patients for home-based monitoring. In 2004, research studies reported that the total healthcare technology market was about $380 million [45]. In 2007, it was predicted that the tele-home care market alone would grow three times to be in billions by 2010 [13] and this prediction has been actually validated [47].

In this work, two common types of PHT solutions were chosen: (1) noncommercial academic solutions (prototypes) and (2) industrial applications (commercial solutions). A comparison of these approaches has been carried out against certain features, i.e., security enabled, mobility aware, and integration support [33]. Then it proposed an alternative set of criteria, which is used to contrast salient contemporary for PHT solutions that have been developed lately.

The rest of this chapter is organized as follows. Section 6.3 surveys related work. Section 6.4 classifies the existing PHC systems as commercial and noncommercial PHT solutions and proposes a classification model and proposed comparison criteria used to discriminate between these two categories of PHT solutions. Section 6.5 employs those criteria in a comparative study and reveals an original tabular view of this study. Section 6.6 proposes an integrated and scalable framework for remote health monitoring. Section 6.7 is describing a novel algorithm for disease detection. Section 6.8 describes the implementation. Section 6.9 concludes the paper with a summary and some future suggested investigations.

6.3 Related Work

Remote health monitoring technologies are fundamentally implemented using sophisticated systems equipped with multiple sensors that are able to acquire, process, and transmit medical data through a secured communication channel to a central processing unit for analysis and diagnosis.

The fundamental architecture of smart health monitoring system consists of a user interface, some clinical devices, a transmission medium, and supporting software and hardware. The healthcare technology user interface is achieved through a healthcare technology workstation [22]. The healthcare technology workstation may be a simple personal computer (PC) or a personal digital assistant (PDA) or a mobile device. The user can interface for instance through a telephone pad, mouse, touch screen, remote control, joystick, and voice recognition system [18]. Clinical devices are connected to the healthcare technology workstation for local healthcare providers to capture patient vital signs or other clinical data, such as images and sounds.

It has been suggested that such systems may not require the patients to expressly engage with any medical devices. The abstraction is achieved through embedding sensors such as in mattresses, toilets, kitchen appliances, and clothing. Such embedded sensors can sense sleep pattern, body weight, body temperature, pulse rate, and so forth [10, 22]

Currently, researchers are conducting experiments on advanced tele-sensing modalities that employ the Doppler radar technique to gather scattered vital signs from throughout the body [63]. These systems can gather multiple clinical parameters and are able to operate autonomously without disturbing the lives of the patients. Various integral components of remote health monitoring system include a widely deployed wireless network and advanced computing technologies [5]. Various monitoring systems have been proposed for monitoring different clinical parameters such as ECG, blood glucose, blood pressure, and body temperature.

Application-based ECG monitoring systems include CardioNet's MCOT™, LifeSync's Wireless ECG System, AUBADE, ICER system, and Pinmed's Pelex-04. Among these, CardioNet's MCOT™ has the ability to enable the physician to be alerted and take action promptly when required [29]. While most of these require the sensors to be disposed on patient's body and involves connections to a device using leads, LifeSync's Wireless ECG System is unique and has the edge for being able to acquire and transmit data wirelessly [39]. AUBADE does propose a basic framework for communication of clinical data over a network; however, it does not provide abilities for full-scale integration between different stakeholders involved in healthcare. Some sensor-based ECG monitoring systems include Toumaz's Sensium™ Life Platform TZ2050 and Shimmer's Wireless ECG Sensor [14, 26, 28, 40]. Soteras Wireless ViSi Mobile™ System provides a mobile-based ECG monitoring system [23, 36, 43]. The ICER system is quite different among these systems, where it evaluates the severity level of parameters related to local hazard and vulnerability by using fuzzy expert system approach; furthermore, the ICER system provides recommendation on needed capabilities for managing each local parameter [30, 31]. The adopted methodology in the ICER system is used in this research.

Analogous monitoring systems have been proposed for other clinical parameters. Entra Health Systems' MyGlucoHealth, vitalo Ltd's SmartLAB® Global Glucose Monitor - D10419-SmartLAB Global, and FORA's D15b BG plus BP Monitor (Bluetooth Version) are examples of sensor-based blood glucose monitoring while AT&T's Diabetes Manager®, AgaMatrix's iBGStar blood glucose monitoring system, and LG's KP8400 cell phone with blood glucose monitor are some key mobile-based blood glucose monitoring systems [6–8, 17, 20, 35, 38, 41, 48, 54–56]. Similar specific monitoring systems have also been proposed for monitoring blood pressure and body temperature, among others.

While myriads of individual clinical parameters monitoring systems have been proposed in the literature, researchers have been making efforts to integrate different sensing modalities to develop systems that are capable of monitoring more than one clinical parameter. Some examples of such systems include A&D's CP-1 THW Complete Wireless Software Connected Health Monitoring System, CareMatrix's CareMatrix Wellness System (CWS), HoneyWell's HomMed Genesis™ DM Remote Patient Care Monitor, Bosch Healthcare's Health Buddy System, and Viterion's Viterion 100 Health Monitor & Viterion 200 TeleHealth Monitor. Most of these cited examples are configured to monitor blood glucose and blood pressure levels in addition to a few other clinical parameters.

Due to the emerging technology in the smart health monitoring nowadays, the energy consumption and usage can be optimized. A recent proposed system called Advanced Metering Infrastructure (AMI) presents new application for monitoring e-health power consumption [19].

While intensive research efforts are being made in remote health monitoring technologies, several challenges are still remaining. First and foremost, high initial cost is one of the strongest barriers to widespread adoption of remote health monitoring technologies. The equipment currently costs several thousands of dollars, and successful adoption entails bringing down the setup costs. There are currently no clear guidelines related to reimbursement of costs through health insurance.

Second most important barrier is the lack of industry standardization in the field of remote health monitoring. Remote health monitoring technologies invariably involves combination of diverse set of devices. In absence of standardization and regulatory guidelines, interoperability of individual components constituting the system poses a challenge. Other barriers related to data transmission and data security. Remote health monitoring necessitates reliable wireless telecommunication infrastructure, which may not be available especially in rural areas. Data security of private medical data of subjects under monitoring presents another challenge.

Another concern frequently voiced by healthcare professionals is that the continuous monitoring systems would lead to excessive influx of information, especially in institutional settings, where clinical data from a large number of subjects would be captured. In the absence of intelligent processing systems, the large amount of data would eventually require large-sized teams just to handle the large information influx and may actually increase the workload.

One of the main reasons that underlies many of the challenges described above is the monolithic architecture-based approach to build remote health monitoring system. In order to overcome several challenges and barriers to adoption of currently available remote health monitoring systems, the proposed framework introduced this research to cover most of the abovementioned challenges and to provide an integrated and scalable architecture that enables customization and flexibility in individual implementations based on the patient profile. It also provides a framework for achieving interoperability between different monitoring products and, thereby, enables reduction in costs, promotes standardization and interoperability, and ensures reliable and secure data transmission [11]

6.4 Classification of Pervasive Healthcare Systems

This section surveys existing pervasive healthcare solutions developed both in research/academia and in healthcare industry sectors. These solutions differ from their ability to expose and use their services, as well as their ability to support integration with other heterogeneous healthcare systems. In addition, they may also differ in the degree of which these solutions can provide high scalability solutions.

6.4.1 Research-Based Solutions

In the current era of technology, pervasive healthcare applications make it possible to provide effective and efficient medical homecare to patients in need. However, these pervasive applications may face several challenges and issues. These issues may involve patient's mobility, network connectivity, and the limited resources of mobile devices and sensors used to collect and transmit sensory data. Therefore, several noncommercial and prototype solutions may play an efficient role in addressing these issues and challenges. Several wireless and networking technologies are implemented in the field of healthcare in order to ensure the integrity of medical applications. Embedded devices such as bioshirts are considered as appropriate and efficient prototype solutions to incorporate innovative technologies and ensure proactive healthcare services. These applications allow healthcare specialists to get real-time and timely information from constantly analyzed and monitored health state data. Furthermore, noncommercial applications may also allow the users to seek alternative diagnosis by communicating the information regarding their healthcare conditions with several specialists [37].

6.4.2 Industrial/Commercial Applications

There has been an extensive set of pervasive healthcare solutions from the industry sector. Technology has also influenced the field of medicine and healthcare through a significant development support. Recent innovations in wireless transmission and biosensor technology have triggered the concept of potential convergences between healthcare and telecommunications. According to recent research and literature surveys, an extremely large number of healthcare organizations and hospitals are trying to adopt and implement commercial and industrial applications of PHTs in order to ensure a high level of confidentiality and integrity of their working procedures. Some of the industrial or commercial applications which are implemented by healthcare organizations may include developments such as wireless polymap systems, biometric wristwatch, connected shimmer, wireless ECG system, mobile cardiac outpatient telemetry (MCOT), and 2Net platform of wireless health solutions. All these commercial applications may significantly help the physicians in monitoring the daily activities and health conditions of patients. These solutions rely on the utilization and the integration of preprogrammed sensors [60].

6.4.3 Comparison Criteria

Table 6.1 incorporates two types of pervasive healthcare solutions that are used to address the issues and challenges of pervasive healthcare. These solutions may include noncommercial solutions (prototypes) and industrial applications (commercial products). In order to understand the efficiency of these applications, Table 6.1 shows some important features, which may or not be supported by the investigated applications. These features are non-intrusiveness, security, mobility awareness, integration support, and context awareness. Some of these features are briefly discussed in [60]. Next, subsections focus on the description of these criteria to demonstrate their relevance in contemporary PHT solutions.

6.4.3.1 Non-intrusive Pervasive Healthcare

Non-intrusive healthcare solutions are characterized by their ability to be used without disturbing the normal life and activities of a patient under observation. They may include sensors used to monitor vital signs, mobile devices used to collect sensory data, communication protocols used to transmit these data, as well as the complexity level when interfacing with patients. They also might include any physical disturbance of the patient's body such as medical examination where the areas of the body are not sensibly probed and the skin is not damaged beyond normal conditions. Non-intrusive devices allow the physicians to diagnose illnesses from their initial stages [37] to avoid further severe consequences.

Table 6.1 Classification of pervasive healthcare solutions [2, 3]

Technology Type	Name	Manufacturer	Classification criteria				
			Non-intrusive	Security enabled	Mobility aware	Support integration	Context-aware
Prototype based solutions (Non-commercial)	LiveNet	–	Yes	No	Yes	No	Yes
	AUBADE	–	Yes	No	Yes	Yes	No
	Bio-Shirt	–	Yes	No	Yes	No	No
	CP-1THW Wireless Software for Health Monitoring System	A&D	Yes	No	No	Yes	No
	DiaSend	–	Yes	Yes	Yes	Yes	Yes
	Numera	–	No	No	Yes	No	No
	Carematix Wellness System (CWS)	CareMatrix	Yes	No	Yes	Yes	Yes
	Health Buddy System	Bosch HealthCare	Yes	Yes	Yes	Yes	No
	e-bra/ e-vest	–	Yes	Yes	Yes	No	No
	MagIC	–	Yes	Yes	Yes	No	No
	LifeShirt™	–	Yes	No	Yes	No	No
	BioHarness™ BT	Zephyr Technology Corporation	Yes	Yes	Yes	No	No
	MyGlucoHealth	Entra Health Systems	Yes	No	Yes	Yes	Yes
	Biosign	Biosign Technologies Inc.	Yes	Yes	Yes	No	No

(continued)

Table 6.1 (continued)

Technology Type	Name	Manufacturer	Classification criteria				
			Non-intrusive	Security enabled	Mobility aware	Support integration	Context-aware
Existing industrial/ commercial applications	2Net Platform	QualComm	Yes	Yes	Yes	Yes	Yes
	MCOT (Mobile CardiacOutpatient Telemetry)	CardioNet	Yes	No	Yes	Yes	Yes
	Wireless ECG System	LifeSync	Yes	No	No	Yes	No
	ShimmerConnect	Shimmer	Yes	Yes	No	No	No
	Biometric wristwatch	Hitachi	Yes	Yes	Yes	No	Yes
	Equivital™ personal devices	Equivital	Yes	Yes	Yes	Yes	Yes
	Polymap Wireless "Polytel System	A&D	Yes	Yes	Yes	No	No
	MedStar	CyberNetMedical	Yes	Yes	Yes	Yes	Yes
	HomMed Genesis™ DM Remote Patient Care Monitor	HoneyWell	Yes	No	No	Yes	Yes
	Viterion 100 TeleHealth Monitor	Viterion	No	Yes	Yes	No	Yes
	VitelNet (Mobile Health Monitoring)	VitelNet	No	No	Yes	Yes	Yes
	Medapps (Web & Cloud based solutions & Products)	Medapps	Yes	Yes	Yes	Yes	Yes
	Mindray (Patient Monitoring System)	Mindray	Yes	Yes	Yes	Yes	Yes
	Mercury Parkinson's disease Epilepsy	–	Yes	No	Yes	Yes	Yes
	CodeBlue: Wireless Sensors for Medical Care	–	Yes	No	Yes	Yes	No
	Aubade	–	Yes	No	Yes	Yes	No
	Pelex-04	Pinmed	Yes	Yes	Yes	No	No
	Sensium™ Life Platform TZ2050	Toumaz	Yes	Yes	Yes	Yes	No

6.4.3.2 Security-Enabled Devices in Pervasive Healthcare

In the field of medicine and healthcare, it is necessary for healthcare organizations to ensure the confidentiality and integrity of patients' information. Thus, security-enabled devices may play an effective role in securing the information of patients from any disclosure or misuse. In PHTs, security-enabled devices may greatly help healthcare organizations to protect and secure sensitive private information in order to improve and enhance the quality of patient care. These devices may also help healthcare organizations in reducing risks and vulnerabilities regarding their confidential information while ensuring the availability and integrity of that information [37, 60].

6.4.3.3 Mobility-Aware Devices in Pervasive Healthcare

Mobility-aware devices are playing a vital role in ensuring the effective and appropriate healthcare operations. These devices allow healthcare organizations to stay updated about the medical conditions of patients, as well as about their health-related activities anywhere and anytime, in local and even global contexts [9, 60]. This can be allowed thanks to mobile devices that are connected to backend systems via wireless, 3G, or 4G networks.

6.4.3.4 Integration Support Across Heterogeneous Pervasive Healthcare Systems

In heterogeneous pervasive healthcare, integration support is an important feature to support communication and interoperation of PHTs to provide a high level of collaboration involving healthcare professionals. This communication interoperability and professional collaboration may greatly help patients to receive superior diagnosis and quality treatments [37]. Standard protocols will allow full integration of heterogeneous healthcare systems to offer a complete set of services independently of the underlying infrastructure and used architecture.

6.4.3.5 Context-Aware Devices in Pervasive Healthcare

Context-aware devices may play an effective role in ensuring the integrity and confidentiality of the patients' information [16]. Context awareness refers to "any information that can be used to characterize the situation of an entity" [25].

Healthcare organizations manage and collect contextual data, both unstructured and structured from a variety of different sources. Context awareness reveals indicators about location and other situational patterns of mobile users and may provide hints to medical practitioners about prevailing trends. Context-aware PHTs in hospitals can also integrate related data of patients and medical staff such as location, staff duty time into the clinical workflow to quickly locate appropriate staff and equipment to increase patient's satisfaction [21]. These types of PHTs may help healthcare organizations in collaboratively archiving, delivering, managing, and creating the data required for the operational requirements in healthcare centers, as described above [11, 37, 60]

6.4.4 Discussion and Analytics

Based on the comparative study shown in Table 6.1 which contrasts existing pervasive healthcare solutions from academia and industry, it has been noticed the following relationships and patterns between commercial and noncommercial pervasive solutions and some common features as described below:

- Most of the compared solutions (either noncommercial or commercial) were designed to be non-intrusive, i.e., among more than 30 solutions that were studied, only three systems have failed to characterize non-intrusiveness, one from academia (*Numera*) and two other solutions from the industry sector. This shows that the convenience and non-disturbance of a patient's daily activity are basic rules for any attempt to introduce PHT solutions.
- Mobility awareness is strongly integrated for both types of solutions. More than 90% of noncommercial and about 80% of commercial models are supporting mobility. The few exceptions, for instance, Wireless ECG or *ShimmerConnect*, are relatively more processing power-aware than the other solutions and require continuous data transmission to a base station. In general, the wide adoption of mobility awareness feature in most of PHT solutions reflects the support telecommunication technologies such as mobile devices, Wi-Fi, and 3G/4G networks to offer novel healthcare monitoring services to patients anywhere and anytime.
- More than 70% of the noncommercial prototypes are not context-aware though, while more than 60% of the investigated commercial solutions are context-aware. This shows that the academic research is focusing on more challenging features such as non-intrusiveness and mobility rather than context awareness and security. This trend may be attributed to the fact that these solutions are developed for demonstration, proof of concept or testing purposes to server research ends. Hence, they are not widely introduced for community use. While commercial products are facing market competition to launch full-loaded devices with features that are essential to users, such as integration with other appliances and high security.
- Security-enabled feature is evidenced in about 50% of both commercial and noncommercial solutions. This shows that the importance of preserving patients' data privacy and security is less insignificant to PHT developers. However, this is a weakness that will hinder future expansion of PHTs and has to be addressed in any future solutions to enhance the integrity and confidentiality of patient's data.

- More than 70% of commercial prototypes support integration, while less than 50% of noncommercial support integration too with other devices or solutions. As mentioned earlier, academic research varies from the commercial production and therefore focuses on key features such as non-intrusiveness and mobility. However, integration is also a discriminating factor that characterizes a complete healthcare solution with wider range of services from different systems.
- Few solutions are somehow robust and implemented the five key features subject of comparison. Examples of these solutions include DiaSend and CareMatix from the noncommercial solution group, while 2Net/QualComm, Equivital, MedStar, MedApps, and Mindray recorded the five features and were equipped with all features.
- An interesting fact found while studying the listed pervasive solutions is that none of them wasn't equipped with at least one or two of the five features examined in this study. This shows that the selected features in this comparison are key features for any successful pervasive healthcare system.

The above features can increase the efficiency of PHT solutions. However, there is an innovative trend to integrate advanced smart data-intensive processing, elaborate analytics, and intelligent power management to sustain next generation PHT solutions.

6.5 Key Properties of Comprehensive Pervasive Healthcare Solution

In this section, the current properties of pervasive healthcare solutions that were identified and analyzed in the previous section are extended, with novel properties that might characterize the next-generation PHT solutions and eventually add value. These properties are mainly related to the following: smart behavior, data intensive management, and intelligent power conservation.

6.5.1 Smart Pervasive Healthcare

Smart healthcare solution refers to an end-to-end architecture that intelligently implements unique features such as: self-adaptation of services in different contexts and reacts proactively to critical situations. They behave autonomously to respond to different conditions and requirements. The following are some features of these prospected PHT solutions:

- *Self-adaptive*: healthcare systems should adapt to different patient environments, different disease types, and different network conditions.
- *Proactive behavior*: the system should take the necessary actions to avoid any severe or resource shortage conditions. Such actions might include the following: (1) send only urgent information when the mobile network is loaded, and (2) switch to standby mode and switch all activities when a patient is in an ideal condition.

- *Autonomous behavior*: the system should implement a set of intelligent processes that will be triggered to respond to different situations, analyze contextual parameters, execute a set of prescribed actions, and collect new contextual information to be used for future similar situations.

6.5.2 Data Intensive Management in Pervasive Healthcare

Pervasive healthcare generates a huge amount of data resulting from executing different types of operations, including monitoring and gathering health-related data. Health data processing and analytics and their interpretation reveal patterns and trends, which are crucial for health professionals. These processes require high-performance data centers, powerful servers, and advanced data analytics tools. Therefore, a pervasive healthcare solution could use the evolving Cloud infrastructure, platform, and services to guarantee high-performance, scalable, and reliable healthcare services. Advanced processing and analytics tools could be used for intensive health data processing, for example, *NoSQL* and *Hadoop* platforms [15, 27], respectively.

6.5.3 Intelligent Power Conservation

Since pervasive healthcare relies on mobile devices to provide services to mobile patients, these devices are susceptible to rapid power drainage. This is a real challenge since this will induce disruption of services if the device's battery is fully drained. However, in an urgent situation, crucial health data should be communicated to physicians or emergency caregivers, which might have severe consequences if not timely communicated. Therefore, intelligent algorithms should be implemented within the mobile to reduce processing tasks at the mobile device and maybe delegate some processing to a backend infrastructure, in addition to disabling all processes running on the background and are not needed [24].

6.5.4 Social Network Integration

In general, pervasive healthcare has greatly improved, to provide continuous monitoring as an essential part of future healthcare systems. The chapter summarized and contrasted some salient pervasive healthcare technologies and surveyed their capabilities and shortcomings. It also identified a set of criteria used to undertake our comparative study. Most of these products were released in the first decade of this century and therefore the research is constricted to that time period. Some of the investigated solutions are already operational though on mobile platforms. They appear easier to use and can ready to be integrated into the user's life making personal care and maintenance significantly easier than before [53]. A number of challenges are still open though in this area, for which further research should be developed. At this paper some future prospects to shape the future of PHT solutions have been identified.

6.6 Integrated and Scalable Framework for Remote (Health) Monitoring: "IS-arm"

6.6.1 Overview Description

The proposed healthcare monitoring system is for the aged care at home with round-the-clock health monitoring in a non-intrusive manner. It helps the elderly to survive independently at their homes by assisting them in their regular day-to-day activities and their activities of daily living (ADL) to make their life easier, more entertaining, and pleasant through consistent interactions across services.

The proposed solution provides an integrated smart technologies infrastructure that monitors the vital signs of the patient at home round the clock. It provides elderly with instantaneous feedback based on their current medical conditions as well as communicates with their physicians for advices or calls in emergency at contingency situations. The proposed system consists of sensors and actuators to provide continuous health monitoring for elderly and helps them to carry out their daily routines individually. It also measures the ability of normal activities of aged care and how active or normal an elderly person and improves the lives of the elderly while they are at home.

The system will enable physicians to support aged community by providing advice, diagnoses, and treatments. This will minimize the number of visits to the hospitals, which will reduce the cost of elderly care and, at the same time, reduce the burden on healthcare providers. In addition, it will allow access to advanced and specialized care to communities in remote locations with no access to specialized medical centers in near vicinity. This will result in minimizing duration of hospital stay or the need for intensive caretaking at home.

The proposed system will provide a state-of-the-art platform to monitor health of the elderly and disabled people round the clock in a smart-home environment seamlessly and non-intrusively. The system can address the problem of providing an expensive home-care visits or establishing a dedicated elderly centers by government authorities. A smart-home for healthcare monitoring is a cost-effective, environment-friendly, and sustainable system to take care of this important component of the society. The key objectives of the proposed system are listed below:

1. Provides an integrated and customizable platform implemented using set of sensors and actuators within the subject's home environment supporting medical data acquisition in a seamless and non-intrusive manner without disturbing the normal daily activities of the subjects.

2. Provides healthcare professionals with access to the medical data acquired at home supporting the healthcare professionals to monitor the medical conditions of the subjects remotely.
3. Generates real-time alerts for healthcare professionals to enable timely and effective intervention depending on medical condition of the subject under monitoring.
4. Provides real-time advices related to the treatment and medication for the elderly depending on their health condition through suitable display devices.
5. Provides real-time nutrition advice related to diet programs, sport practices, nutrition plans to enable preventive health management.
6. Enables time and cost-efficient administration of healthcare services to elderly resident at home or old-age care centers.
7. Facilitates early discharge of non-vital cases to retain their normal life at home in a reliable and safe environment while enabling continued monitoring of health condition for the desirable lengthy time.
8. Provides health authorities with access to medical data of elderly suffering from chronic diseases for a long-term period which helps in data analytics and finding patterns regarding chronic diseases in the country and other statistics for health-related issues.

The key differentiator of the proposed system is the modular architecture with clearly defined interfaces between different modules and standardization of communication protocols, which enables scalability and interoperability.

6.6.2 New Proposed Framework

The framework proposed in Fig. 6.2 describes the monitoring of different clinical parameters for the elderly. It involves data acquisition, processing, and analysis of the acquired data using an artificial intelligence-based diagnostic engine. It renders the most appropriate recommendations to the concerned users in suitable display, i.e., visual display, screen, and voice command. While the framework exploits artificial intelligence to automate processing of large volumes of inflowing data, it also provides flexibility of manual intervention. A medical expert may initiate manual intervention on his accord or in response prompted by the system for doing so. At the beginning, the patient vital signs will be captured by various non-intrusive and wearable sensors and will be sent to the data acquisition modules. The data then will be filtered from the noise and invalid values and stored in the database. Various data will also be obtained from other sources such as knowledge database and social network. If there is an emergency case, such as falling, the intervention module will be triggered immediately.

The expert system will examine the obtained data from the patient and check the predefined rules for matching cases and will take the appropriate action accordingly. The action will be in a form of advises, recommendations, or generating new rules

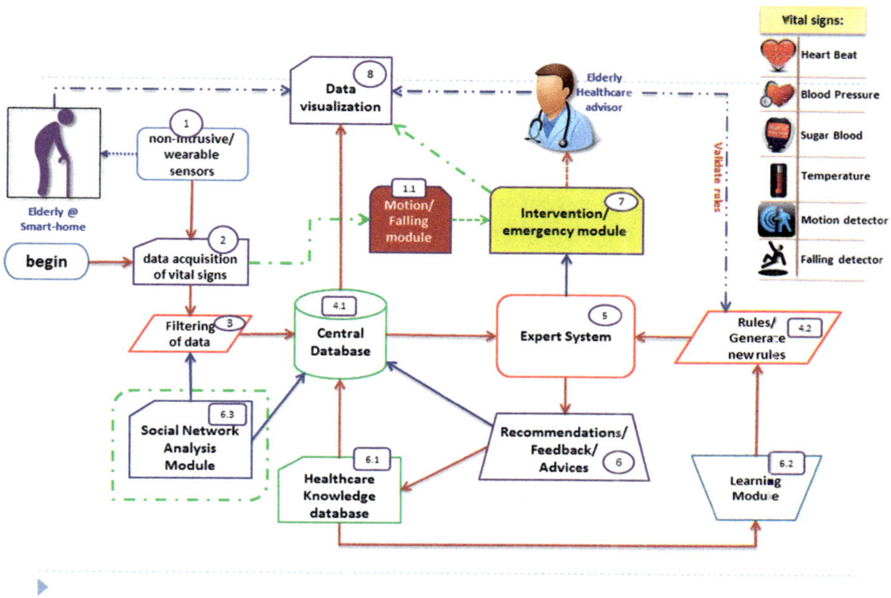

Fig. 6.2 Framework for remote health monitoring in smart-home [3]

in case it was not defined in the knowledge base through the learning module. To ensure rule validation and the system accuracy, no new rules will be added without human expert validation. The system supposed to display the information to both the patient and the physician simultaneously, such as warning, alarms, or medical advices.

The main modules of our proposed framework are listed below:

1. *Data Acquisition Module*: includes a network of sensors for acquiring clinical data of monitored subjects in their homes. Various examples of clinical data acquired through this module include heartbeat rate, blood pressure, sugar blood, temperature, motion, and falling detectors.
2. *Data Filtering Module*: filters the acquired data to eliminate noise, errors, or invalid values using different filtering techniques/algorithms including low-pass filters, high-pass filters, and noise removal/cancellation that can be used to pre-process sensory data before processing.
3. *Diagnostic Engine*: or expert system is implemented to effectively substitute a human medical expert. The engine is interfaced with data repository containing, for instance, a set of rules related to diagnostics, alert generation, and prompting manual intervention. Bayesian network for instance can be used for data classification. However, a novel algorithm for real-time diagnosing medical conditions (diseases) was developed using new mathematical expressions ("indicator value"). Further details about the algorithm will be explained in the next section.

4. *Publication Module*: or recommendation/advice module this module will interface with the diagnostic engine and publish advice and/or recommendations to suitable target devices (monitored subjects, responsible healthcare professional, or both).
5. *Intervention/Emergency Module*: deals with urgent and emergency cases where the captured signs were identified by the diagnostic engine as extremely abnormal. It will generate an urgent warning to the responsible healthcare professionals as well as trigger other actions such as triggering a request for dispatching ambulance to the subject's location. Additionally, the module will allow the medical staff to review and validate or update the final advice for the patient.
6. *Learning Module*: this module will be configured to update the data repository with new rules whenever required. It learns from the new recommendations/feedback provided by the medical expert and the previous experiences. Then, it will continuously improve the knowledge base captured using data classification technique such as "Bayesian classifier" to generate accurate and valid medical advice and, therefore, result in the best possible decision.
7. *Data Visualization Module*: it is responsible for rendering the recommendation/advice generated by the diagnostic engine or provided by the medical expert through the manual intervention module for both the elderly patient and the healthcare staff.
8. *Motion/Falling Module*: deals with emergency situations and consists of detecting fall incidents. It has been separated from the other medical conditions since it requires an immediate attention and high processing. It has been mapped directly to the intervention/emergency module such that necessary action is taken immediately, such as calling an ambulance and alerting the responsible healthcare professional.
9. *Social network analytics module*: this module will collect data from social networks, filter this data, and analyze it to get some relevant insights that will help in the monitoring. Analysis resulted will be used also to enrich the knowledge base with new knowledge and possibly derive new rules.

Some other components are supportive components that complement the roles of the above modules, for examples:

10. *Central Database*: it is used to store the medical data from various sources, which encompasses data from the sensors, the Knowledge-Base module, Feedback/Advices module, and external databases such as "Social Networks".
11. *Knowledge Database*: it uses data from various sources, for example, data acquired from the sensors, patient's medical history information, and knowledge from other external databases such as data collected from the "*social networks*." We will explain the concept of acquiring medical knowledge from the social network in the following section.
12. *Generating New Rules*: this process is related to the learning module; if a new medical condition is detected, then a new rule will be generated after it has been validated by the human expert.

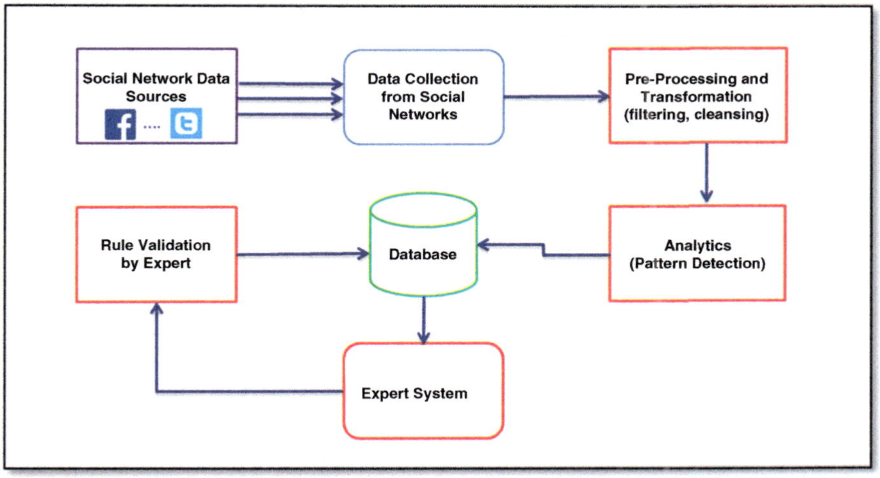

Fig. 6.3 Detailed view of the social network health sensing module

6.6.3 *Social Network Sensing Module*

This module is responsible of collecting health-related data from social network and doing some pre-processing and transformation activities (e.g., filtering, restructuration) on this data in order to derive some relevant health-related information and augment the knowledge base with this data. Data analytics could result in patterns detection, rule validation, and eventually new rule deduction. Figure 6.3 describes the Social Network module's components and the sequence of interactions among them:

1. Collect data from the social networks (i.e., Facebook and Twitter) as row data.
2. Filter the data to remove the unwanted or invalid data (or information), and keep only the relevant structured health-related data in specific format (i.e., XML, ERD).
3. Analyze the health data collected from the social networks and filtered in steps (1) and (2) and derive patterns and new rules.
4. Store the data in the central database.

6.7 Novel Algorithm for Disease Detection

Self-diagnosis of diseases becomes very important nowadays especially with the application of Artificial Intelligence techniques. Self-diagnosis will allow for early disease detection and for an appropriate and prompt treatment. Furthermore, due to the increasing spread of diseases and their corresponding symptoms nowadays, it becomes impossible for a physician to recall all symptoms and medical conditions for all kinds of diseases. Hence, it is introduced here an innovate algorithm that could be incorporated

```
Algorithm 1 DISEASE SEARCH ALGORITHM
Input: smarEhealthSensor // Vital signs database table
Output: ehealthIndex , detectedDisease
procedure LookupTableAlgorithm(smarEhealthSensor )
    eHealthIndex ← 0
    for all row ∈ smarEhealthSensor do
        val ← row[primaryValue] // vital sign value
        wf ← row[weightFactor]/100 // weight factor
        vsMin ← row[vitalSignMin] // min normal range
        vsMax ← row[vitalSignMax] // max normal range
// we include vital sign if its value is outside normal range
        if (val >=vsMin && val <=vsMax)
          subIndex ← 0
        else
          subIndex ← val * wf
        endif
        eHealthIndex ← eHealthIndex+subIndex // update ehealth index
    end for
    for all row ∈ smarEhealthSensor do
        dMin ← row[dMin]
        dMax ← row[dMax]
        if (eHealthIndex >=dMin && eHealthIndex <=dMax)
// a disease is detected add it to disease string
          detectedDisease ← detectedDisease.row[dName]
        endif
    end for
end procedure=0
```

Fig. 6.4 Disease search algorithm [1]

with a web-based platform to support any medical expert system in diagnosing the diseases based on the vital signs values acquired from the wearable sensors on a real-time basis. The algorithm uses a single and a unique "indicator" value to search in a lookup table for the predefined corresponding disease. The existing expert systems are the sequential or "serial" search algorithms. The latter incorporates inference rule method for disease diagnosis, which is a very sophisticated process that requires high resources in terms of computational time and energy consumption. The process of detecting the diseases using the new algorithm is formally specified in Fig. 6.4.

6.8 Implementation

In this section, monitoring system implementation is described. First, describe the system technical and nonfunctional requirements. Second, the monitoring system's components and key technologies used to implement them are depicted. Then, the set of health condition rules have been developed and are used for monitoring couple of health conditions (Fig. 6.5). Finally, it presents the implementation of the

Rules	Examples

Blood Sugar (BS)	IF (Fasting **Blood_Sugar** is **>= 200**) THEN (**FBglucose** is very *high*)
	IF (Fasting **Blood_Sugar** is **>= 125**) THEN (**FBglucose** is *high*)
	IF (Fasting **Blood_Sugar** is **50-70**) THEN (**FBglucose** is *low*)
	IF (Fasting **Blood_Sugar** is **>=70**) THEN (**FBglucose** is *normal*)

Blood Pressure (BP)	IF (**SysToLic** is 90-120) AND (**DiasToLic** is 60-80) **THEN** (BP is *normal*)
	IF (**SysToLic** is <90) AND (**DiasToLic** is <60) **THEN** (BP is *low*)
	IF (**SysToLic** is >120) AND (**DiasToLic** is >80) **THEN** (BP is *High*)

Falling	IF (**Falling_Sensor** is **= ON**) AND (Voice_Command = Respond) THEN (False_Alarm is = ON)
	IF (**Falling_Sensor** is **= ON**) AND (Voice_Command = No_Respond) THEN (Un_Conscious is = ON)
	IF (Un_Conscious is = ON) AND (Call_Check is = Respond) THEN (False_Alarm is = ON)
	IF (Un_Conscious is = ON) AND (Call_Check is = No_Respond) THEN (Call_Emergency)

Motion	IF **(Motion_Timer is= OFF >= 20 min) AND (Bed_Sensor** is **= OFF) THEN Motion_Module is = ON**
	IF (**Motion_Module** is **= ON**) AND (**Call_Check** is **= Respond**) THEN (False_Alarm is = ON)
	IF (**Motion_Module** is **= ON**) AND (**Call_Check** is **= No_Respond**) THEN (**Caregiver_Call**
	IF (**Motion_Module** is **= ON**) AND (**Caregiver_Call** is = No_Response) **THEN** (Call_Emergency)

Falling & Motion	IF (**Falling_Sensor** is **= ON**) AND (**Motion_Timer** is**= OFF >= 10 sec**) THEN (Call_Emergency)

Fig. 6.5 Sample set of diagnostic and monitoring rules

health monitoring system including both the web application and mobile application used to report monitoring data. A sample set of rules that may be configured in the data repository is provided in Fig. 6.5.

6.8.1 System Technical and Non-functional Requirements

The following are some technical requirement our monitoring system is going to support as well as some nonfunctional properties it is supporting.

13. *Application Server*: the system is hosted on Glass Fish application server and RabbitMQ servers running on a Linux platform in a cloud environment and connecting to MySQL Database servers. Various technologies like JSP, Servlet, EJB, MDB, and JDBC were used to develop features of our monitoring system. All communication with client complies with public HTTPS, TCP/IP communication protocol standards.
14. *Client*: users will be able to access the system through mobile application as well as via web applications browsers. HTM5 is used to support interoperability across mobile applications and provide high flexibility and dynamic features.

In terms of nonfunctional requirements, our system is developed to support the following features.

15. *Security*: access rights will be granted to any user accessing the application-landing page. Only administrator user can add or remove other creators and perform other administrative tasks.
16. *Persistence*: we use relational database for data persistence; data management rules were enforced to ensure consistency and accuracy of data.

17. *Performance*: there is no particular constrains related to system performance. It is anticipated that the system should respond to any request under standard database and web server script timeouts. Also system performance can depend on available hardware, network and Internet connection capabilities.

6.8.2 Prototype Implementation

The monitoring system is implemented using the following technologies and devices. Table 6.2 provides a brief description of these.

As mentioned previously, the healthcare monitoring system after collecting the data from various sources (sensors, knowledge base, social networks, etc.) will be checked against the set of rules stored in the system database. If the captured sensor value is within the abnormal range of a specific medical condition, then a matching case will be found, and a set of actions (i.e., recommendations) will be generated and displayed to the patients and/or the medical staff. All these rules are set and validated by the medical human expert in advance [12].

6.8.3 Mobile Application Implementation

The mobile application was developed on Android platform and has been deployed on different mobile devices including smartphone and tablet. Table 6.3 depicts different views of our monitoring system including medical staff view, patient view, and both the web application views and the mobile application views.

Table 6.2 System components and technologies

System	Description and use	Examples
Database server	Servers used to store data acquired by sensors, generated rules, medical advices, and recommendations	– MySQL – Oracle
Visualization devicestion devices	Display information to patient and physician based on user role (Sys Admin, physicians, elderly, nutritionist, etc.)	– Tablet, smart phones – LCD, dashboard, printout
Processing / application servers	Hosted locally (or as webserver) to process data and analyze conditions using expert system	– iCloud servers (hosting the expert system, process recommendations, and learning module)
Sensor devices	Set of sensors to capture different parameters from patient at home, i.e., wearable sensors	– Zephyr HxM (heartrate) for Android and iPhone – MyTech (blood pressure) – iBGStar (blood glucose)
Diagnostic engine	Expert system consists of set of rules and conditions based on medical case. Then it processes the rules and generates advices or triggers intervention emergency module if needed	Expert system is developed using Java-based engine

Table 6.3 System views for monitored subjects and healthcare professionals

<table>
<tr><td colspan="2">(a) Medical staff web application view</td></tr>
<tr><td>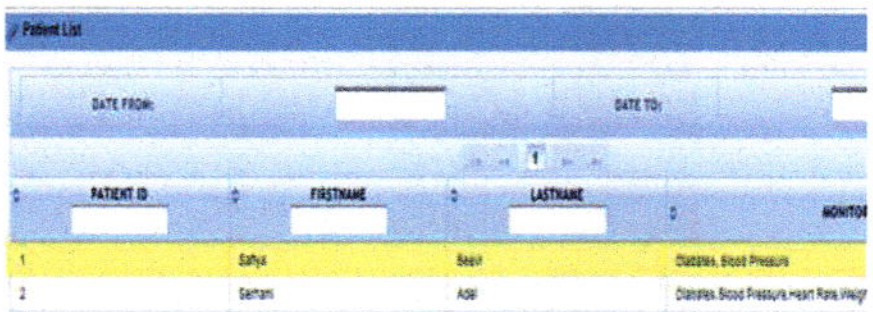
Real-time monitor for patient medical records</td><td>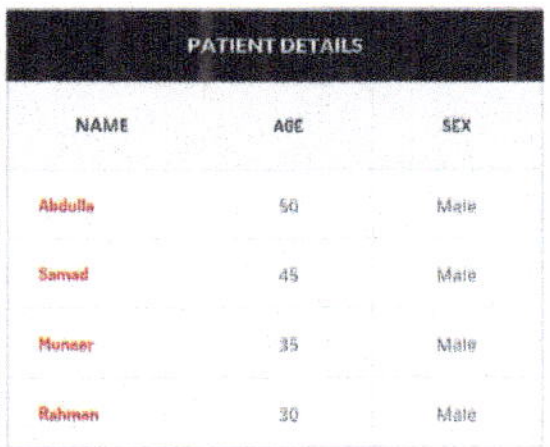

List of patients allocated to specific physician based on the medical case</td></tr>
<tr><td colspan="2">Figures above are examples of captured data and how it is processed and analyzed simultaneously to supply medical staff with up-to-date data and patient's history to provide the best medical advice</td></tr>
<tr><td>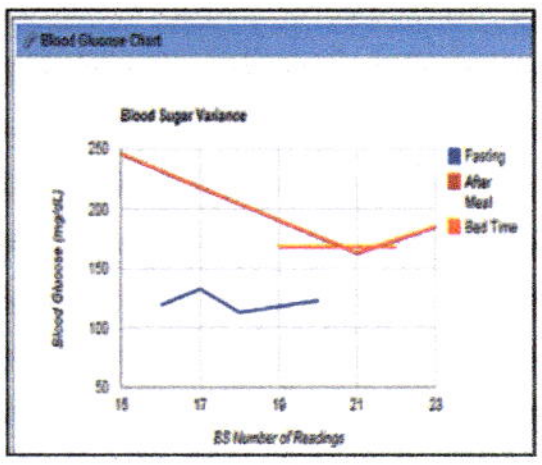
Blood glucose monitoring screen</td><td>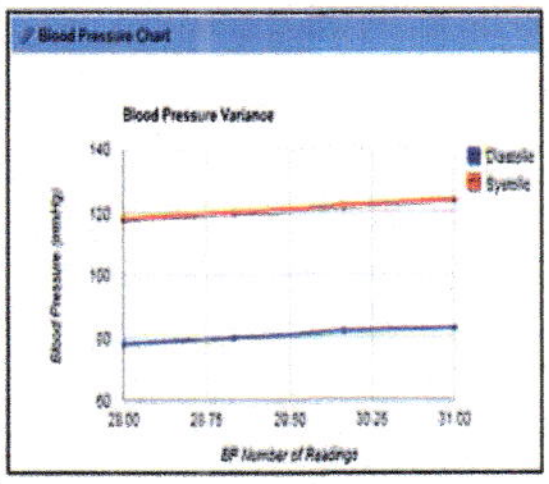
Blood pressure monitoring screen</td></tr>
<tr><td colspan="2">(b) Patient mobile application view</td></tr>
<tr><td colspan="2">The following are examples of how the data captured by the system will be processed and analyzed simultaneously to provide an up-to-date medical status for the patient</td></tr>
<tr><td>

Smart-home eHealth system home page
Features two login options:
Physician view
Patient view</td><td>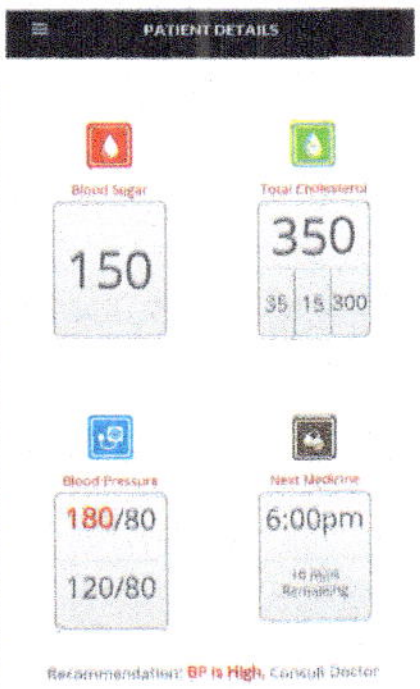

Simple and clear messages shown to patients</td></tr>
</table>

(continued)

Table 6.3 (continued)

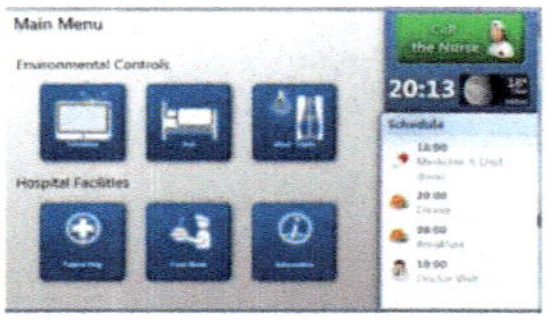 The eHealth system uses mobile devices as the target devices for visualization. Sample screenshots of a mobile-based application to be used with the proposed system have been provided	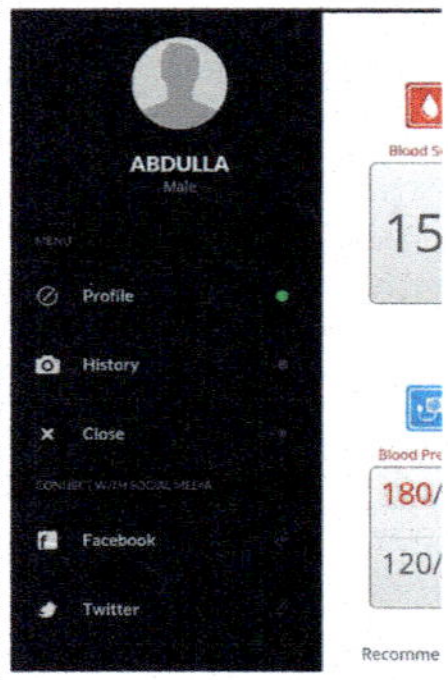 Patient view: user profile and other tools with simple view
(c) Physician mobile view	
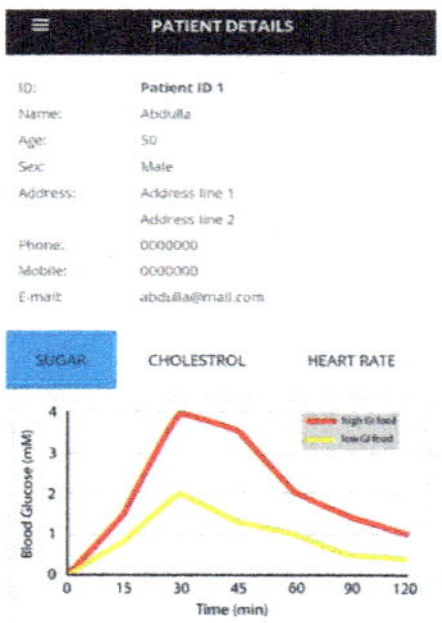 Physician view: patient's blood sugar records	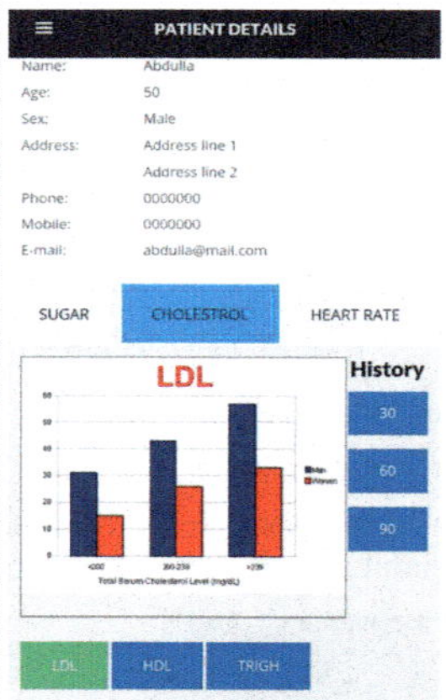 Physician view: patient's cholesterol level records
Physician view: patient's heart rate records	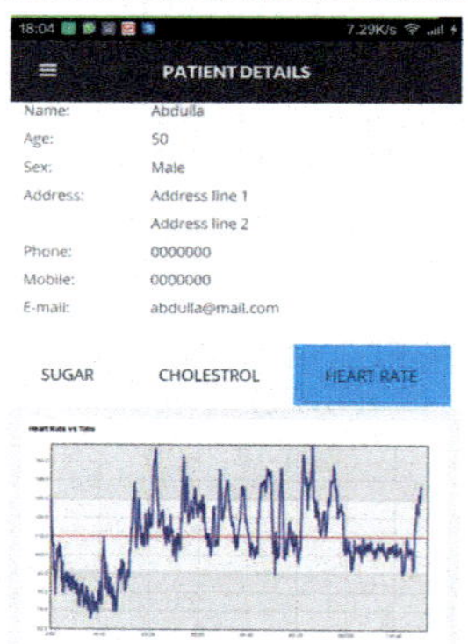

6.8.4 Discussion of Results

Based on the above sample of monitoring results and after experiencing different monitoring scenarios, our system proves its efficiency in detecting degradation of health condition after reporting abnormal readings.

For a considerable period of monitoring of 100 patients, the system was able to detect health deterioration with detection accuracy most of the time close to 100%. By scaling the number of monitored patients with different diseases (blood pressure, cholesterol, and blood sugar), the system remains stable and performs very well. No major performance issues or delay in communication and data transfer was recorded; only slight performance deterioration was reported due to network bandwidth and sensor calibration. For vital signs, blood sugar, cholesterol level, and heart rate, the monitoring results are always accurate and reflect the real situation. For failing and motion monitoring, the monitoring results are mostly appropriate; however, the accuracy in this situation is explained by the fact that some patient's movement might not be related to health deterioration but to infrequent physical practices the patient is used to do at home.

6.9 Conclusion and Future Work

As the world population ages and the old-age support ratios diminish, innovative and smart healthcare administration is imperative. An important cornerstone of healthcare future administration is the adoption of remote health monitoring technologies. While significant research and development work is currently being undertaken, several barriers related to costs, industry standardization, regulatory frameworks, user acceptance, data privacy and security, among others are yet to be overcome. This proposed monitoring system exploits various emerging technologies including biosensors, mobile technologies, and communication media to develop reliable, efficient, and complete solution, which easily can scale with a number of users, sensors, and homes. Thus, the proposed system addresses several key barriers related to widespread adoption of remote health monitoring technologies.

As future work, we are planning to extend our monitoring model to cope with varying number of vital signs and learn the correlation between them. We are also planning to implement and integrate social network module to retrieve data from social network with sensory data to offer better insights. We also plan to evaluate our model on a large-scale scenario, involving a large number of patients and considering different health situations.

References

1. Al Ahmad M., Al-Hemairy, M., & Amin, A. inventor. (2016). *A Novel Algorithm for fast disease detection based on vital signs*. USA Patent Application no. US 62/377,223. Filed on 19 August 2016.
2. Al-Hemairy, M., Serhani, A., Atif Y., & Amin, S. (2013). Classification of Pervasive Healthcare Systems. In *Developments in eSystems Engineering (DeSE), 6th International Conference on eSystems Engineering* (pp. 105–110). Abu Dhabi.

3. Al-Hemairy, M., Serhani, M. A., Amin, S., Alahmad M., & Hijji M. (2016). Integrated and scalable architecture for providing cost-effective remote health monitoring, Developments in eSystems Engineering (DeSE). In *2016 9th International Conference on eSystems Engineering*, Liverpool.
4. Anderson, S., & Wittwer, W. (2004). Using bar-code point-of-care technology for patient safety. *Journal for Healthcare Quality, 26*(6), 5–11. doi:10.1111/j.1945-1474.2004.tb00527.x.
5. Anderson, K., Valenzuela, E., & Wittwer, G. (2011). Wine export shocks and wine tax reform in Australia: Regional consequences using an economy-wide approach*. *Economic Papers: A journal of applied economics and policy, 30*(3), 386–399. doi:10.1111/j.1759-3441.2011.00124.x.
6. Andre, D., & Teller, A. (2005). Health. Care. Anywhere. Today. *Studies in Health Technology and Informatics, 118*, 89–110.
7. Anliker, U., Ward, J. A., Lukowicz, P., Troster, G., Dolveck, F., Baer, M., Keita, F., Schenker, E. B., Catarsi, F., Coluccini, L., Belardinelli, A., Shklarski, D., Alon, M., Hirt, E., Schmid, R., & Vuskovic, M. (2004). AMON: A wearable Multiparameter medical monitoring and alert system. *IEEE Transactions on Information Technology in Biomedicine, 8*(4), 415–427. doi:10.1109/titb.2004.837888.
8. Banerjee, S., Steenkeste, F., Couturier, P., Debray, M., & Franco, A. (2003). Telesurveillance of elderly patients by use of passive infra-red sensors in a 'smart' room. *Journal of Telemedicine and Telecare, 9*(1), 23–29. doi:10.1258/135763303321159657.
9. Bardram, J. E. (2005). Activity-based computing: Support for mobility and collaboration in ubiquitous computing. *Personal and Ubiquitous Computing, 9*(5), 312–322. doi:10.1007/s00779-004-0335-2.
10. Bardram, J., & Christensen, H. (2007). Pervasive computing support for hospitals: An overview of the activity-based computing project. *IEEE Pervasive Computing, 6*(1), 44–51. doi:10.1109/mprv.2007.19.
11. Benharref, A., & Serhani, M. A. (2014). Novel cloud and SOA-based framework for e-health monitoring using wireless biosensors. *IEEE Journal of Biomedical and Health Informatics, 18*(1), 46–55. doi:10.1109/jbhi.2013.2262659.
12. Benharref, A., Serhani, M. A., & Nujum, A. R. (2014, August). Closing the loop from continuous M-health monitoring to fuzzy logic-based optimized recommendations. In *2014 36th Annual International Conference of the IEEE Engineering in Medicine and Biology Society* (pp. 2698–2701). IEEE.
13. Blobel, B. (2007). Comparing approaches for advanced e-health security infrastructures. *International Journal of Medical Informatics, 76*(5–6), 454–459. doi:10.1016/j.ijmedinf.2006.09.012.
14. Bohn, J., Coroamă, V., Langheinrich, M., Mattern, F., & Rohs, M. (2004). Living in a world of smart everyday objects – Social, economic, and ethical implications. *Human and Ecological Risk Assessment: An International Journal, 10*(5), 763–785. doi:10.1080/10807030490513793.
15. Borthakur, D. (2007). Hadoop. In *Hadoop*, Retrieved 4 September 2013 from http://hadoop.apache.org.
16. Bottazzi, D., Corradi, A., & Montanari, R. (2006). Context-aware middleware solutions for anytime and anywhere emergency assistance to elderly people. *IEEE Communications Magazine, 44*(4), 82–90. doi:10.1109/mcom.2006.1632653.
17. Camarinha-Matos, L. M., & Afsarmanesh, H. (2004). TeleCARE: Collaborative virtual elderly care support communities. *The Journal on Information Technology in Healthcare, 2*(2), 73–86.
18. Cattamanchi, A., Smith, R., Steingart, K. R., Metcalfe, J. Z., Date, A., Coleman, C., Marston, B. J., Huang, L., Hopewell, P. C., & Pai, M. (2011). Interferon-gamma release assays for the diagnosis of latent tuberculosis infection in HIV-infected individuals: A systematic review and meta-analysis. *Journal of Acquired Immune Deficiency Syndromes, 56*(3), 230–238. doi:10.1097/qai.0b013e31820b07ab.
19. Chalmers, C., Hurst, W., Mackay, M., & Fergus, P. (2015, October). Smart health monitoring using the advance metering infrastructure. In *Computer and Information Technology; Ubiquitous Computing and Communications; Dependable, Autonomic and Secure Computing; Pervasive Intelligence and Computing (CIT/IUCC/DASC/PICOM), 2015 IEEE International Conference on* (pp. 2297–2302). IEEE.
20. Chen, C. I., Liu, C. Y., Li, Y. C., Chao, C. C., Liu, C. T., Chen, C. F., & Kuan, C. F. (2005). Pervasive observation medicine: The application of RFID to improve patient safety in obser-

vation unit of hospital emergency department. *Studies in Health Technology and Informatics, 116*, 311–315.
21. *Cisco context-aware healthcare solution* (n.d..) Available at: http://www.21.com/web/strategy/healthcare/Context-Aware_for_Healthcare.html. Accessed 15 November 2016.
22. Coronato, A., & Pietro, G. D. (2010). Formal specification of wireless and pervasive healthcare applications. *ACM Transactions on Embedded Computing Systems, 10*(1), 1–18. doi:10.1145/1814539.1814551.
23. Cruz-Correia, R. J., Vieira-Marques, P. M., Ferreira, A. M., Almeida, F. C., Wyatt, J. C., & Costa-Pereira, A. M. (2007). Reviewing the integration of patient data: How systems are evolving in practice to meet patient needs. *BMC Medical Informatics and Decision Making, 7*(1). doi:10.1186/1472-6947-7-14.
24. De Jager, D., Wood, A. L., Merrett, G. V., Al-Hashimi, B. M., O'Hara, K., Shadbolt, N. R., & Hall, W. (2011, May). A low-power, distributed, pervasive healthcare system for supporting memory. In *Proceedings of the First ACM MobiHoc Workshop on Pervasive Wireless Healthcare* (p. 5). ACM.
25. Dey, A. K. (2001). Understanding and using context. *Personal and Ubiquitous Computing, 5*(1), 4–7. doi:10.1007/s007790170019.
26. Ducatel, K., Bogdanowicz, M., Scapolo, F., Leijten, J., & Burgelman, J. C. (2001). *Scenarios for ambient intelligence in 2010*. Office for official publications of the European Communities.
27. Edlich (2009). NoSQL. *In NoSQL*. Retrieved 1 September 2013 from http://nosql-database.org.
28. Hansen, T. R., Bardram, J. E., & Soegaard, M. (2006). Moving out of the lab: Deploying pervasive technologies in a hospital. *IEEE Pervasive Computing, 5*(3), 24–31. doi:10.1109/mprv.2006.53.
29. Hayes, A. F. (2014). *Introduction to mediation, moderation, and conditional process analysis: A regression-based approach*. New York: Guilford Publications.
30. Hijji, M., Amin, S., Iqbal, R., & Harrop, W. (2013, December). A Critical Evaluation of the Rational Need for an IT Management System for Flash Flood Events in Jeddah, Saudi Arabia. In *Developments in eSystems Engineering (DeSE), 2013 Sixth International Conference on* (pp. 209–214). IEEE.
31. Hijji, M., Amin, S., Iqbal, R., & Harrop, W. (2015). The significance of using "expert system" to assess the preparedness of training capabilities against different flash flood scenarios. *Lecture Notes on Software Engineering, 3*(3), 214–219. doi:10.7763/lnse.2015.v3.193.
32. Katz, J., & Rice, R. (2009). Public views of mobile medical devices and services: A US national survey of consumer sentiments towards RFID healthcare technology. *International Journal of Medical Informatics, 78*(2), 104–114. doi:10.1016/j.ijmedinf.2008.06.001.
33. Kenny, L. (2006). *Exploring the business and social impacts of pervasive computing*. Zurich: Swiss Re.
34. Lankton, N. K., & Wilson, E. V. (2007). Factors influencing expectations of e-health services within a direct-effects model of user satisfaction. *e-Service Journal, 5*(2), 85–112. doi:10.2979/esj.2007.5.2.85.
35. Lee, T., & Mihailidis, A. (2005). An intelligent emergency response system: Preliminary development and testing of automated fall detection. *Journal of Telemedicine and Telecare, 11*(4), 194–198. doi:10.1258/1357633054068946.
36. Lukowicz, P., Kirstein, T., & Troster, G. (2004). Wearable systems for health care applications. *Methods of Information in Medicine-Methodik der Information in der Medizin, 43*(3), 232–238.
37. Magedanz, T., Karmouch, A., Pierre, S., & Venieris, I. (2005). *Mobility aware technologies and applications: Second international workshop, MATA 2005, Montreal, Canada, October 17–19, 2005, Proceedings* (Vol. Vol. 3744). Berlin/Heidelberg: Springer.
38. Mahoney, D. F., & Tarlow, B. (2006). Workplace response to virtual caregiver support and remote home monitoring of elders: The WIN project. *Studies in Health Technology and Informatics, 122*, 676.
39. Mattila, E., Korhonen, I., & Saranummi, N. (2007). Mobile and personal health and wellness management systems, In *Pervasive computing in healthcare* (pp. 105–134). Boca Raton: CRC Press.
40. Bardram, J., Mihailidis, A., & Wan, D. (2007). *Pervasive computing in healthcare*. Boca Raton: CRC Press.
41. Morris, M. E. (2005). Social networks as health feedback displays. *IEEE Internet Computing, 9*(5), 29–37. doi:10.1109/mic.2005.109.

42. Munnelly, J., & Clarke, S. (2007, March). ALPH: A domain-specific language for crosscutting pervasive healthcare concerns. In *Proceedings of the 2nd workshop on Domain specific aspect languages* (p. 4). ACM.
43. Muras, J.A., Cahill, V., & Stokes E. K. (2006). In *2006 pervasive health Conference and workshops "a taxonomy of pervasive healthcare systems"*.
44. Nachman, L., Baxi, A., Bhattacharya, S., Darera, V., Deshpande, P., Kodalapura, N., Mageshkumar, V., Rath, S., Shahabdeen, J., & Acharya, R. (2010, May). Jog falls: A pervasive healthcare platform for diabetes management. In *International Conference on Pervasive Computing* (pp. 94–111). Springer Berlin Heidelberg.
45. Office of technology policy an overview of e-waste policy issues (2006). Available at: http://www.bvsde.paho.org/bvsacd/cd57/recycling/intro.pdf. Accessed 15 November 2016.
46. Olguin, D. O., Gloor, P. A., & Pentland, A. (2009, April). Wearable sensors for pervasive healthcare management. In *2009 3rd International Conference on Pervasive Computing Technologies for Healthcare* (pp. 1–4). IEEE.
47. Pai, F.-Y., & Huang, K.-I. (2011). Applying the technology acceptance model to the introduction of healthcare information systems. *Technological Forecasting and Social Change, 78*(4), 650–660. doi:10.1016/j.techfore.2010.11.007.
48. Philipose, M., Fishkin, K. P., Perkowitz, M., Patterson, D. J., Fox, D., Kautz, H., & Hahnel, D. (2004). Inferring activities from interactions with objects. *IEEE Pervasive Computing, 3*(4), 50–57. doi:10.1109/mprv.2004.7.
49. Rafalimanana, H., & Lai, M. (2013). *World population ageing 2013*. New York: United Nations, Department of Economic and Social Affairs. Population Division.
50. Riva, G. (2005). *Ambient intelligence: The evolution of technology, communication and cognition towards the future of human-computer interaction* (Vol. Vol. 6). Amsterdam: IOS Press.
51. Saha, D., & Mukherjee, A. (2003). Pervasive computing: A paradigm for the 21st century. *Computer, 36*(3), 25–31. doi:10.1109/mc.2003.1185214.
52. Satyanarayanan, M. (2001). Pervasive computing: Vision and challenges. *IEEE Personal Communications, 8*(4), 10–17. doi:10.1109/98.943998.
53. Schaechinger, U., Rockelein, W., Perk, A., Asbach, P., & Nerlich, M. (2003). NOAH – A mobile emergency care system. *Studies in Health Technolgy and Informatics, 97*, 147–158.
54. Scheffler, M., & Hirt, E. (2005). Wearable devices for telemedicine applications. *Journal of Telemedicine and Telecare, 11*(5), 11–14. doi:10.1258/1357633054461994.
55. Schrooyen, F., Baert, I., Truijen, S., Pieters, L., Denis, T., Williame, K., & Weyn, M. (2006). Real time location system over WiFi in a healthcare environment. *Journal on Information Technology in Healthcare, 4*(6), 401–416.
56. Sixsmith, A., & Johnson, N. (2004). A smart sensor to detect the falls of the elderly. *IEEE Pervasive Computing, 3*(2), 42–47. doi:10.1109/mprv.2004.1316817.
57. Thuemmler, C., Buchanan, W., Fekri, A. H., & Lawson, A. (2009). Radio frequency identification (RFID) in pervasive healthcare. *International Journal of Healthcare Technology and Management, 10*(1/2), 119. doi:10.1504/ijhtm.2009.023731.
58. Tu, Y.-J., Zhou, W., & Piramuthu, S. (2009). Identifying RFID-embedded objects in pervasive healthcare applications. *Decision Support Systems, 46*(2), 586–593. doi:10.1016/j.dss.2008.10.001.
59. Varshney, U. (2007). Pervasive healthcare and wireless health monitoring. *Mobile Networks and Applications, 12*(2–3), 113–127. doi:10.1007/s11036-007-0017-1.
60. Varshney, U. (2009). *Pervasive healthcare computing: EMR/EHR, wireless and health monitoring*. Boston: Springer Science & Business Media.
61. Washburn, S. C., & Hornberger, C. A. (2008). Nurse educator guidelines for the management of heart failure. *The Journal of Continuing Education in Nursing, 39*(6), 263–267. doi:10.3928/00220124-20080601-10.
62. World Population Ageing Report (2013) Available at: http://www.un.org/en/development/desa/population/publications/pdf/ageing/WorldPopulationAgeing2013.pdf. Accessed 1 October 2016.
63. Ziefle, M., & Rocker, C. (2010). Acceptance of pervasive healthcare systems: A comparison of different implementation concepts, pervasive computing technologies for healthcare (PervasiveHealth), 2010. In 4th International Conference on No Permissions.

Chapter 7
Technology Implementation Case Studies: Lincus Software as a Service

Adie Blanchard, Faye Prior, Laura Gilbert, and Tom Dawson

7.1 Introduction

Health and social care systems worldwide are facing increasing pressures, driven by limited resources, an ageing population and the increasing prevalence of long-term conditions [73]. In England, the last available figures indicate that 15 million people (a quarter of the population) have at least one long-term condition [69]. These figures are of concern, since this population are frequent users of health care and social care services, accounting for 50% of all general practitioner appointments, 70% of hospital bed days and 70% of total health and social care expenditure [69]. This burden to health and social care systems is unsustainable, particularly given that the number of people living with multiple long-term conditions will rise over the next decade [31].

Evidence of this burden is already reflected through several key problem areas with service efficiency, such as extended waiting times for accident and emergency admittance [52], general practitioner appointments [33], referral to treatments [53] and discharge from hospital [55]. The pressures have also been coupled with increasing financial restrictions, making it difficult for health and social care services to deliver within their means [51, 69]. As a result, there has been an increasingly urgent need to enhance service efficiency in order to deliver high-quality care at a lower cost [7].

A. Blanchard • F. Prior • L. Gilbert • T. Dawson (✉)
Rescon Ltd, 2 The Kilns, Penn Croft Farm, Crondall, UK, GU10 5PX
e-mail: tom.dawson@rescontechnologies.com

M. Dastbaz et al. (eds.), *Technology for Smart Futures*,
DOI 10.1007/978-3-319-60137-3_7

7.1.1 The Role of Technology

Technology has been implemented in health and social care services to help reduce the increasing gap between supply and demand for resources [45], whilst simultaneously providing an opportunity to improve services and outcomes at a reduced cost [6]. As this strain increases, it is accepted that technology, both software and hardware, will have a greater role to play in the management of health and wellbeing in both healthy individuals and those with long-term conditions [28, 41].

New technologies provide many potential advantages for organisations including improved efficiency, quality of care, health outcomes and provision of new services [74]. Technology can also transform how individuals engage with services and help them to manage their own health and wellbeing [66]. However, despite reported benefits, technology implementation remains challenging, with a lack of acceptance and uptake often stalling the benefits that could be gained from its use ([6], [16]; [40]).

The majority of research in the area of health and social care technology implementation thus far has focused on electronic health records (EHRs) which provide a digital version of a person's medical history. Though the potential benefits have been well documented, adoption has been slow [34, 38, 67]. Reported barriers to uptake have included design and technical concerns, cost of purchase, maintenance costs, resistance from clinicians, interoperability concerns, unclear return on investment and digital exclusion for multiple reasons [34, 49]. One example of a challenging EHR implementation in the UK is that of Cambridge University Hospitals NHS Foundation Trust in 2013 [70]. Extensive service disruption was reported shortly after implementation leading to a negative audit by the UK's Care Quality Commission (CQC) which reported eight areas requiring improvement. Three years later these issues were largely resolved, with benefits reported to include integration of patient-related information, cost savings in staff time through eliminating the need to retrieve paper notes and freeing up of clinic appointments as a result of the ability to review notes virtually.

Another technology opportunity worth addressing is telecare, which involves the use of technology to support delivery of health and social care to individuals within the home or wider community [4], for example, fall sensors to aid independent living in the elderly [61]. Telecare can improve cost-effectiveness and responsiveness of services, minimise the need for travel and provide reassurance to individuals in their homes [60].

Additionally, telehealth allows the remote exchange of clinical data (e.g. blood pressure readings) from an individual to health professionals, supporting the diagnosis and management of health conditions [61]. Telehealth can improve access to services at a lower cost, though the benefits may only be visible over a long period of time [19]. Although there are significant demonstrated benefits relating to both telehealth and telecare, barriers to implementation are evident and include threats to independence and a required level of technical competence [61].

In the UK, the Department of Health [68] conducted a Whole System Demonstrator randomised control trial involving 6191 patients (3030 with diabetes, heart failure or COPD) and 238 GP practices to identify telehealth and telecare capabilities. Early indications showed that when correctly used telehealth could reduce accident and emergency visits, emergency admissions, total bed days spent in hospital and risk of mortality [63, 68]. Despite this potential further analysis found no significant health economic benefits of the telehealth interventions utilised within in the Whole System Demonstrator. The study examined acceptability, effectiveness and cost-effectiveness of telehealth in 845 randomised individuals receiving telehealth compared to 728 receiving normal care. There were found to be no improvements in quality of life despite higher costs in the telehealth group. It was concluded that telehealth (as it stood in 2008–2009 at least) was not a cost-effective addition to standard support and treatment [30]. In addition to this, no significant reductions in service use were found when using telecare as implemented in the Whole Systems Demonstrator when compared with a control group [64]. Technology has, however, dramatically advanced since this study was completed, as have the approaches to implementing technologies, so it is important to consider current capabilities, and where the future is leading, when considering the potential health economic benefits of telehealth.

Whilst telehealth focuses on extending the care reach of health professionals, there is limited research relating to implementation of self-management technologies, such as mobile health applications, which can facilitate self-care and positive behavioural change to improve self-management and health outcomes [56] at a low cost [48]. The increasing ownership and popularity of mobile devices in addition to their progressively improving technological capabilities means they provide great potential as a platform for improving health management [18]. Though evidence relating to their impact remains sparse [42], supporting evidence is growing [72]. Despite the potential benefits to be gained from smartphone technology, implementation can be challenging and impacted by many factors such as usability, data security, funding, lack of evidence relating to cost-effectiveness, digital literacy and access to technology [46, 48, 72].

7.1.2 Barriers to Technology Implementation

With substantial advances in health technology, and the increase in funding set aside for future digital projects [66], there is a growing need to better understand the factors that contribute to the translation of development of a sound technology to successful implementation and ongoing use [56]. An area of understanding is that of the barriers to technology implementation, of which a considerable amount of literature has been published in the setting of health and social care.

It is evident that many barriers can impede adoption and ongoing use of new technologies [1]. Therefore, they must be anticipated and addressed throughout technology implementation to maximise success. Implementation can be problematic due to several interrelated factors [11], with a combination of technical, social, organisational, ethical, financial and legal factors identified as the key barriers to successful implementation [1, 11, 32, 38, 56, 61].

7.1.2.1 Usability

The usability of technology can be a barrier to implementation [11, 16]. Interfaces can be difficult to navigate, rely upon a high level of literacy or be available in limited languages. These can be blockers to widespread use, in particular for those with low literacy levels or learning disabilities [8], further extending health and social inequalities in these populations [24].

Usability issues can also arise when health and social care technologies are developed with minimal engagement from users, therefore not being user-friendly or providing the user-experience as anticipated by the end user [10, 14, 21]. User involvement can therefore be beneficial to codevelop the technology [29]. Perceived ease of use has also been found to impact intention to use the technology [40], though it is considered to have a weaker impact than usefulness [15, 20]. Stability and reliability of technology is also associated with effective implementation [13].

Familiarity and competency with technology can also impact adoption by health and social care staff, particularly given the time constraints of staff, which may limit the time available for the learning and familiarity process [20]. It is known that adoption can be slowed if it takes more time to learn how to effectively use a system [27]. This highlights the importance of intuitive user-friendly interfaces which are easy to understand, to increase the likelihood of use [45] and implementation success [57]. This is particularly important given the case that technology is often implemented in health and social care to help reduce workload; therefore perception of increased workload can lead to negative attitudes towards the system [71], and poorly designed technology can result in user frustration [70].

7.1.2.2 Usefulness

An extensive body of literature suggests that perceived usefulness is important for successful adoption [15, 20, 35] and intention to use technology [9, 32], though this has been questioned [40]. Nevertheless, the evidence suggests that the benefits of usage must be outweighed by the effort of using the technology if it is to be adopted by the user [15].

7.1.2.3 Interoperability

Interoperability is critical for the sustainability of technologies in health and social care systems [70]. A high level of interoperability is necessary to increase adoption and effective use of technologies [12]. Any new technology must be able to integrate with existing workflows, infrastructures and other existing technologies such as information technology systems [13], particularly given the future expectations for technologies to exchange information [70]. Implementation can therefore be severely and negatively impacted if a technology lacks interoperability with existing systems [2, 37].

7.1.2.4 Training and Support

As with any new practice introduced in to an organisation, and due to the uniqueness of new health and social care technologies, training and support for both staff and end users can be vital to implementation [20, 26] and can improve satisfaction with the technology [13]. Poor timing of training [44], lack of training or training that is not relevant to the individual roles of users [13] can impede success. Therefore, tailored, flexible and hands-on training delivered just prior to the start of implementation can be beneficial [13]. However, even with appropriate training, digital literacy may still pose a barrier to technology use [58]. To limit this barrier, it can be useful to identify those who may need additional support [13].

Unsatisfactory technical support has also been identified as a barrier to adoption [6]. Consequently, ongoing support, particularly early on in the implementation process, can be beneficial to sustain momentum and ensure technology use [38].

7.1.2.5 Organisational Change

Implementation of new technology in health and social care often requires organisational change [62] and can impact on working practices [10]. This can be met with high resistance at an individual and organisational level [16]. Resistance to adoption has been noted to have many potential causes including inconvenience and disruption to current workflows, unhelpful features and security and reliability concerns [47, 50]. Technology that is perceived as inadequate, or interfering with values, aspirations and roles is also likely to be resisted [11]. Poor compatibility with current work processes can also increase resistance to adoption [20].

Effective leadership that promotes technology implementation can help to reduce resistance to change [44]. At a senior level, management attitude [71] and support [45] have been found to impact success. Buy-in from senior management is vital to ensure appropriate resourcing of implementation [12]. However, leaders at any level can impact implementation [14]. ‘Champions’ who make personal commitments to the success of technology can transmit enthusiasm to staff [44] and help overcome resistance within the organisation [14].

Resistance to organisational change can be actively addressed by identifying core values, understanding stakeholder concerns and needs, creating a vision and a need for change and being responsive to organisational stress resulting from change [38]. Therefore, ongoing involvement of all stakeholders is crucial from the initial stages of implementation, particularly to increase perceived value of the technology [11, 21, 38], and address workflow concerns [58]. It is also beneficial if the perceived benefits of the technology are shared by its users [20], which can be facilitated by providing evidence supporting its use [14]. In addition, stakeholder expectations must be managed to what is realistically achievable in relation to technological capabilities, outcomes and timescales for implementation [12]. However, to realise the potential of a new technology, organisations must be willing to invest in the resourced required for implementation [14].

7.1.2.6 Process of Implementation

The process of implementation can be as important as the technology itself and can have a significant impact on success. 'Big bang' implementations, where the technology is implemented quickly and used immediately, tend to be resisted, perhaps since they fail to address barriers such as sufficient training and organisational acceptance [10]. 'Big bang' implementation can also lead to large scale disruption caused by the occurrence of unexpected issues [70]. In contrast, incremental implementation is more likely to be beneficial, particularly for larger organisations with more complex processes [45]. It is also useful for organisations to reduce workloads during the implementation period to provide users with additional time to learn how to use the technology [11].

As health and social care is a dynamic and ever-changing environment, an active change process taking on board user feedback is likely to be required [14]. The technology itself must therefore be readily adaptable to the changing needs and contexts of the organisation [11]. Without adaptation, technology may be a poor fit for the organisation, which can lead to increased resistance [14]. Flexibility in the design is crucial [5]. Knowing when and how to adapt an aspect of a technology to better fit workflows can be facilitated by senior staff and champions through redesign of workflows, training and support and addressing issues [11].

7.1.2.7 Cost of Implementation

The cost of technology and the costs associated with implementation are an important barrier, particularly if there is no existing evidence for return on investment [14, 45]. As with any innovation, there are many uncertainties surrounding implementation [58]. For this reason, running proof of concept pilots can be a beneficial strategy to give organisations the opportunity to test and reflect upon the use of the technology [14].

7.2 Case Studies: Background

Whilst there is extensive literature surrounding the barriers to technology implementation in health and social care, there are few practical lessons that provide real-world implementation guidance. The following case studies will discuss the barriers and practical challenges faced, and lessons learned, whilst implementing a 'software as a service' application within the health and social care sector.

The software as a service application to which these case studies relate is known as 'Lincus'. Lincus is a tool for recording and monitoring health and wellbeing information over time and provides both self-care and shared care capabilities. It is available to use via a web browser or can be downloaded as a companion application on iOS and Android mobile devices.

Lincus was initially developed to overcome the barriers associated with contemporary history taking, including memory, recall [39] and communication [65]. It allows individuals to quantify and record subjective and objective measures of health and wellbeing through the use of pictorial surveys (Fig. 7.1).

Events can be logged, enabling the impact they may have had on health and wellbeing to be examined. Nutrition, physical activity and multiple clinical measurements can also be recorded. Over time, the information entered provides a history of health and wellbeing, which can be communicated to stakeholders using a variety of data visualization formats (Fig. 7.2).

Lincus pilot trials have demonstrated usability across a diverse and challenging range of user groups. During these trials, users have reported numerous benefits, including improved health and wellbeing, identification of previously undiagnosed conditions and enhanced engagement and communication with service providers. However, the primary objectives of the pilot trials were feasibility and usability, and these benefits were secondary outcomes which could not be generalised as robust conclusions about the effectiveness of Lincus. Despite the positive primary and secondary outcomes of these trials, many implementation barriers were and continue to be faced. Lessons learned from successes and failures from each case study were used to inform future implementations, refining and improving the technology itself (iterative technology development), as well as the implementation process.

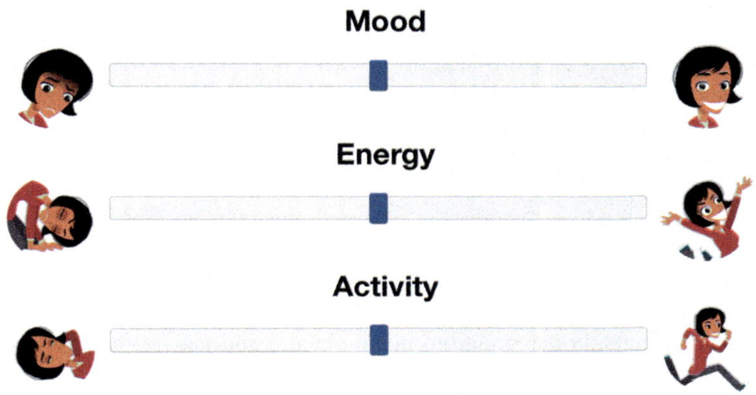

Fig. 7.1 Lincus pictorial surveys

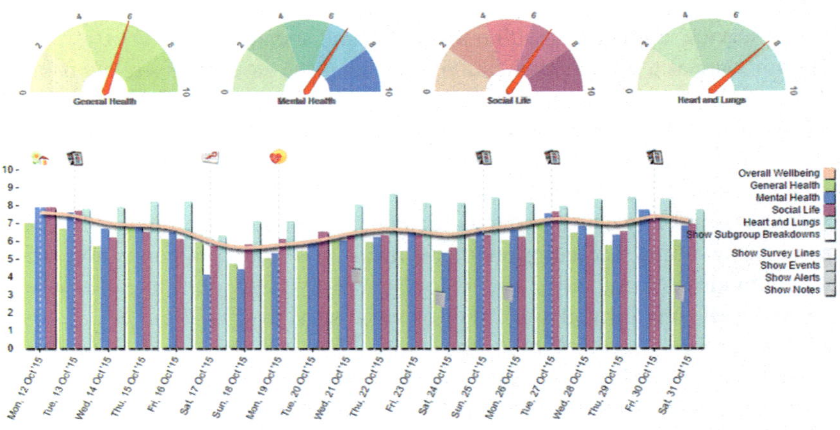

Fig. 7.2 Lincus results visualisation

7.3 Case Study 1: Supporting Individuals with Multiple and Complex Needs

7.3.1 Trial Summary

The first usability pilot trial of Lincus was conducted with a service that supports individuals with multiple and complex needs at a risk of homelessness [54]. This was a three-month usability trial, which began after considerable co-development with the organisation and stakeholders to customise the tool to specific needs.

The objectives of the pilot trial were to assess the usability of the Lincus system, establish how effectively the effect of events on health and wellbeing could be recorded and examined, progressively develop Lincus to meet organisational requirements and create learning points for future developments.

The trial included 12 individuals with complex needs who were living in supported accommodation and two support workers. The support workers helped the individuals to complete five separate surveys on Lincus every day for 4 weeks, thereby acting as a mechanism to facilitate communication between support workers and service users. Five surveys were provided which questioned the subjects' mental health, their housing/homelessness situation, their general health, any alcohol/substance abuse and their offending. Events that occurred and interventions performed were also recorded on Lincus. Surveys, along with events, were codeveloped prior to the trial with subject matter experts to ensure that they were relevant to the population.

Over the course of the trial, general improvements were reported in overall wellbeing of participants, home situation, mental health and general health. However, due to the small size of the trial, the results were not statistically convincing; therefore firm conclusions about the study impact and relationship between wellbeing and events could not be made. However, it gave insight on a case-by-case basis into how the information recorded on Lincus could be used by the organisation to make links between events and wellbeing amongst service users. This was used in the study to help the organisation identify interventions that were beneficial or caused potential harm to individuals to inform and improve on quality of care. The primary outcome from this trial was behaviour change both for the service users and their support staff. This led to Lincus receiving National Institute for Health and Care Excellence (NICE) recognition, as a tool which facilitates behaviour change [54].

Further benefit reported by the support workers included participants feeling more listened to and engaged with as a result of using Lincus. The length of completing the surveys was predicted to be less than 2 min, though in practice the completion took longer than anticipated (around 8 min), due to the responses to the questions stimulating further discussion between the support worker and service user. As a result, more support sessions were scheduled due to issues arising during these discussions. It was also reported that service users felt more supported as a result of using the technology. The support workers also stated improved engagement and job satisfaction. Lincus use also provided an auditable log of care provision and the impact of that care provision from a service user perspective. In one case, it was used to review wellbeing and events leading up to an individual being evicted from the programme, which evidenced the actions taken by the organisation to support the individual in attempt to prevent the events that led to eviction.

7.3.2 Co-development

The initial co-development of the system with the organisation was a crucial step in facilitating acceptance of the technology by staff and ensuring usability and usefulness [29, 71]. Through this process it was identified that the system would be used in a way other than it was originally designed for, for example, that the surveys would be administered by support workers to facilitate discussion with service users. This highlights the potential for disconnect between the expectations of the technology developer and the consumer [12], which can be overcome by intensive engagement between both parties prior to, and throughout the implementation process.

The co-development and engagement process also increased the likelihood of the system being successfully implemented, as evidenced by the smooth running of the trial from start to finish. However, it must be noted that the existing culture of openness to technology and positive change also played a major factor in the trial's success. As such, ongoing involvement of stakeholders in development and deployment almost certainly facilitated the perceived value of the technology [11, 21, 38].

7.3.3 Usability and Usefulness

Support workers found the system intuitive to use, and participants responded well to it, suggesting that the usability and usefulness barrier was prevented by co-development [11]. However, digital literacy with hardware and software was identified as a potential issue during the trial, including familiarity in interpreting data. However, identifying this issue helped tailor future training and support.

Some service users became more involved in the process than others. This resulted in some participants being keen to be surveyed more regularly, whilst others preferred to complete the surveys only two to three times per week. This highlighted the importance of allowing participants some flexibility around the frequency of surveying and types of surveys completed.

It is known that technologies must be flexible in design and readily adaptable to the changing needs of an organisation [5, 11]. Therefore, engagement with support workers was sustained throughout the trial, with continued co-developed adaptations made to the system [14] to meet additional requirements. This was particularly beneficial to ensure ongoing use of the technology.

7.3.4 Training and Support

Provision of training which was tailored to the users needs prior to implementation ensured staff were well trained and comfortable working with the technology prior to implementation to help facilitate the trial [20]. By the end of the trial, the support workers were in a good position to provide advice and training to others. In the future this could help to provide additional support within the organisation and enable them to provide advice to the technology providers on how the technology can be used or developed to better fit workflows [11]. However, a key area for training outlined at the end of the trial included verbal questioning techniques to gain accurate information and build trust between support workers and service users and recognise an individual's preference for relevancy and frequency of surveys. As a result, these areas would be addressed in future training.

7.3.5 Future Work

It was not until 3 years after the initial pilot that Lincus was again used as a mobile assessment tool in this organisation, despite the multiple benefits that had been demonstrated. The delay was mainly due to communication, commissioning and procurement barriers, initially due to a management change in the organisation. After 18 months, communication was re-established and a procurement pathway was found; however a change to this pathway caused further delays in the process. It was then decided, due to the funding source and commissioning barriers, to procure a solution through a public tender process, causing further delays due to tender publication, application and the subsequent interview process. Once the contract had been awarded to procure Lincus, contract negotiation was a further barrier to starting the project. These sorts of delays to small- and medium-sized enterprises (SMEs) with typically limited cash flow and turnover can heavily impact on company resources. Such delays in commissioning post pilot studies could, and have in the past, led to company closure, particularly if there are limited funds coming from other sources.

Following procurement as a mobile assessment tool, Lincus was used to record and monitor health and wellbeing information for individuals with multiple and complex needs, using a 15 research-driven questions survey developed as an evaluation tool for universal population comparisons [59]. Further continued collaboration with the group is expected to further tailor the tool to best suit the needs of the service users and those that support them.

7.3.6 Summary

This case study highlights the importance of co-development and long-term engagement to develop a technology that is usable and useful for an organisation. The outcomes and positive experience from the initial has led to future collaboration and technology development.

7.4 Case Study 2: Improving the Performance of Individuals with Multiple Long-Term Conditions

One Precious Life [22] was an ambitious project that would treat individuals in the UK with long-term conditions as high-performance athletes would be treated. The project was based on the service principles of preparing athletes for elite competition, including training, rehabilitation following illness or injury, being coached to be better than before and proactively addressing causes rather than the symptoms.

The technology aspect of the One Precious Life project incorporated three main features. The first was updates to the Lincus platform, making it more accessible to the general public and adding a number of features to be used as part of a person-centred coaching programme for individuals diagnosed with multiple long-term conditions. The second was the provision of a friendly and accessible online information hub for a number of health and wellbeing articles and access to relevant e-books. The third was the development of an Android application which connected to an activity tracker to assist positive behavioural change.

As part of the project, life coaches were recruited and trained for 3 months to learn how to apply the principles of high-performance athlete training and services to those with long-term conditions. They were deployed into the community to improve the physical, mental, social and emotional performance of individuals diagnosed with multiple long-term conditions, utilising the technologies and providing a personalised service rather than a one-size-fits-all programme. Eight participants were recruited over a period of 3 months. Participants were provided with access to their life coach on Lincus, where the coach could also provide advice and support remotely. The coaches taught the participants how to use the technologies, including the Lincus platform and integrated activity tracker, whilst educating them on how to improve their health and wellbeing.

Multiple changes were made to the Lincus platform to make it more accessible and appealing [45, 57], including interface redesign, messaging, coaching and alert functionality, integration of the activity tracker to enable wireless upload of activity data from the devices provided, improved data visualisation and modularisation. Simple nutrition and activity tracking were also developed to help individuals track, understand and improve these lifestyle behaviours. The technology developers continued to adapt and modify the technologies in accordance with feedback on user experience.

At completion of the trial, all of the participants said they would recommend the programme, stating that they had benefited from the intervention, that their quality of life had improved and that they would continue to make improvements to their lifestyle based on what they had learnt. Some participants expressed disappointment at the conclusion of the trial and requested to maintain contact with the coaches. Provision of the activity tracker also led to self-reported positive behavioural change, with results suggesting that use of the activity tracker enhanced motivation to be more active.

Though the exact model was not commissionable at the time, variants of it have been picked up by Rescon's commercial and customer partners. The project provided both the funding and a catalyst to develop technologies and implementation strategies. It gave a greater capacity to support individuals to improve their health through accessible education, self-tracking and other eCoaching tools. The technologies developed from this project continue to be integrated within the Lincus platform and used when a customer need has been identified.

7.5 Case Study 3: Supporting the Care of People with Learning Disabilities – Initial Trial and Wider Roll Out

7.5.1 Trial Summary

As part of an ongoing collaboration, Lincus was adapted and trialled with a learning disabilities charity, initially to assess usability for individuals with learning disabilities. A 17-week pilot trial for 11 service users supported by 16 support workers was conducted to support people with severe learning disabilities living in supported accommodation [23].

Prior to the adaptation of the system, meetings were held with subject matter experts to develop survey questions and approach, with Lincus further developed for the trial over 10 weeks. The selection of survey questions to be used in the trial resulted in the creation of short surveys for general health (Fig. 7.3), mental health, social life and emotional health. Each survey consisted of four to five questions with simple words or phrases, supported by pictorial representations to enable the

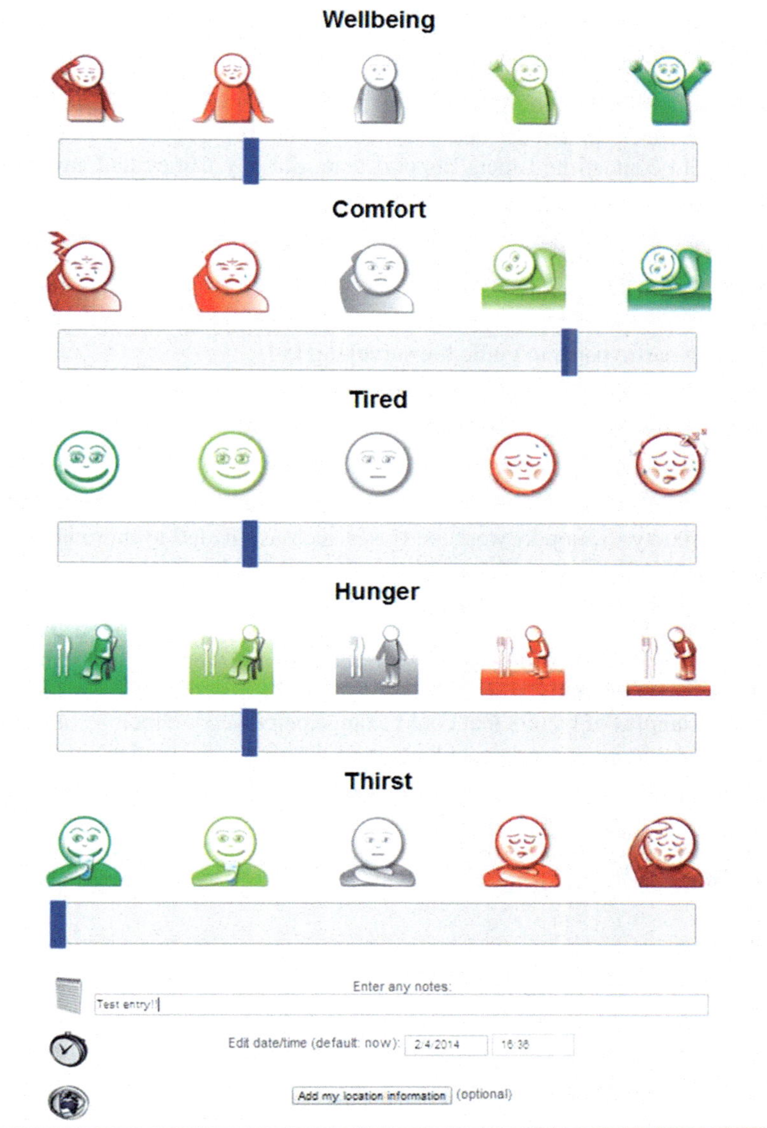

Fig. 7.3 General health (short) survey

individual to communicate their view of their wellbeing in certain areas of their lives. During the trial, support workers would assist service users to report information on Lincus using touch screen tablets, though service users were encouraged to enter their response directly where possible. As well as providing a detailed history for each user over time, it would enable the impact of events on the wellbeing of an individual or group to be measured.

Staff attended a training day prior to the start of the pilot where staff were given a demonstration on how to use Lincus. The training day also provided an opportunity for staff to raise questions or concerns. Once the trial had started, the Lincus support team provided ongoing support and contacted support workers throughout the trial to ensure continued use and issue resolution [38]. They were also informed that they could contact the Lincus support team directly if they had any issues or concerns to reinforce support [38].

For the first month of the trial, support workers were asked to complete the surveys with the service users they supported twice daily. After this period, surveys were completed as frequently as support workers felt necessary or when service users asked to complete them. This followed the lesson learned from the case study 1, in allowing participants to guide the surveying [54].

Through the pilot the usability of Lincus was demonstrated as a tool for communicating perceived overall health and wellbeing for those with learning disabilities living in supported accommodation. Usability varied within the group, suggesting that it may be of greater benefit to certain subgroups. In one case, an individual asked to use Lincus when she was feeling anxious, despite never previously identifying feelings of anxiety to support workers. Cases such as this led to more buy-in from support workers who witnessed the technology really making a difference.

The overall reported wellbeing, mental and emotional health of participants was observed to improve during the trial (statistical analysis not reported due to low numbers). Many factors could have contributed to the improvements during the trial, which may be directly or indirectly related to the use of Lincus or completely unrelated. Examples of factors that could change perceived wellbeing could include service users receiving more targeted attention during the trial independently of the actual technology or improvements in the weather which occurred during the trial period [36]. Support workers reported that using Lincus was a positive experience, which improved engagement with some service users. Following the outcomes from the initial small-scale trial, Lincus was widely deployed throughout the charity.

7.5.2 Staff Engagement

Staff engagement was the strongest barrier encountered when implementing Lincus during the trial and wider implementation. This was potentially due to resistance to change [16] and the perception of increased workload [71]. Lack of engagement was evidenced by a reduction in survey completion during the mid-stage of the initial trial (Fig. 7.4). Though this inadvertently highlighted the use of Lincus as a tool to potentially assess staff engagement, it suggested lack of integration with current workflows and perceived utility of staff using the technology. Both are known to impact the success of new technology implementations [20].

As a result, more focus was placed on better integration of Lincus into existing workflows through service and technology redesign [11]. This was facilitated by senior management [45], providing feedback from support workers in relation to

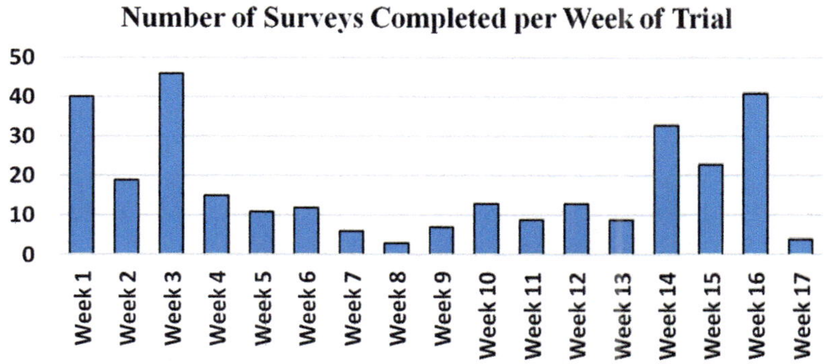

Fig. 7.4 Number of Lincus surveys completed per week of the learning disabilities trial

changes that would be beneficial for themselves and the service users they support. To gain further comments, after the pilot, a feedback button was implemented into Lincus to enable all users to report positive or negative feedback about the system, ensuring ongoing communication of user needs [12].

Lack of engagement was further evidenced in the pilot as no events were recorded by support staff, despite initial demonstration, mutual agreement on utility and regular follow-up. Consequently, the impact of events could not be assessed, which was a key objective of the pilot trial. This may have been down to multiple factors including: inadequate training [13, 44], lack of time [71], no perceived incentive to use the tool [14], poor communication from management [14], lack of understanding about the tool [14, 71], technical issues [11] or digital literacy [11]. The system was consequently adapted after the trial to automatically ask users if they wished to add an event after completing a survey, to consciously prompt logging of events. This adaption, along with communication from senior staff on the importance of logging events, resulted in improved and consistent event logging.

Since effective leadership and support are important for staff engagement ([14, 44, 45]), management support was more actively utilised after the trial, with senior staff asserting the need for support workers to utilise the events feature. The Lincus system was further developed after the pilot to monitor and quantify staff engagement, allowing management and administrative users to generate user reports on survey completion rates and the amount of events recorded to reflect staff engagement. This has been found to be a useful human resource management tool.

Though these benefits were recognised by senior staff, many did not take the time to log in to monitor both use and also the wellbeing of the individuals the organisation supported. After discussion it was found that actively logging into the system was a barrier to using the system, due to the high workload of senior management. As a result, a new feature was developed in Lincus that would parse reports from the system to an individual's email account with desired (depersonalised) information at a frequency that suited them. This example of improved workflow integration resulted in better adoption and use of the tool by senior management, which has been crucial to longer term success and resourcing.

Perceived usefulness was targeted to improved staff engagement, as it is critical for successful adoption of new technologies [15, 20, 35]. Dissemination of the benefits of using Lincus helped to gain 'buy-in' from staff during wider implementation, as they were aware of the benefits of using the tool within their organisation. This resulted in some staff taking greater ownership of the technology and becoming champions, which could transmit enthusiasm [44]. Furthermore, the benefits of using the technology became evident to staff during the trial, with several positive outcomes translating into success stories which were widely shared, this resulted in staff empowerment and eagerness to engage with the platform. The most notable case from the pilot trial with this user group was that of an individual being diagnosed with a previously undetected long-term condition (arthritis) as a result of using Lincus to help them communicate.

Lack of engagement also occurred at an organisation level due to resource constraints [14]. This was evident in a lack of staff, limited time and availability of equipment. Continued engagement with management helped to manage this barrier to make the system beneficial and easy to use for administrators as well as support staff and service users [11]. Ongoing phone and email support was provided to management staff to encourage continued use of the system [38].

7.5.3 Staff Training

To overcome the barrier of unfamiliarity with new technology and to improve acceptance [20] prior to the pilot, staff engaged in a full training day to become competent in the use of Lincus. To maximise the usage of the system, staff were actively encouraged during the training and the trial to log events, so that they could further investigate the impact they may have had on the perceived health and well-being of end users. Despite this, as previously discussed, no events were recorded during the first trial. This led to training during wider roll out placing greater emphasis on events and why they are beneficial to record.

Given the range of digital literacy levels amongst support workers, barriers were also faced with some staff having difficulties using the software and laptops/tablets. Though this could not be mitigated completely, specific and thorough staff training along with support to use the system was identified as an important factor for reducing barriers relating to variable digital literacy [20]. This was also further mitigated by recommending and training on specific tablets for use by service providers. Consequently, a further full training day was provided to all support staff, and in a 2-day train the trainer course was made available to management staff in order to train them to a level so that they were able to teach others. This helped to improve support available to users from within the organisation, and as a result of the train the trainer course, the organisation had a number of staff members who could deliver training internally. Ongoing phone and email support alongside user guides were also made available to ensure staff were provided with adequate support [38].

Since the initial trial, Lincus has been implemented incrementally throughout the organisation. This provided an opportunity to deliver sufficient training and resolve any teething issues before it was implemented more widely. Future application of Lincus would ideally utilise more rigorous training and management of staff implementation.

7.5.4 System Design

Considerable effort has been made from the initial trial and beyond to improve end user engagement, including customisable interfaces, regular updates to improve usability and new functionality. Early stakeholder engagement and consideration of workflow have been critical throughout the collaboration to increase perceived usefulness of the technology [11, 21, 38]. Feedback from users helped to develop the technology [12] and improve implementation. For example, some service users found having five pictorial icons for each survey question to be confusing (Fig. 7.3). Therefore, survey questions were adapted during the pilot to include only two pictorial icons that depicted the highest and lowest representations of the question (Fig. 7.5). Following this, two icons have been used for all surveys.

In addition, some users were overwhelmed when all survey questions were presented on the same page. Therefore, an accessibility view was developed to provide the option of showing one question at a time to help users focus on individual questions, which was particularly important for this user group. Lincus was also optimised for tablet use, as it became evident that this platform was preferred by many users during the trial.

To ensure flexibility of the platform [5, 11], Lincus was developed as a modular system, allowing the platform to be configured for the organisation's specific needs. This enabled the addition or removal of features, which provide a tighter, more focused system. Modularisation can help to prevent users from feeling overwhelmed or confused when there are too many features available to them in a new and unfamiliar technology platform.

Usability has continued to be improved by working closely with the organisation to meet the needs and requirements of all stakeholders [11, 21, 38]. As part of the ongoing collaboration, secure and logged video and messaging communication was integrated into the platform to improve communication for service users who are supported by the organisation.

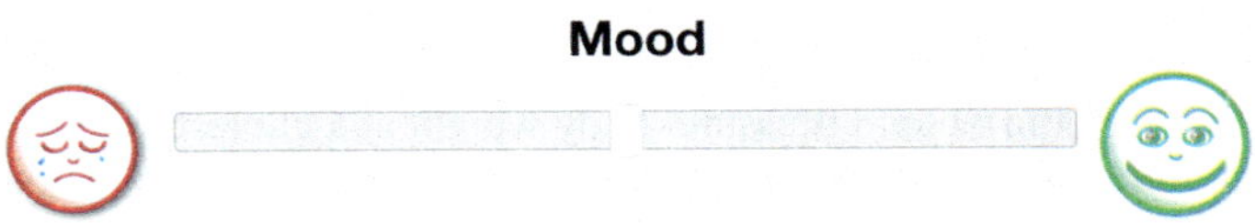

Fig. 7.5 Lincus survey question adapted to include only two pictorial icons

7.5.5 *Technical Issues*

Throughout wider implementation of Lincus following the pilot trial, technical issues have arisen from poor or intermittent internet connectivity. Though this was due to faulty or limited internet connections within the organisation, it led to frustration, resistance and negative attitudes towards using the technology, which negatively impacted implementation. It was vital to work closely with the organisation to ensure users would have stable internet connectivity prior to use of the technology. As a result, the organisation put measures in place to improve internet connectivity where it was identified as an issue. Provision of strong and reliable internet connectivity has improved the Lincus user experience considerably.

7.5.6 *Summary*

The development of a strong culture of collaboration has resulted in long-term commissioning of the solution with expansion from a person-centred recording platform to a complete care management package.

7.6 Case Study 4: Implementation Without Stakeholder Communication – A Medical Group and a City Council

7.6.1 *Medical Group*

In 2015 Lincus was commissioned by a medical group to support patients with diabetes. The plan for the trial was to focus on the initial use of Lincus at the patient's annual or biannual diabetes review. It was agreed with the senior medical group stakeholders that aspects of long-term blood glucose control, cholesterol and blood pressure would be assessed by the practice nurse and could be inputted along with patient-reported outcome measures relating to diabetes into the Lincus tool. It was proposed that this initial use and endorsement of Lincus would lead to continued use, improved engagement from patients with their condition and hence better long-term outcomes [6, 74].

It was proposed that prior to or during this practice nurse consultation, patients would be encouraged and supported to download the Lincus Companion App on Android or iOS mobile devices, where the wellbeing and diabetes survey could be completed during the appointment. Patients clinical measurements including HbA1c (glycated haemoglobin), blood pressure, body weight and cholesterol could also be entered, providing an opportunity for further discussion around health and wellbeing. This would reinforce the notion of feeling supported and helped to improve communication between the practice nurse and patients.

Patients would have the option to continue using the Lincus app at no cost to track and improve their health and wellbeing, with a view to improving multiple disease markers [3]. They would also have the option to log in to the web platform to access more self-management features. Whether they continued to use Lincus or not, the platform could be used as a person held record which could be brought with them to the next clinic review to stimulate discussion and reflection of how their health and wellbeing may have changed. This provided a tool for discussion around issues such as lifestyle change and medication adherence, which are known to improve health outcomes [17, 25].

Despite numerous engagements being held with the medical group at the date of the writing of this chapter, there was no real clinical progress with the trial as no patients have been recruited. Particular resistance was met from the only one general practitioner and one practice nurse who turned up to the first engagement (despite many being invited) who had workload, workflow and utility concerns, issues that have previously been identified in other clinical settings [38, 71].

The medical group had commissioned the most basic use of Lincus, which may have contributed to the lack of engagement, potentially due to the technology being undervalued or the most basic version not meeting the perceived needs of the practitioners. It also illustrates that if not enough resource is focused into implementation of a technology, including staff time for training and education, then the likelihood of failed implementation is greater [14].

In this case, a top-down commissioning process [41] that does not involve all stakeholders as early as possible in the process has also been a barrier to implementation. Staff at the medical group had recently been through the process of implementing a digital technology, which was decommissioned after a short period of time. This led to negative views about implementing and using a new technology, as staff felt that this was a disruptive process which would occur yearly.

After several months of communications, a further engagement has been scheduled. This engagement with staff will focus on improving collaboration and perceived usefulness of the technology [15, 20, 35]. It is hoped that this will result in a better outcome for the use of Lincus, with real benefits for both patients and also the clinical practitioners.

7.6.2 City Council

Lincus was also tailored for a 12-month project with a city council to support 300 individuals across a range of services including supported living, care homes and young people during transition. In this trial, the city council (commissioning authority) would identify a number of service providers to use Lincus to identify personal triggers for early intervention and would also support improved and timely access to services and information, improved quality of services, and improved self-help and lead to more effective and efficient organisations.

As for the previous example, Lincus was commissioned by senior management without the early inclusion of the providers who would be using the technology the most. This led to lack of engagement at the beginning of the project due to both lack of ownership in the project by providers and their own resource constraints [14], which meant the start date was revised from the original date to ensure benefit to all parties. With more time and engagements, providers gained a better understanding of how the system works and how it might benefit them. As a result of these engagements, providers became increasingly keen to use the platform. It also changed the way in which the platform was to be used for the project, with an increased focus on shared care rather than self-care alone, being the original commissioned work.

After the contract began, the city council also found difficulty in identifying potential service providers to use Lincus, which slowed implementation. Support was provided to the authority during this process to help identify and engage with service providers that could benefit from Lincus use. In addition, access to hardware in order to use Lincus was also a barrier to implementation for service providers, as the funding did not extend to hardware provision. This meant service providers had to already have access to laptops or tablets. After the revised start of the trial, staff and service users reported positive feedback when using the tool. However, outcomes have yet to be identified from this trial.

7.6.3 Summary

Lessons learned as a result of both of these trials is the need to include those that will work with the solution early on in the commissioning process. This strategy would increase the likelihood of stakeholder engagement throughout technology implementation. Organisations that are too busy, overworked and do not have time to engage and see the value in the tool were evidence of failures in these deployments. Throughout the implementation process, it is vital to keep engaging the organisation and collaboratively exploring how the technology can be used to prevent failed implementation. Organisations must be aware of the resource required to adopt a new technology, including engagements, education and training [14]. In addition, it would be beneficial to have a better understanding of what providers need from the platform to help target use of the technology in the early stages of implementation. Future implementations would therefore benefit from a full analysis of all stakeholder needs, including them in the technology design and implementation process as early as possible.

7.7 Case Study 5: Leveraging Existing Services to Codevelop a Solution

Lincus has been used as the underlying technology to deliver a programme that was sponsored by a clinical commissioning group, in collaboration with a community outreach organisation. The project aim was to identify high blood pressure in individuals living in a city, whilst educating people on the importance of blood pressure, leveraging both the Lincus technology and an existing service provider network. Blood pressure screening clinics were set up around the city over a 6-month period by the service provider. A tailored website was also developed for the project in close consultation with the provider to provide educational information on blood pressure and access to a blood pressure checker to allow people to see how their blood pressure compared to NHS advice. Each person that was screened at the clinic was given the opportunity to register on Lincus to allow them to track and improve their health and wellbeing.

The programme was highly successful, with more than 1000 people screened by the community outreach organisation at the time of writing, and is a great example of technology and service integration through collaborative working, starting from the moment of conception of the tender approach. Many individuals with high blood pressure were identified as a result of the screening. A key learning point from this programme is to work with a provider partner early on leveraging existing services to codevelop a solution that will change lives at scale.

7.8 Summary

There have been many barriers faced and lessons learned throughout the case studies outlined in this chapter, which have helped to inform improvement of future technology and deployment implementations. Whilst it is important to get the technology right in the first instance through engagement and collaboration with stakeholders [11, 21, 38], it is vital to ensure that it is used to its full potential through effective implementation including integration into existing systems, services and workflows [13, 14].

Developing easy to use technology can help to increase acceptance and reduce barriers during implementation [15, 40]. Whilst new technology can take time for organisations to implement and learn, an intuitive design can facilitate implementation by minimising the need for formal training ([45, 57]). For this reason, Lincus has been developed with a modular design to target the needs of specific organisations without providing users with an overwhelming amount of functionality. However, whilst technology needs to be easy to use at the front end, it also needs to be useful [15, 20, 35]. Working closely with stakeholders at an early stage can help improve perceived usefulness [11, 21, 38] and subsequently increase successful adoption [15, 20, 35].

Staff must be trained appropriately to use the technology [20]. When implementing Lincus, staff training days were not only useful for organisations but also for Lincus trainers to gather and implement feedback. Training can be a useful time to identify core values, create a vision for change [38] and convey the benefits of use to different stakeholders to increase perceived usefulness [15, 20]. Ongoing support provided to organisations and users throughout Lincus implementation has helped to identify and overcome issues particularly relating to digital literacy and technical issues [20], which has helped to improve stakeholder experience [38].

Initial barriers particularly relating to resistance to change can be broken down at the very start of the implementation process through engagement with all stakeholders [12, 45], which is essential when a top-down commissioning process has been used. It can also facilitate system design at an early stage [11]. The support of senior staff is also crucial throughout implementation [43–45]. To gain support from senior staff, the technology needs to target the objectives of the organisation with optimal resource [21].

However, it is long-term engagement strategies which are vital to gain trust, increase the usability of the technology [14], improve implementation [57] and sustain technology use over time. It is also important to explore how people want to engage with digital services [66] to align technology implementation with needs and expectations [11].

Successful implementation can require both technical and adaptive changes which require ongoing engagement from those implementing the changes and the individuals who are using the technology [70]. The start of implementation is just the beginning of this process [13]. For this reason, flexible design is vital to meet the changing needs of an organisation [5, 11]. This can also facilitate a long-term collaboration as evidenced in a number of case studies in this chapter, which can act as a catalyst to further develop the technology capabilities.

The process of implementation must also be considered to improve success. Lincus has been implemented incrementally to resolve organisation and technical issues prior to wider implementation, also limiting resistance and disruption that can result from 'big bang implementations' [10, 70]. This has helped to successfully implement Lincus from small-scale pilots to wider use. Incremental implementation has led to some individuals within organisations becoming champions of the technology. Peers championing the use of a previously unheard of technology has facilitated wider uptake and generated positive attitudes towards the technology.

A further important consideration is that organisations must be aware of the resource required to implement a new technology and must be willing to dedicate the required resource, including training, education and time [14]. This was evident in case study four where lack of resource heavily impacted technology implementation.

Most importantly, a holistic approach to implementation considering actual and potential barriers can facilitate successful and sustained technology adoption [45]. A number of case studies outlined in this chapter highlight the success of the tool as a result of ongoing co-development, engagement and collaboration. Whilst usefulness and usability can be key factors in implementation success [15, 20], the technology must align with organisational needs and workflows [11], and involvement of all stakeholders is paramount [12].

Future work will focus on continued improvements to the usability of Lincus, alongside close collaboration with key stakeholders. Perceived usefulness and uptake is likely to be improved by further integrating the system into existing organisation workflows through collaboration [20].

7.9 Conclusion

The case studies outlined in this chapter demonstrate that every implementation is different and unique to each organisation. Therefore, implementation strategies must reflect this. In any case, a collaborative approach with tight communication is imperative. Working closely with an organisation and its stakeholders on an ongoing basis is key to address potential and actual barriers to improve implementation and technology uptake. A collaborative learning process is essential, with flexible system design to allow for adaptations that must be made to reflect changing needs and requirements of organisations.

The barriers identified in these case studies and lessons learned as they were overcome can be used to facilitate future implementations of technologies focused on improving care. There is an opportunity for future research in this area to explore holistic whole ecosystem approaches to the implementation and development of new health and social care technology.

References

1. Anderson, J. G. (2007). Social, ethical and legal barriers to e-health. *International Journal of Medical Informatics, 76*(5), 480–483.
2. Anderson, G. F., Frogner, B. K., Johns, R. A., & Reinhardt, U. E. (2006). Health care spending and use of information technology in OECD countries. *Health Affairs, 25*(3), 819–831.
3. Barlow, J., Wright, C., Sheasby, J., Turner, A., & Hainsworth, J. (2002). Self-management approaches for people with chronic conditions: A review. *Patient Education and Counseling, 48*(2), 177–187.
4. Bayer, S., Barlow, J., & Curry, R. (2007). Assessing the impact of a care innovation: Telecare. *System Dynamics Review, 23*(1), 61–80.
5. Beuscart-Zéphir, M. C., Anceaux, F., Crinquette, V., & Renard, J. M. (2001). Integrating users' activity modeling in the design and assessment of hospital electronic patient records: The example of anesthesia. *International Journal of Medical Informatics, 64*(2), 157–171.
6. Buntin, M. B., Burke, M. F., Hoaglin, M. C., & Blumenthal, D. (2011). The benefits of health information technology: A review of the recent literature shows predominantly positive results. *Health Affairs, 30*(3), 464–471.
7. Care Quality Commission. (2015). *The state of health care and adult social care in England.* [Online]. Available at: http://www.cqc.org.uk/sites/default/files/documents/6227_cqc_soc2009-10_tag.pdf. Accessed November 2016.
8. Cashen, M. S., Dykes, P., & Gerber, B. (2004). eHealth technology and internet resources: Barriers for vulnerable populations. *Journal of Cardiovascular Nursing, 19*(3), 209–214.
9. Chismar, W.G., & Wiley-Patton, S. (2002). Test of the technology acceptance model for the internet in pediatrics. In *Proceedings of the AMIA Symposium* (p. 155). American Medical Informatics Association.
10. Cresswell, K., & Sheikh, A. (2009). The NHS Care Record Service (NHS CRS): Recommendations from the literature on successful implementation and adoption. *Journal of Innovation in Health Informatics, 17*(3), 153–160.
11. Cresswell, K., & Sheikh, A. (2013). Organizational issues in the implementation and adoption of health information technology innovations: An interpretative review. *International Journal of Medical Informatics, 82*(5), e73–e86.

12. Cresswell, K., Coleman, J., Slee, A., Williams, R., & Sheikh, A. (2013a). Investigating and learning lessons from early experiences of implementing ePrescribing systems into NHS hospitals: A questionnaire study. *PloS One, 8*(1), e53369.
13. Cresswell, K. M., Bates, D. W., & Sheikh, A. (2013b). Ten key considerations for the successful implementation and adoption of large-scale health information technology. *Journal of the American Medical Informatics Association, 20*(e1), e9–e13.
14. Damschroder, L. J., Aron, D. C., Keith, R. E., Kirsh, S. R., Alexander, J. A., & Lowery, J. C. (2009). Fostering implementation of health services research findings into practice: A consolidated framework for advancing implementation science. *Implementation Science, 4*(1), 1.
15. Davis, F. D. (1989). Perceived usefulness, perceived ease of use, and user acceptance of information technology. *MIS Quarterly, 13*, 319–340.
16. Edmondson, A. C., Bohmer, R. M., & Pisano, G. P. (2001). Disrupted routines: Team learning and new technology implementation in hospitals. *Administrative Science Quarterly, 46*(4), 685–716.
17. Forouzanfar, M. H., Alexander, L., Anderson, H. R., Bachman, V. F., Biryukov, S., Brauer, M., Burnett, R., Casey, D., Coates, M. M., Cohen, A., Delwiche, K., et al. (2015). Global, regional, and national comparative risk assessment of 79 behavioural, environmental and occupational, and metabolic risks or clusters of risks in 188 countries, 1990–2013: A systematic analysis for the Global Burden of Disease Study 2013. *The Lancet, 386*(10010), 2287–2323.
18. Free, C., Phillips, G., Galli, L., Watson, L., Felix, L., Edwards, P., Patel V., & Haines, A. (2013). The effectiveness of mobile-health technology-based health behaviour change or disease management interventions for health care consumers: A systematic review. *PLoS Medicine, 10*(1), e1001362.
19. Gagnon, M. P., Duplantie, J., Fortin, J. P., & Landry, R. (2006). Implementing telehealth to support medical practice in rural/remote regions: What are the conditions for success? *Implementation Science, 1*(1), 1.
20. Gagnon, M. P., Desmartis, M., Labrecque, M., Car, J., Pagliari, C., Pluye, P., Frémont, P., Gagnon, J., Tremblay, N., & Légaré, F. (2012). Systematic review of factors influencing the adoption of information and communication technologies by healthcare professionals. *Journal of Medical Systems, 36*(1), 241–277.
21. van Gemert-Pijnen, J. E., Nijland, N., van Limburg, M., Ossebaard, H. C., Kelders, S. M., Eysenbach, G., & Seydel, E. R. (2011). A holistic framework to improve the uptake and impact of eHealth technologies. *Journal of Medical Internet Research, 13*(4), e111.
22. Gilbert, L., & Dawson, T. (2014). *One precious life: Technology report*. Rescon Internal Publication. https://lincus.rescontechnologies.com/.
23. Gilbert, L., Blanchard, A., Barton, J., & Dawson, T. (2014). *Lincus: An adaptation and pilot usability study on individuals with learning disabilities*. Rescon Internal Publication. https://lincus.rescontechnologies.com/.
24. Glied, S., & Lleras-Muney, A. (2008). Technological innovation and inequality in health. *Demography, 45*(3), 741–761.
25. Grant, R., Adams, A. S., Trinacty, C. M., Zhang, F., Kleinman, K., Soumerai, S. B., Meigs, J. B., & Ross-Degnan, D. (2007). Relationship between patient medication adherence and subsequent clinical inertia in type 2 diabetes glycemic management. *Diabetes Care, 30*(4), 807–812.
26. Gruber, D., Cummings, G. G., Leblanc, L., & Smith, D. L. (2009). Factors influencing outcomes of clinical information systems implementation: A systematic review. *CIN: Computers, Informatics, Nursing, 27*(3), 151–163.
27. Hall, B.H., & Khan, B. (2003). *Adoption of new technology* (No. w9730). National Bureau of Economic Research.
28. Ham, C. (2010). The ten characteristics of the high-performing chronic care system. *Health Economics, Policy, and Law, 5*(01), 71–90.
29. Hartswood, M., Procter, R., Rouncefield, M., & Sharpe, M. (2000). Being there and doing IT in the workplace: A case study of a co-development approach in healthcare. In *Proceedings of the 2000 Participatory Design Conference*.

30. Henderson, C., Knapp, M., Fernández, J. L., Beecham, J., Hirani, S. P., Cartwright, M., Rixon, L., Beynon, M., Rogers, A., Bower, P., Doll, H., Fitzpatrick, R., Steventon, A., Bardsley, M., Hendy, J., Newman, S. P; Whole System Demonstrator evaluation team. (2013). Cost effectiveness of telehealth for patients with long term conditions (Whole Systems Demonstrator telehealth questionnaire study): Nested economic evaluation in a pragmatic, cluster randomised controlled trial. *British Medical Journal, 346,* f1035.
31. House of Commons Health Committee. (2014). *Managing the care of people with long-term conditions.* [Online]. Available at: http://www.publications.parliament.uk/pa/cm201415/cmselect/cmhealth/401/401.pdf. Accessed November 2016.
32. Hu, P. J., Chau, P. Y., Sheng, O. R. L., & Tam, K. Y. (1999). Examining the technology acceptance model using physician acceptance of telemedicine technology. *Journal of Management Information Systems, 16*(2), 91–112.
33. Ipsos MORI. (2015). *GP patient survey – National summary report.* [Online]. Available at: http://gp-survey-production.s3.amazonaws.com/archive/2015/July/July%202015%20National%20Summary%20Report.pdf. Accessed November 2016.
34. Jha, A. K., DesRoches, C. M., Campbell, E. G., Donelan, K., Rao, S. R., Ferris, T. G., Shields, A., Rosenbaum, S., & Blumenthal, D. (2009). Use of electronic health records in US hospitals. *New England Journal of Medicine, 360*(16), 1628–1638.
35. Jimison, H., Gorman, P., Woods, S., Nygren, P., Walker, M., Norris, S., & Hersh, W. (2008). *Barriers and drivers of health information technology use for the elderly, chronically Ill, and underserved.* Evidence Reports/Technology Assessments, No. 175.
36. Keller, M. C., Fredrickson, B. L., Ybarra, O., Côté, S., Johnson, K., Mikels, J., Conway, A., & Wager, T. (2005). A warm heart and a clear head the contingent effects of weather on mood and cognition. *Psychological Science, 16*(9), 724–731.
37. Kemper, A. R., Uren, R. L., & Clark, S. J. (2006). Adoption of electronic health records in primary care pediatric practices. *Pediatrics, 118*(1), e20–e24.
38. Keshavjee, K., Bosomworth, J., Copen, J., Lai, J., Kucukyazici, B., et al. (2006). Best practices in EMR implementation: A systematic review. *American Medical Informatics Association Annual Symposium Proceedings, 2006,* 982.
39. Kessels, R. P. (2003). Patients' memory for medical information. *Journal of the Royal Society of Medicine, 96*(5), 219–222.
40. Ketikidis, P., Dimitrovski, T., Lazuras, L., & Bath, P. A. (2012). Acceptance of health information technology in health professionals: An application of the revised technology acceptance model. *Health Informatics Journal, 18*(2), 124–134.
41. Kings Fund. (2008). *Technology in the NHS. Transforming the patient's experience of care.* [Online]. Available at: https://www.kingsfund.org.uk/sites/files/kf/Technology-in-the-NHS-Transforming-patients-experience-of-care-Liddell-Adshead-and-Burgess-Kings-Fund-October-2008_0.pdf. Accessed November 2016.
42. Kumar, S., Nilsen, W. J., Abernethy, A., Atienza, A., Patrick, K., Pavel, M., Riley, W. T., Shar, A., Spring, B., Spruijt-Metz, D., & Hedeker, D. (2013). Mobile health technology evaluation: The mHealth evidence workshop. *American Journal of Preventive Medicine, 45*(2), 228–236.
43. Leonard-Barton, D., & Deschamps, I. (1988). Managerial influence in the implementation of new technology. *Management Science, 34*(10), 1252–1265.
44. Lorenzi, N. M., & Riley, R. T. (2000). Managing change. *Journal of the American Medical Informatics Association, 7*(2), 116–124.
45. Ludwick, D. A., & Doucette, J. (2009). Adopting electronic medical records in primary care: Lessons learned from health information systems implementation experience in seven countries. *International Journal of Medical Informatics, 78*(1), 22–31.
46. Luxton, D. D., McCann, R. A., Bush, N. E., Mishkind, M. C., & Reger, G. M. (2011). mHealth for mental health: Integrating smartphone technology in behavioral healthcare. *Professional Psychology: Research and Practice, 42*(6), 505.
47. Mair, F. S., May, C., Finch, T., Murray, E., Anderson, G., Sullivan, F., O'Donnell, C., Wallace, P., & Epstein, O. (2007). Understanding the implementation and integration of e-health services. *Journal of Telemedicine and Telecare, 13*(suppl 1), 36–37.

48. Martin, T. (2012). Assessing mHealth: Opportunities and barriers to patient engagement. *Journal of Health Care for the Poor and Underserved, 23*(3), 935–941.
49. McGinn, C. A., Grenier, S., Duplantie, J., Shaw, N., Sicotte, C., Mathieu, L., Leduc, Y., Légaré, F., & Gagnon, M. P. (2011). Comparison of user groups' perspectives of barriers and facilitators to implementing electronic health records: A systematic review. *BMC Medicine, 9*(1), 1.
50. Mitzner, T. L., Boron, J. B., Fausset, C. B., Adams, A. E., Charness, N., Czaja, S. J., Dijkstra, K., Fisk, A. D., Rogers, W. A., & Sharit, J. (2010). Older adults talk technology: Technology usage and attitudes. *Computers in Human Behavior, 26*(6), 1710–1721.
51. NHS England. (2013). *The NHS belongs to the people: A call to action*. [Online]. Available at: https://www.england.nhs.uk/wp-content/uploads/2013/07/nhs_belongs.pdf. Accessed November 2016.
52. NHS England. (2016). *A and E attendances and emergency admissions*. [Online]. Available at: https://www.england.nhs.uk/statistics/wp-content/uploads/sites/2/2016/06/Monthly-AE-Report-August-16.pdf. Accessed November 2016.
53. NHS England Analytical Services. (2016). *Referral to treatment (RTT) waiting times statistics for consultant-led elective care*. [Online]. Available at: https://www.england.nhs.uk/statistics/wp-content/uploads/sites/2/2013/04/RTT-Annual-Report-2015-16-v3_final.pdf. Accessed November 2016.
54. NICE Shared Learning Database. (2013). *Lincus Trial: YMCA Liverpool pilot study for individuals with multiple and complex needs*. [Online]. Available at: https://www.nice.org.uk/sharedlearning/lincus-trial-ymca-liverpool-pilot-study-for-individuals-with-multiple-and--complex-needs. Accessed November 2016.
55. Operational Information for Commissioning. (2016). *Delayed Transfers of Care Statistics for England 2015/16*. [Online]. Available at: https://www.england.nhs.uk/statistics/wp-content/uploads/sites/2/2013/04/2015-16-Delayed-Transfers-of-Care-Annual-Report-1.pdf. Accessed November 2016.
56. Or, C. K., Karsh, B. T., Severtson, D. J., Burke, L. J., Brown, R. L., & Brennan, P. F. (2011). Factors affecting home care patients' acceptance of a web-based interactive self-management technology. *Journal of the American Medical Informatics Association, 18*(1), 51–59.
57. Øvretveit, J., Scott, T., Rundall, T. G., Shortell, S. M., & Brommels, M. (2007). Improving quality through effective implementation of information technology in healthcare. *International Journal for Quality in Health Care, 19*(5), 259–266.
58. Poon, E. G., Blumenthal, D., Jaggi, T., Honour, M. M., Bates, D. W., & Kaushal, R. (2004). Overcoming barriers to adopting and implementing computerized physician order entry systems in US hospitals. *Health Affairs, 23*(4), 184–190.
59. Prior, F., & Dawson, T. (2016). Development of a holistic health economic evaluation tool leveraging patient self-report. *IEEE*. In Press.
60. Rogers, A., Kirk, S., Gately, C., May, C. R., & Finch, T. (2011). Established users and the making of telecare work in long term condition management: Implications for health policy. *Social Science and Medicine, 72*(7), 1077–1084.
61. Sanders, C., Rogers, A., Bowen, R., Bower, P., Hirani, S., Cartwright, M., Fitzpatrick, R., Knapp, M., Barlow, J., Hendy, J., & Chrysanthaki, T. (2012). Exploring barriers to participation and adoption of telehealth and telecare within the whole system demonstrator trial: A qualitative study. *BMC Health Services Research, 12*(1), 1.
62. Southon, F. C. G., Sauer, C., & Dampney, C. N. G. K. (1997). Information technology in complex health services. *Journal of the American Medical Informatics Association, 4*(2), 112–124.
63. Steventon, A., Bardsley, M., Billings. J., Dixon. J., Doll, H., Hirani, S., Cartwright, M., Rixon, L., Knapp, M., Henderson, C., Rogers, A., Fitzpatrick, R., Hendy, J., Newman, S; Whole System Demonstrator Evaluation Team. (2012). Effect of telehealth on use of secondary care and mortality: Findings from the Whole System Demonstrator cluster randomised trial. *British Medical Journal, 344*, e3874.
64. Steventon, A., Bardsley, M., Billings, J., Dixon, J., Doll, H., Beynon, M., Hirani, S., Cartwright, M., Rixon, L., Knapp, M., Henderson, C., Rogers, A., Hendy, J., Fitzpatrick, R., & Newman,

S. (2013). Effect of telecare on use of health and social care services: Findings from the Whole Systems Demonstrator cluster randomised trial. *Age and Ageing, 42*(4), 501–508.
65. Sudore, R. L., Landefeld, C. S., Pérez-Stable, E. J., Bibbins-Domingo, K., Williams, B. A., & Schillinger, D. (2009). Unraveling the relationship between literacy, language proficiency, and patient–physician communication. *Patient Education and Counseling, 75*(3), 398–402.
66. The Kings Fund. (2016). *A digital NHS? An introduction to the digital agenda and plans for implementation*. [Online]. Available at: https://www.kingsfund.org.uk/sites/files/kf/field/field_publication_file/A_digital_NHS_Kings_Fund_Sep_2016.pdf. Accessed November 2016.
67. Thiru, K., Hassey, A., & Sullivan, F. (2003). Systematic review of scope and quality of electronic patient record data in primary care. *British Medical Journal, 326*(7398), 1070.
68. UK Department of Health. (2011). *Whole system demonstrator programme*. [Online]. Available at: https://www.gov.uk/government/uploads/system/uploads/attachment_data/file/215264/dh_131689.pdf . Accessed November 2016.
69. UK Department of Health. (2012). *Long term conditions compendium of information: 3rd edition*. [Online]. Available at: https://www.gov.uk/government/publications/long-term-conditions-compendium-of-information-third-edition. Accessed November 2016.
70. UK Department of Health. (2016). *Making IT work: Harnessing the power of health information technology to improve care in England*. [Online]. Available at: https://www.gov.uk/government/uploads/system/uploads/attachment_data/file/550866/Wachter_Review_Accessible.pdf. Accessed November 2016.
71. Ward, R., Stevens, C., Brentnall, P., & Briddon, J. (2008). The attitudes of health care staff to information technology: A comprehensive review of the research literature. *Health Information and Libraries Journal, 25*(2), 81–97.
72. Whittaker, R. (2012). Issues in mHealth: Findings from key informant interviews. *Journal of Medical Internet Research, 14*(5), e129.
73. World Health Organization. (2013). *Global action plan for the prevention and control of NCDs 2013–2020*. [Online]. Available at: http://www.who.int/nmh/publications/ncd-action-plan/en/. Accessed November 2016.
74. York Health Economics Consortium. (2009). *Organisational and behavioural barriers to medical technology adoption* (pp. 1–144). Coventry: York Health Economics Consortium.

Part III
Technological Challenges for "Smart Futures", Evaluation and Monitoring

Chapter 8
Environmental Responsibility Assessment using Belief Rule Based Inference

Anar Bazarhanova, Ah-Lian Kor, and Colin Pattinson

8.1 Introduction

Turning sustainable development into action and taking control over consequences of not doing so became a central issue of the twenty-first century. A large body of data concerning environmental problems is claiming to be results of unsustainable consumption practices of industrialized world in a large scale [1, 2]. In recent years, organizations have become increasingly interested in commitment to environmental issues. Environment is one of the pillars of the sustainability triangle [3], along with economic and social dimensions. The definition of Environmental Responsibility can be defined as the obligation that a company has to operate in a way that protects the environment. This research is focused on assessing the Environmental Responsibility level of an organization.

Large companies are usually legally bound to prevent their activities from contributing to water discharge, CO_2 emissions to the atmosphere, waste management, and soil and noise contamination [4]. ICT is believed to have a great potential in solving these problems. In Sobotta's book [2], many experts debate on how ICTs can support an organization in reducing CO_2 emissions, saving energy, and optimizing resource utilization – thus becoming greener and developing toward a more environment-friendly society.

Due to legislation pressure and increase of community awareness, a variety of environmental management systems, standards, and tools are being developed and used in order to assist companies to become more environmentally friendly. Each of them has its own particular benefits and advantages, but there is no indication of

A. Bazarhanova (✉)
Department of Computer Science, Aalto University, Espoo, Finland
e-mail: anar.bazarhanova@aalto.fi

A.-L. Kor • C. Pattinson
School of Computing, Creative Technologies and Engineering, Leeds Beckett University, Leeds, UK

M. Dastbaz et al. (eds.), *Technology for Smart Futures*,
DOI 10.1007/978-3-319-60137-3_8

which of them is better for the company's current state. The primary focus of an enterprise's environmental management depends on which industrial sector it is in. Companies might take a proactive approach to implementing environmental practices based on specific ISO standards relevant to their industry in order to reduce the environmental impact of their activities. Nevertheless, this research concentrates on a more generic and aggregated perspective of defining the Environmental Responsibility of a company.

Environmental Responsibility level is a very abstract concept, and measuring it in an absolute manner is not feasible. Attitude surveys provide many kinds of useful information, and environmentally friendly behavior has often been studied successfully, but neither method truly reveals much about environmental performance assessment in organizations [5].

8.1.1 SMEs and Sustainability

Small-, micro-, and medium-sized enterprises make up more than 90% of estimated total number of business sector bodies in the EU [6] and could contribute up to 70% of all industrial pollution [7]. Mostly, large enterprises and corporations are legally bound to incorporate CSR policies and follow internationally recognized environmental standards to secure sustainability in their operations. A compelling amount of research has been conducted, and voluntary industry initiatives evolved, such as Eco-Management and Audit Scheme (EMAS), Environmental Management System (EMS), and ISO 14001 standard, as means to develop systematic approaches in improving environmental performances of enterprises. Hence, smaller enterprises are usually exempted from those standards due to lack of financial, human resources and time. The research in the field of EMS adoption among SMEs has revealed other obstacles such as low awareness, absence of pressure from customers, poor information quality from advisors, and skepticism in benefit gaining [8]. That emphasizes the need to provide small- and medium-sized enterprises with an easy to access and comprehendible, attractive financial saving mechanism to reduce their footprint and optimize operations in a sustainable way [8–10].

How to measure the Environmental Responsibility level of SMEs? Which is the recommending path that companies should follow toward environmental performance excellence? This research addresses these questions. Therefore, the research aim primarily focuses on the development of a novel assessment and decision support model to help companies evaluate their current state followed by recommendations of behavioral and operational best practices to enhance their Environmental Responsibility level. This paper demonstrates the feasibility of the Belif Rule Based (BRB) approach in the assessment of enterprise's level commitment to environmental issues.

In order to address the research aims stated above, the research work has been implemented according to the steps described in Fig. 8.1:

The work commenced with relevant background work and literature review for the problem statement, and the BRB Expert Systems theory has been chosen for the ER assessment methodology. After the toolkit development, the BRB model has

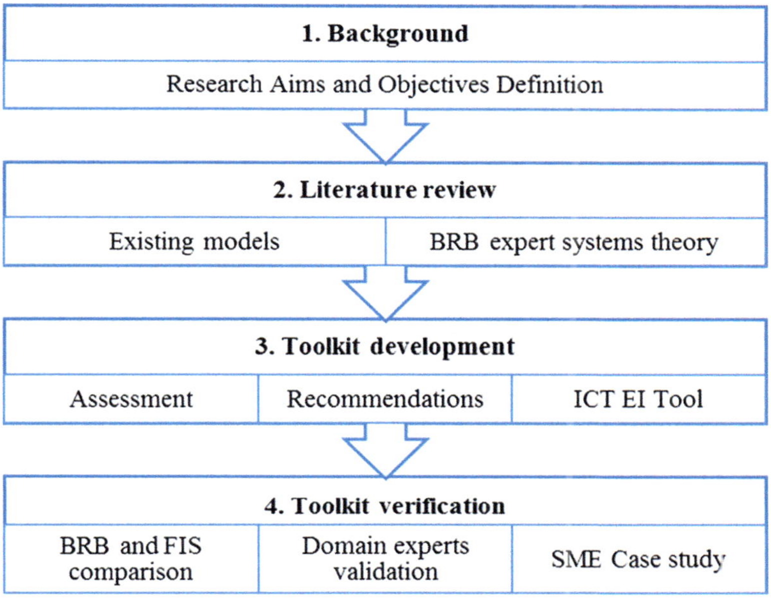

Fig. 8.1 Research plan

been compared and validated with Fuzzy Inference System theory [11], verified by domain experts, and a case study with a selected SME has been conducted (outside the scope of this chapter, but details could be found in [34]).

The remaining of the paper is structured as follows. Section 8.2 presents the existing corporate environmental impact assessment models. Section 8.3 provides an overview of the theoretical basis of Belif Rule Based methodology applied in research. BRB inference validation and performances in comparison with a Fuzzy Inference System (FIS) theory are then presented in Sect. 8.4. A final section concludes the paper with findings, limitations of this research, as well as potential avenues for future research.

8.2 Literature Review

8.2.1 Existing Environmental Assessment Models

Various evaluation approaches and models for the assessment of companies' environmental impact have evolved. Some of the most well known are SURF Green ICT Maturity Model [12], Sustainability Maturity Model from Industrial Research Institute [13], Sustainability Management Maturity Model of FairRidge Group, Systematic Action Plan from Fachgruppe Green IT [14], UK HM Government

Green ICT Maturity Model [15], Green IT Readiness Framework [16], and SustainaBits Framework and Rating System for Sustainable IT [17] by using which organizations may benefit and raise their awareness in environmental issues.

Works dedicated to data center assessment and greening operations have been intentionally excluded from this literature review. Most of the models surveyed are research-related models which require minimum knowledge on Green ICT domain and are in formats of scientific works, tables, and publications or are abstract and conceptual [16, 17], mitigating the chances to be adopted by nonacademic organizations. Some models [12, 13] include an actual assessment by assigning scores per categories, but are not applicable for small- and medium-sized enterprises. Most of the models focus on eliminating negative impacts of ICT infrastructure, whereas SMEs need a simple, comprehensive, easy-to-use, and access tool for an assessment of their level of Environmental Responsibility.

It is evident from literature review that Green ICT and ICT for greening domain fundamentals in a corporate context need a proper classification and standardization, recognized both by industry and academia. Categorization inconsistencies are demonstrated in models above and expected to be even more diversified among those which were not identified, skipped, or missed. Also, assessment systems do not address qualitative reviews and adaptations toward targeted user groups [14–16]. Environmental responsibility level assessment is a multidimensional, observational process that requires a more rigorous reasoning approach to handle uncertainties and imprecisions and, at the same time, be positive perspective oriented.

Environmental Responsibility (ER) assessment is characterized by a number of identified factors which are qualitative in nature and can be assessed based on human judgment. Thus, a general ER assessment problem for SMEs could be addressed without a detailed and rigorous audit conducted by affiliated authorities. Such an approach would be able to handle uncertainties, vagueness, and fuzziness. Assessment models presented above follow mostly traditional approaches in Green readiness assessment and reasoning, which are incapable of producing accurate ER level results. Expert systems are widely used to deal with knowledge-based decision support systems. Thus, the Belif Rule Based approach with its ability to infer uncertain knowledge in the domain of Environmental Responsibility has been applied in this research.

8.2.2 Belif Rule Based Expert Systems

Expert system development involves the deployment of an appropriate inference rule engine. Knowledge representation systems are mainly used to support human decision-making and can be transformed into rule-based schemes, which are easy to perceive, understand, and deploy [18]. These schemes which express different types of knowledge are usually constructed in the formats of if-then rules, which are widely deployed in the areas of artificial intelligence, decision support, and expert systems. Belif Rule Based Expert Systems consist of two parts: knowledge base and inference engine, which are used to derive conclusions from rules, either established by experts with domain-specific knowledge, historical data, or observation facts provided by users. That is to say, inference engine is a core algorithm of the Belif Rule Based (BRB) Expert System, and the following section will examine available reasoning patterns and justify the selection of forward chaining inference-based rule engine.

As an alternative to rule engines, one can consider data-driven design methods, using "lookup tables," database manipulations where scripts are updated on the fly, or hand-coded if-then-type cases in the application. A primary purpose of a rule engine is the separation of the business and system logic, so that rules can be easily maintained without intervention into the application logic, code recompilation, etc. Moreover, rules are stored in an external file and encoded into a human-readable format, ensuring that nontechnical experts will be able to collaborate. Logic and data separation is a good OOP pattern, also ensuring loose coupling parameter for service-oriented architecture (SOA)-based application types. Having a separate file containing all rules supports the knowledge centralization aspect, which is even further strengthened by the availability of a wide range of integrated development environment (IDE) plug-ins.

Rule engines have great potential in reducing application maintenance cost, because reasoning makes a clear separation between the logic and data, i.e., separating the application source code (not modified) from the logic code (modified if logic is changed).

An inference engine from the practical view is a piece of software that helps to derive logical conclusions from a set of facts and user observations. There are two types of inference engine logics: forward and backward chaining. Forward chaining (or data driven) is a method that starts with the available information and uses rules to extract more data, as the input data determines which rules are to be used [19]. While in backward chaining (also goal driven), an inference engine would iterate all rules until it finds the one with a consequent, matching the requirement.

8.2.3 *Inference Engines*

There is a big number of business logic rule engines available in the market, most of which are open source and show impressive performance indicators, but each of which is dedicated to address specific problems. The selection of a suitable engine for this research is based on the following criteria: no cost for a complete version (i.e., open source), Java and IDE integration, extensive documentation, and acceptable inference engine performance.

Jess [20] is considered as one the fastest rule engines for Java platform, which offers a direct manipulation and interaction with all Java objects. Lisp-similar description language Jess uses an enhanced version of a Rete pattern matching algorithm. Nevertheless, the last version of Jess 7 has been released in 2007 which makes Jess less likely to meet research needs.

Drools [21] is an open-source JBoss and Red Hat Inc. Business Rule Management System (BRMS). It offers several editing and managing tools along with high-performance execution. It also provides Eclipse IDE plug-in for core development and Drools Flow graphical modeling editor.

OpenRules [22] is another Business Decision Management System (BDMS) that provides a number of tools for rule-based decision system development, which requires less developer support. The main strength of OpenRules is that it allows to import and edit rules in MS Excel, Word, or Google Docs formats, which makes it attractive to nontechnical domain experts due to ease of its operation. A complete version of OpenRules includes an Eclipse plug-in that enables debugging, web service deployment, and integration with any Java or .NET applications. Similar to Jess, a full version of OpenRules requires a license with a nominal fee. CLIPS [23] is an acronym for C Language Integrated Production System, a software tool for building expert systems. CLIPS itself is written in C language. CLIPS Java API (CLIPS JNI) distribution is also available, but using CLIPS brings an additional overhead with version support and IDE integration due to the latest Java version incompatibility.

This is a brief description of selected engines on different measurement criteria; however, it should be noted this list is not exhaustive because many other engines have not been discussed in this section. A comparative analysis above reveals that JBoss Drools is an engine that fits all selection criteria and has been chosen to be deployed in this research.

8.3 Assessment Methodology

8.3.1 Belif Rule Based Knowledge Representation and Inference Procedures

Constructed rule-based expert systems based on human knowledge are considered the most visible and fastest-growing branch of artificial intelligence (AI) according to Sun's work [24]. There are several common types of knowledge propositions in rule-based systems: Boolean for the concepts which have either true or false values, fuzzy set of propositions for non-clearly defined concepts, or an attribute as a variable having a set of possible values it can take. In this research, the possibility of defining Environmental Responsibility in an organization by a list of actions that will lead to more efficient and sustainable performance is proposed. However, it is recommended that the assessment results be accompanied by a more rigorous and continuous audit of the company environmental performance. For the purpose of this research, Boolean and fuzzy knowledge proposition sets are used, for example:

Prioritization of using eco-labelled equipment will lead to savings on energy consumption is more deterministic rather than probabilistic, and it is derived from conclusions established by experts and observation facts provided by statistics.

There are many types of uncertainties in real-world decision support systems such as vagueness, imprecision, and ambiguity [25], because each knowledge proposition attribute can be described as "high," "medium," and "low" or "true" and "false." The whole concept of ER assessment for a company is a fuzzy, scalable, and continuous (i.e., could be in a continuous continuum from 0% to 100%) concept, due to infeasibility to obtain precise input data, which will cause inaccuracy in an evaluation process. As it is described earlier, an inference is a reasoning procedure to derive conclusions from a knowledge base. In a forward chaining algorithm, an inference starts iteratively searching for the pattern match of an input and an if-then clause. When a match is found, it fires the if-then clause followed by triggering an action. However, forward chaining mechanism is not equipped with uncertainty handling. Therefore, decision is made to deploy the forward chaining and elements of belief degrees with a hybrid knowledge representation inference scheme to accommodate uncertainties. For example, in the Hossain's measles diagnosis paper [25], belief distribution is described as follows:

$$Rk : \text{If}\left(\text{Fever is'Medium'}^\wedge \text{ Rash is'High'}^\wedge \ldots\right)$$

$$\text{Then measles diagnosis probability is}\left\{(\text{High, } 0.90),(\text{Medium, } 0.10),(\textit{Low}\text{, } 0.00)\right\}, \quad (8.1)$$

Proposition in (Eq. 8.1) states: belief degree is 90% that the condition is "high," 10% that it is "medium," and 0% that it is "low". Moreover, input variables involved in inference may not be of the same type. They might be expressed quantitatively and qualitatively and could be different both in type and range. To summarize, there is a need to deploy a hybrid inference schema with forward chaining under uncertainty to provide mathematical handling of various input data types and uncertainty handling.

First step in building the knowledge base of a BRB system is to identify relevant antecedent attributes, types of uncertainties, and corresponding weights. These then form a generic domain knowledge representation scheme using belief structure. Secondly, a rule base is constructed on the basis of a belief structure, which apprehends nonlinear causal relationships of rules. In a complete Belif Rule Based scheme, input for each antecedent variable is transformed with a set of available referential values. This distribution describes the degree of each antecedent being activated [26].

Suppose N is a set of distinctive referential values for an antecedent attribute $x_i(i = 1 \ldots T)$ represented by

$$H(x_i) = \{H_{i,n}, n = 1..N_i\} \quad (8.2)$$

where $H_{i,n}$ denotes the n_{th} evaluation value for an attribute x_i. Correspondingly, the belief distribution of x_i can be defined as

$$S(x_i) = \{(H_{i,n}, \alpha_{i,n}), n = 1..N_i\} \quad (8.3)$$

where $\alpha_{i,n}$ is the belief degree to which x_i is assessed to evaluation degree $H_{i,n}$, and

$\alpha_{i,n} \geq 0 \, and \sum_{n=1}^{N_i} \alpha_{i,n} \leq 1.$

The belief degree is considered to be complete when it is equal to 1 and incomplete when less than 1. Any data type, even without uncertainties can be transformed into evaluation belief distribution [27].

Belif Rule Based schema (conjunctive Boolean expression) is defined as follows:

$$\text{If } x_1 \, is \, A_1^k \wedge x_2 \, is \, A_2^k \wedge \ldots \wedge x_{Tk} \, is \, A_{Tk}^k, \quad (8.4)$$

$$\text{Then} \{(D_1, \beta_{1,k}), (D_2, \beta_{2,k}) \ldots (D_n, \beta_{n,k}),\}$$

$$\text{where} \sum_{n=1}^{N} \beta_{n,k} \leq 1$$

with a rule weight θ_k and attribute weight $\delta_{1,k}, \delta_{2,k} \ldots \delta_{Tk,k}, k \in \{1 \ldots L\}$.

Here, $x_1, x_2 \ldots x_{1Tk}$ denote the antecedent variables in the k_{th} rule. These attributes belong to the set of antecedent variables $X = \{x_i; i = 1 \ldots T\}$ in which each element takes a value from an array of finite sets $A = \{A_1 \ldots A_t\}$. The vector $A_i = \{A_{i,n} : n = 1 \ldots N_i = |A_i|\}$ is defined as the set of referential attributes for antecedent variable x_i. In the k_{th} rule, A_i^k represents the referential value corresponding to i_{th} antecedent variable. T_k denotes the total number of antecedent attributes in the k_{th} rule; $\beta_{n,k}$ is a belief degree to which D_n is assumed to be consequent, taking into account the logical relationship of the k_{th} rule: $Fk: x_1 \, is \, A_1^k \wedge x_2 \, is \, A_2^k \wedge \ldots \wedge x_{Tk} \, is \, A_{Tk}^k$. If $\sum_{n=1}^{N} \beta_{n,k}$ the k_{th} rule is said to be complete and incomplete otherwise. In exceptional and extreme cases where $\sum_{n=1}^{N} \beta_{n,k}$ it denotes total ignorance on the consequent variable. It should be mentioned that $D = \{D_n; n = 1..N\}$ can act either as a firing action or a concluding statement [28]. For example, in case of ER assessment:

> R_k: If the *use* of ecolabelled equipment is high *and* switch *off and* standby policy is medium *and* standards compliant strategy is adoption is high},

$$\text{Then } ER \text{ level is} \{(\text{good}, 0.7), (\text{average}, 0.2), (\text{fair}, 0.1), (\text{poor}, 0)\}, \quad (8.5)$$

where belief distribution representation for ER is considered good with 70% of confidence, 20% for average, and 10% sure that ER level is fair. In general it is expressed as

$$\left(A_1^*, \varepsilon_1\right) \wedge \left(A_2^*, \varepsilon_2\right) \wedge \ldots \wedge \left(A_T^*, \varepsilon_T\right) \quad (8.6)$$

where ε_1 is a degree of belief corresponding to antecedent A_i^* of the i_{th} variable $i = 1 \ldots T$ which reflects uncertainty of data and T is the total number of input attributes. And,

$$A\left(A_i^*, \varepsilon_i\right) = \left\{\left(A_{i,j}, \alpha_{i,j}\right); j = 1..J_i\right\}, i = 1..T \quad (8.7)$$

where $A_{i,j}$ is the i_{th} referential value of the i_{th} attribute and $\alpha_{i,j}$ is a degree to which A_i^* belongs to $A_{i,j}$. The total degree $\alpha_k$with input match of A^k antecedent in the k_{th} rule is calculated by:

$$\alpha_k = \varphi\left(\left(\delta_{k1}, \alpha_{k1}\right) \ldots \left(\delta_{kT_k}, \alpha_{Tk}^k\right)\right), \quad (8.8)$$

where φ is an aggregation function for T_k antecedents in k_{th} rule and δ_{k_1} $\left(i = 1 \ldots Tk\right)$ is the weight of the i_{th} antecedent variable. An aggregation function for subjective probabilities generation is "$\bigwedge$" operator, i.e., $\varphi_{sum}(a, b) = a + b - ab$ [29]. Particularly, the consequent part of a rule is true if only all antecedent variables meet the rule conditions, so the following weighted multiplicative aggregation function is used:

$$\alpha_k = \prod_{i-1}^{T_k} \left(\alpha_i^k\right)^{\overline{\delta_{ki}}}$$

$$\text{where } \overline{\delta_{ki}} = \frac{\delta_{ki}}{\max_{i=1\ldots Tk}\left\{\delta_{ki}\right\}} \quad (8.9)$$

For the formula above $0 \le \alpha_k \le 1, \alpha_k = 1\, if\, \alpha_i^k = 1$ for all $i = 1 \ldots T_k$ and $\alpha_k \ge 0,\ if\, \alpha_i^k = 0$ for any $i = 1 \ldots T_k$ [27].

Also, a consequent value is linearly dependent on an antecedent variable weight, where one attribute may contribute more than another. Unfortunately, in the concept of ER, there is no existing work or justification of assigning different weights for different activities. Thus, in the context of this research, corresponding weights are considered to be equal. Additionally, the inference engine should take into account incompleteness of input data, where antecedent value is not known or partially known. For such cases in this research, inference is being held with worst case scenario for this antecedent, i.e.,

$$\text{If } 0 \le \sum_{i=1}^{N} \overline{\beta_{i,k}} \le 1$$

$$\beta_{i,k} = \frac{\overline{\beta_{i,k}}\left(\sum_{t=1}^{Tk}\left(\tau(t,k)\sum_{j=1}^{Jt}\alpha_{t,j}\right)\right)}{\sum_{t=1}^{Tk}\tau(t,k)} \tag{8.10}$$

$$\text{where}\ \tau(t,k) = \begin{cases} 1, \mathit{for}\ Ut\ \text{in}\ Rk(t = 1\ldots Tk) \\ 0\ \text{otherwise} \end{cases}$$

The formula above saves consequent calculation in cases when not all antecedent variables are involved in a rule inference. Recapitulating again, in an ER calculation, all antecedent attributes are included and using worst cases for incomplete input data. If there is any incomplete data, the lowest possible referential value is assumed in order to compute the consequent. Thus, final consequent variable will be generated by combining each consequent for corresponding antecedent. From J. B. Yang's "Rule and utility based evidential reasoning approach for multi-attribute decision analysis under uncertainties" work [30], $m_{j,I(k)}$ is the combined probability mass degree of belief in D_j, where:

$$m_{j,k} = \omega_k \beta_{j,k}, j = 1\ldots N$$

$$m_{D,k} = \omega_k \left(1 - \sum_{j=1}^{N}\beta_{j,k}\right) \tag{8.11}$$

Suppose $m_{j,I(K)}$ is the combined belief degree in D_j antecedent-belief degree pair and $m_{D,I(L)}$ is the remaining degree. Then, the overall aggregated belief degree β_j *in* D_j is defined as

$$\{D_j\} : \beta_j = \frac{m_{j,I(L)}}{1 - m_{D,I(L)}} \tag{8.12}$$

Thus, the concluding consequent is generated by aggregating L, and input data from vector $A^* = \{A^{*k}, k = 1 \ldots L\}$ number of rules is represented as

$$S(A^*) = \{(D_j, \beta_j);\ j = 1\ldots N\} \tag{8.13}$$

Assuming that $u(D_j)$ is the utility of an individual consequent variable (crisp value), single value converted result is equal to:

$$u(S(A^*)) = \sum_{j=1}^{N} u(D_j)\beta_j \tag{8.14}$$

Lastly, the overall belief degrees are measured by individual antecedent degrees of the k_{th} rule activated by an input which is a building base for the overall output belief degree.

8.3.1.1 Knowledge Base in ER Assessment

As previously mentioned, knowledge base in Belif Rule Based systems is either established by experts with domain-specific knowledge, historical data, or observation facts or statistics. In this research, it is based on an in-depth literature review and experts in Green ICT validation:

Here, V_j denotes the category, e.g., V_1 (equipment procurement compliant with Green ICT guidelines and the optimization of enterprise operations), V_2 (energy performance improvements and monitoring toward the use of alternative energy resources), V_3 (energy-aware network engineering adherence), V_4 (social commitment), and V_5 (waste management). The categories proposed based on the research scope are:

1. Dedicated to small- and medium-sized enterprises
2. Only for SMEs' in-office ICT infrastructure deployment and behavioral best practices in equipment usage
3. SMEs' ICT equipment procurement, usage, and end-of-life treatment life cycle stages
   Thus, the knowledge base for the Environmental Responsibility assessment of SMEs is presented in Table 8.1:

Below is a set of a structured questionnaire with categories of items relevant to the assessment of Environmental Responsibility:

V_1:

1. Does your company follow Green ICT procurement guidelines when ICT equipment is purchased?

Table 8.1 Categories

Category	Antecedents
V_1: equipment procurement compliance with Green ICT guidelines	$A^1_{1,}A^1_{2,}A^1_{3,}A^1_{4,}A^1_{5,}$
V_2: energy performance improvement and monitoring	$A^2_{1,}A^2_{2,}A^2_{3,}A^2_{4,}$
V_3: energy-aware networks engineering adherence	$A^3_{1,}A^3_{2,}A^3_{3,}$
V_4: social commitment	$A^4_{1,}A^4_{2,}A^4_{3,}A^4_{4,}$
V_5: waste management	$A^5_{1,}A^5_{2}$

- Always
- Sometimes
- Never

2. Have you ever used life cycle impact assessment as a product/service purchase criterion?

- Yes
- No

3. Do you prioritize eco labels (e.g., EPEAT, Energy Star, EU Ecolabel, SWAN, etc.)?

- For 100% of equipment (excellent)
- For between 70% and 99% of equipment (good)
- For between 40% and 69% of equipment (fair)
- For less than 40% of equipment (poor)

4. Are you familiar with the use of services that minimize the energy consumption and environmental impact of ICT equipment (e.g., virtualization, optimization, etc.)?

- Extremely
- Moderately
- None

V_2:

5. Have you ever conducted any ICT equipment energy consumption assessment?

- Yes
- No

6. Is the use of switch off and standby modes common in your company?

- Yes for all
- Occasionally
- No

7. Have you installed any power management software in your company ICT equipment?

- Yes
- No

8. Have you followed any systematic approach for energy efficiency improvement (e.g., data collection and data analysis)?

- Always for all
- Sometimes
- Never

9. Does your company use energy from any of these renewable sources? (e.g., solar, wind, geothermal, or biomass energy)?
   - Yes, from at least one
   - No

V_3:

10. Do the following statements apply to your company? "The network infrastructure makes use of equipment that adheres to the latest energy efficiency standards (sleep mode or Energy Efficient Ethernet)."
    - Extremely
    - Moderately
    - Not at all
11. Do the following statements apply to your company? "The number of required IT equipment, functionalities, and quality of service are optimized in order to reduce environmental impact."
    - Extremely
    - Moderately
    - Not at all
12. Do the following statements apply to your company? "Routing is made energy aware and offers possibilities to choose the most energy efficient route instead of the shortest path."
    - Extremely
    - Moderately
    - Not at all

V_4:

13. Has your company adopted any documented reference architecture (with guiding principles for designing new services/products) aimed to minimize environmental impact?
    - True
    - False
14. Does your company have any sustainable development-related training and communication activities for employees?
    - True
    - False
15. Does your company promote the use of audio and videoconferencing facilities to reduce travel?

- True
- False

V_5:

16. Does the following statement apply to your company?

"Ensure a strict implementation of an e-waste policy for the reuse or recycling of ICT equipment to minimize environmental and social hazards after disposal."

- Extremely
- Moderately
- Not at all

17. Does your company have any collection and recovery (e.g., reuse, repairing, remanufacturing) channels (subcontractors) that can reduce the amount of waste sent to landfill?
    - True
    - False

The knowledge base in this research is constructed after an in-depth literature review and critical analyses of existing environmental performance assessment models and primarily guided by the EU Draft Background Report for the development of an EMAS Sectoral Reference Document on "Best Environmental Management Practice in the Telecommunications and ICT Services Sector" [31]. Thus, questions numbered from 1 to 3 relate to the V_1, equipment procurement compliant with Green ICT guidelines and the optimization of enterprise operations; questions numbered from 4 to 8 are for the V_2, energy performance improvements and monitoring toward the use of alternative energy resources; questions numbered from 10 to 12 are for the V_3, energy-aware network engineering adherence; and questions numbered from 13 to 15 are for the V_4, social commitment; and questions numbered from 16 to 17 are for the V_5, waste management.

As it can be seen, the knowledge base has 5 V_j parent categories and 17 A_j^i antecedent attributes. Each category consists of antecedent attributes that comprise a set of questionnaire items that users will need to answer. In order to provide mathematical handling of various input data types and uncertainties, a set of available referential values is described as {(High, 0.0), (Medium, 0.0), (Low, 0.0)}. It is important to mention that the research has followed several iterations in refining the knowledge categorization in the knowledge base: initially, there are 8 independent parent categories, which would lead to 3^8 = 6561 cases of combinations to consider, adding additional complications and overhead. Subsequently, the total number of categories has been streamlined into five categories for simplicity and integrity purposes. The total ER index is calculated (Eq. 8.14), by aggregating $N = 5$ number of parent categories, which in turn consist of $\sum_{i=1}^{N} A_j^i$ aggregation of corresponding antecedents. A_j^i represents the corresponding questionnaire item for each V_j category (Fig. 8.2).

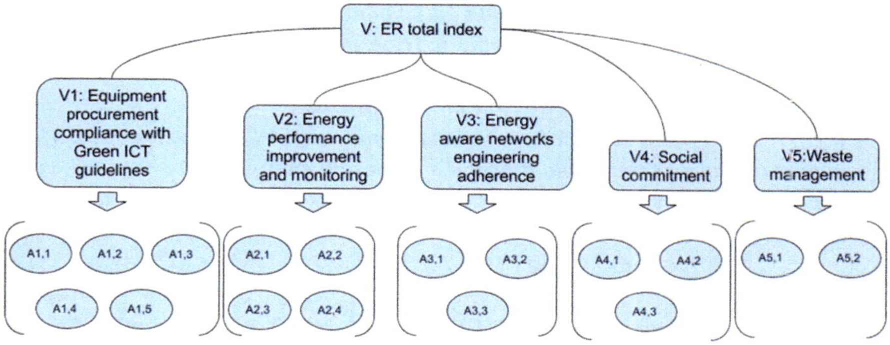

Fig. 8.2 Knowledge base tree

Table 8.2 Rule base matrix

Rule id	Rule weight	If	Then
1	1	V_1 is H & V_2 is H & V_3 is H & V_4 is H & V_5 is H	V is {H} or V is {(H, 1.0), (M, 0.0), (L, 0.0)}
2	1	V_1 is M & V_2 is H & V_3 is H & V_4 is H & V_5 is H	V is {H} or V is {(H, 0.9), (M, 0.1), (L, 0.0)}
3	1	V_1 is L & V_2 is H & V_3 is H & V_4 is H & V_5 is H	V is {H} or V is {(H, 0.8), (M, 0.2), (L, 0.0)}
4	1	V_1 is H & V_2 is M & V_3 is H & V_4 is H & V_5 is H	V is {H} or V is {(H, 0.8), (M, 0.2), (L, 0.0)}
…	…	…	…
243	1	V_1 is L & V_2 is L & V_3 is L & V_4 is L & V_5 is L	V is {L} or V is {(H, 0.0), (M, 0.0), (L, 1.0)}

Having 5 antecedent parent categories with 3 referential values results in 243 total number of rules. A total number of 243 rules is determined based on the number of categories $X = \{x_i; i = 1 \ldots T\}$, *where* $T = 5$ and 3 referential attributes (high, medium, low) $3^5 = 243$. To enumerate all possible combinations, the R language for statistical computing and graphics is used (function–expand.grid(1:3,1:3,1:3,1:3,1:3)). Table 8.2 is the extract of a matrix with 243 inference rules:

Here, {H} is a high, {M} is a medium, and {L} is a low degree of Environmental Responsibility index. Table 8.2 describes two different approaches for producing the total index: ER is {H} with implicit uncertainty handling and ER is {(H, 1.0), (M, 0.0), (L, 0.0)} with explicit uncertainty handling (Eq. 4). It has been decided to keep the weights to be one for all the rules, i.e., assigning the same importance to each rule. Examples of a belief rule taken from Table 8.2 are:

R1: If energy performance improvement and monitoring is High and energy performance improvement and monitoring is High and energy-aware network engineering adherence is High and social commitment is High and waste management is High, THEN ER index is High.

R2: If energy performance improvement and monitoring is Low and energy performance improvement and monitoring is High and energy-aware network engineering adherence is Medium and social commitment is Low and waste management is Medium, THEN ER index is Medium.

R3: If energy performance improvement and monitoring is Low and energy performance improvement and monitoring is High and energy-aware network engineering adherence is Low and social commitment is Low and waste management is Low, THEN ER index is Low.

Here, belief degrees are attached to three referential values and weighted equal. Upon inference completion, the total ER index ($\sum_{i=1}^{N} V_n$) is generated with the following breakdown: initial level for 0–20% range, beginning 20–40%, improving 40–60%, succeeding 60–80%, and leading 80–100% accordingly. The total ER index is displayed without uncertainties in a single deterministic value in percentages, i.e., V is {H} or V is {M} or V is {L}, as it is shown in Fig. 8.3. Additionally, output result of the toolkit developed shows the total ER index score and subcategory score breakdown in %.

An Environmental Responsibility assessment system toolkit is intended to be used by small- and medium-sized enterprises' regular employees or ICT department representatives. Therefore, all aggregated consequent numbers are rounded to the nearest decimal, and end users are notified of the estimated (not precise) figures, due to nature ER index value.

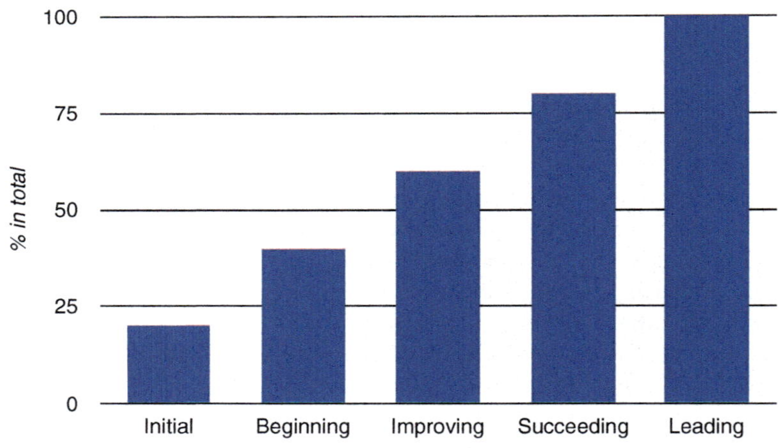

Fig. 8.3 Maturity levels

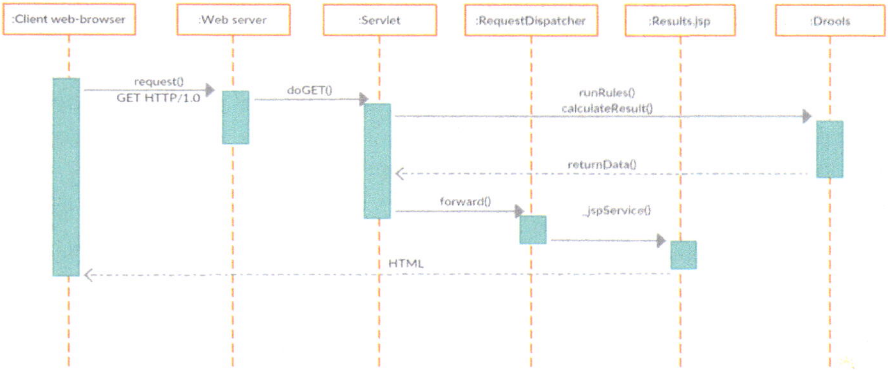

Fig. 8.4 Web-based ER assessment sequence diagram

8.3.1.2 Environmental Responsibility Toolkit

As a proof of concept, a Java web application "Environmental Responsibility Toolkit for SMEs" has been developed (http://demo1-ersme.rhcloud.com), with a JBoss Drools inference engine to provide the reasoning mechanism. The inference engine has been populated with rules described in Table 8.2, where an inference starts iteratively searching for the pattern match of an input and if-then clause. If it is true, the relevant then clause is fired triggering an appropriate action (Fig. 8.4).

The assessment itself encompasses a questionnaire with 17 items to be responded with predefined degrees of uncertainty for each rule, defined in Table 8.2. Upon completion of the questionnaire, the results are automatically analyzed, and the page is displayed with a total ER index score and subcategory score breakdown. Also, based on the results, an individualized set of recommendations to improve the Environmental Responsibility level are presented (outside the scope of this chapter, but details could be found in [34]). Recommendations are based on EU Draft Background Report for the development of an EMAS Sectoral Reference Document "Best Environmental Management Practice in the Telecommunications and ICT Services Sector" [31].

8.4 BRB and FIS Performance Comparison

This section presents validation results of BRB system developed with a fuzzy logic (FL) reasoning-based approach for the Environmental Responsibility assessment. The modeling has been performed using MATLAB Fuzzy Logic Toolbox. Comparison between BRB and fuzzy approach results has been performed and proved the validity of a proposed BRB technique.

In the last two decades, fuzzy logic theory application has increased significantly, which was firstly introduced by L.A. Zadeh [32]. Fuzzy logic (FL) comes from theory of fuzzy sets, where classes of input objects have unsharp boundaries with a certain degree of belief, which in a wider sense can be described as a theory of multivalued logic. FL has been widely deployed in different applications of decision support, industrial process controls, and consumer product selection. A crucial significance of that approach is the use of linguistic variables in describing complex systems, e.g., descriptive words in human-readable and comprehensible format [33].

8.4.1 Fuzzy Logic Design

A basic fuzzy logic system consists of three components: fuzzifier, FIS, and defuzzifier. A fuzzifier finds a mapping between input variable values into a fuzzy set, and defuzzifier does a reverse operation of mapping set of output values into a crisp value. Hence, fuzzy logic is a helpful tool to map an input space to an output space. The primary mechanism for doing this resides in FIS through the list of if-then rules. An architectural view of basic fuzzy system is shown in Fig. 8.5:

There are three steps in a fuzzy inference process: fuzzification, fuzzy rule inference, and defuzzification. The algorithmic steps of a fuzzy logic system life cycle are as follows:

1. Decide on linguistic variables and notations for input and output variables.
2. Determine pertinent membership functions for each input and output variables.
3. Construct knowledge base in rule format.
4. Fuzzification: convert crisp data into fuzzy data sets using membership functions.
5. Fuzzy rule inference: rule evaluation in the rule base and the result set combination.
6. Defuzzification: convert fuzzy output values into crisp data.

The Mamdani-type inference has been deployed for building the model, which expects the output membership functions to be fuzzy sets. The input variables are fuzzified with three linguistic attributes "low," "medium," and "high," as previously defined in the BRB system, with the *gaussmf* membership function (MF) in a range [0 100]. The output variable is described with *trimf* MF with "poor," "average," and "good" attributes. The model consists of three rules, defined in accordance with the BRB system. The Surface Viewer of the MATLAB Fuzzy Logic Toolbox shows the graphical mapping between any two inputs and an output (Fig. 8.6).

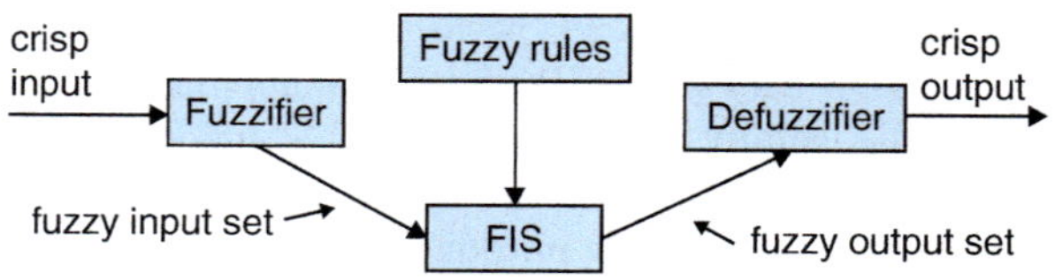

Fig. 8.5 Fuzzy logic system design

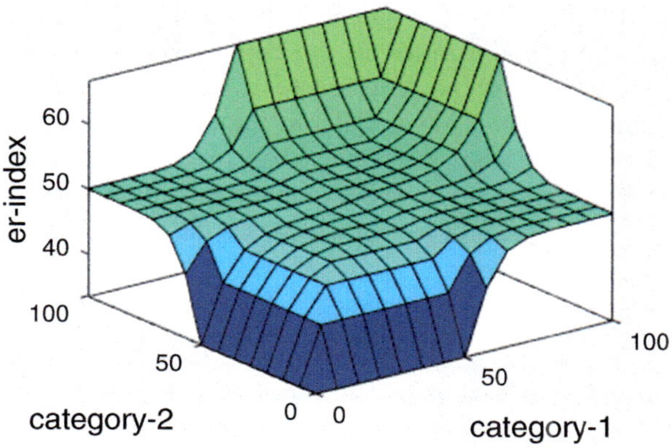

Fig. 8.6 Surface view of ER SME assessment

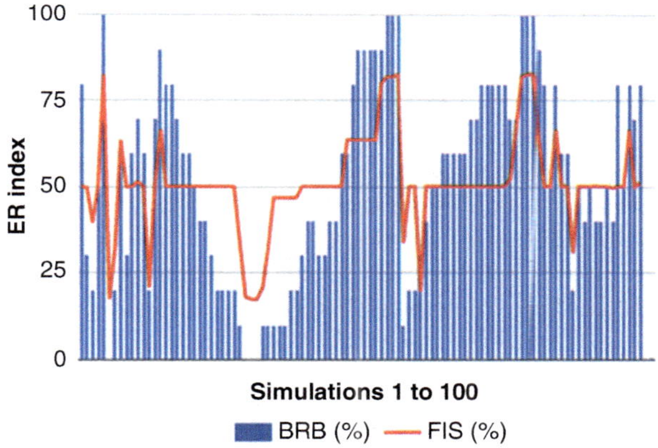

Fig. 8.7 Method comparison

8.4.2 *Results and Comparison*

The Belif Rule Based approach is compared with FIS in this section. In order to verify the validity of the methodology chosen in ER assessment research, 100 simulations with randomized input variables (answers to questions) were carried out. The chart below presents results of the simulations, and an additional mean value is included in comparative analysis as a benchmark (Fig. 8.7).

It is observed that the results of the fuzzy approach are close to the calculated mean value for almost all the experiments. Results of two methods are within the range of the standard deviation. The standard deviation value could be used to calculate the maximum, minimum, and mean values. It is found that FIS performs better for the range of values closer to the mean value (calculated mean value) but not for the maximum nor the minimum values. Per category effect and consequence are not handled in output variables generated based on mean value calculations. For example, when the “Category 1 - Equipment procurement compliant with Green ICT guidelines and the optimization of enterprise operations” accomplishes its maximum value, that implies a positive degree of awareness of an organization on environmental issues; hence, result in BRB in simulation 1 is higher compared to FIS and mean calculations. Also, when at least one of the input variables is equal to zero, BRB approach demonstrates a lesser output variable, which is legitimate: failing to address even one aspect of a problem causes the whole case status unsteady and unreliable.

8.4.3 *Statistical Analysis and Significance Tests*

Significance test has been performed to determine if the BRB approach and FIS theory produce significantly different results. As a first step, we need to test the assumption that the samples come from normal distributions. From the plots below, it is evident that BRB (blue) approximately follows a straight line, indicating approximate normal distribution. The FIS sample (orange colored) shows an increasing departure from normality in the lower tail (Fig. 8.8).

The difference between BRB and FIS is evident. Next, we need to set up hypotheses and evaluate whether the difference between BRB and FIS is significant. In this statistical test, μ_1 is the mean score for BRB, while μ_2 is the mean score for FIS. H_0 is the null hypothesis where μ_1 equals to μ_2, while H_1 is the alternative hypothesis where μ_1 is not equal to μ_2. A two-tail z-test is conducted at a level of confidence of 0.05. Difference between BRB and FIS is evident.

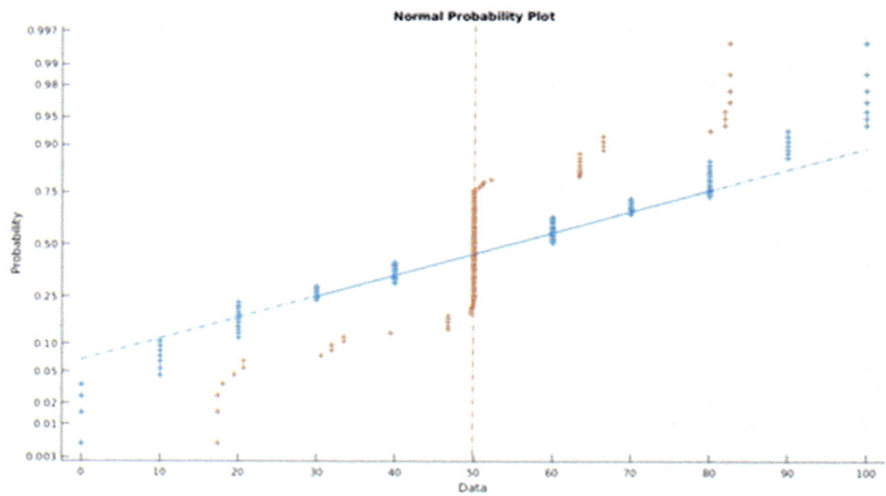

Fig. 8.8 Normal distribution probability for BRB and FIS

Table 8.3 z-Test: two Sample for Means

	BRB	FIS
Mean, μ	52.32	51.01
Known Variance, σ	819.43	194.70
No of observations (n)	99.00	99.00
Hypothesized Mean Difference	0.00	
Z value	0.41	
$P(Z \leq z)$ one-tail	0.34	
z Critical one-tail	1.64	
p-value $(Z \leq z)$ two-tail	0.68	
z Critical two-tail	1.96	

$$H_0: \mu_1 = \mu_2; H_1: \mu_1 \neq \mu_2, \alpha = 0.05 \tag{8.15}$$

Table 8.3 shows a statistical analysis of the two independent data samples. Because both data samples are large (>30), a z-test has been chosen opposed to t-test. The calculated variance values are different for BRB (819.43) and FIS (194.70), and thus, we shall conduct a two-tail z-test for two samples with unequal variances at a level of confidence of 0.05.

The p-value is a probability that measures the evidence against the null hypothesis. The p-value in our research is greater than α, so we fail to reject H0. This means that BRB and FIS approaches do not generate significantly different results. However, as we can see from Table 8.3, the variances of two populations are unequal (BRB is four times higher than FIS). Again, the hypothesis test indicates that there is not enough evidence to reject the null hypothesis that the two-batch z-score is equal to the 0.05 significance level. This failure may reflect normality in the population or it may reflect a lack of strong evidence against the null hypothesis due to the small sample size.

A two-tail z-test for the means of two independent samples with unequal variances investigates whether two independent samples come from normal distributions with unequal variances and the same mean values. For a hypothesis test, quality assurance of two data samples has been tested also using the two-sample Kolmogorov-Smirnov test. The function *kstest2(columnBRB, columnFIS)* returned the value equal to 1. The result rejects the hypothesis that the data in BRB and FIS samples are from the same continuous distribution at the 5% significance level. This confirms our assumption in two approaches of heterogeneity.

To conclude, the BRB approach deployed in this research dedicated to small- and medium-sized enterprises to assess their Environmental Responsibility level generates correlative, well-balanced, and sensible results compared to FIS approach. So the observed difference between the sample means (52.32–51.01) is not convincing enough to say that BRB and FIS approaches generate notably different consequences. Although, the test only performs comparison between mean values and does not focus on such a substantial difference in variances. The most likely explanation of z-test result is that BRB approach, in contrast to FIS, deploys rule-based context-adapted inference procedures for a total consequent variable calculation.

8.5 Conclusions

This research proposes the use of a Belif Rule Based approach to assess an enterprise's level of commitment to environmental issues. Participating companies have to complete a structured questionnaire of the Environmental Responsibility BRB assessment system developed. An automated analysis of their responses (using the Belif Rule Based approach) determines their Environmental Responsibility level. This is followed by a recommendation on how to progress to the next level. The recommended best practices will help promote understanding, increase awareness, and make the organization greener. It is posited that such a system generates well-balanced, sensible, and context-adapted results. The aim of the Environmental Responsibility Assessment System is to help small- and medium-sized enterprises focus on making improvements on more sustainable business operations.

Future work recommendations are drawn from the validation by experts, focus groups, and the target SM, which are: (1) test the usefulness of the assessment in practice and benchmark the results among a pool of similar types of enterprise respondents; (2) include required capital investment amount, payback period, and cost saving for each recommended activity; (3) provide specialized assessment and recommendation road maps for a participating organization's industrial sector (manufacturing, nonmanufacturing); and (4) develop a paid version with more questions and recommendations for more accurate ER assessment estimations, where the users would be able to save their scores, export results, and track their progress.

The key limitations of this study are: (1) assessment does not take into account regional differences among potential participating organizations; and (2) assessment model is not sector-specific. The final limitation extracted from validation sessions is the assessment model applicability to medium-sized enterprises, rather than small and micro organizations. This is due to the type of content and level of questions asked in the assessment process. The assessment model and recommendations would benefit from being evaluated by more Green ICT and sustainability experts, and this could be potentially addressed in the future research. Furthermore, the forward chaining inference algorithm is considered to be less powerful than alternative methods like evidential reasoning, D-S theory or Bayesian networks, etc. The ER assessment model could be improved by integrating and enhancing the current forward chaining logic with uncertainty handling mechanisms.

Acknowledgments This work has been conducted as part of the Erasmus + Master program PERCCOM (Pervasive Computing and Communications for Sustainable Development). Thus, the author would like to thank the Erasmus Mundus program and the European Union for the opportunity to participate in this program and complete this study. The authors are grateful to referees for their feedback comments that have helped to improve the quality of the system.

References

1. Group TC. (2008). *SMART 2020: Enabling the low carbon economy in the information age*. London: Climate Group.
2. Sobotta, A. T., Sobotta, I. N., & Gøtze, J. (2010). Greening it. Adrian Sobotta.
3. Dastbaz, M., Pattinson, C., & Akhgar, B. (2015). *Green information technology: A sustainable approach*. Waltham: Morgan Kaufmann.
4. Marazza, D., Bandini, V., & Contin, A. (2010). Ranking environmental aspects in environmental management systems: A new method tested on local authorities. *Environment International, 36*, 168–179.
5. Harju-Autti, P., & Kokkinen, E. (2014). Open Access Research Article.
6. Structural Business Statistics Overview. (2015). Eurostat. [ONLINE] Available at: http://ec.europa.eu/eurostat/web/structural-business-statistics. (Accessed 5 April 2016).
7. Hillary, R. (1995). Small firms and the environment. Groundwork.
8. Hillary, R. (2004). Environmental management systems and the smaller enterprise. *Journal of Cleaner Production, 12*, 561–569.
9. Biondi, V., Frey, M., & Iraldo, F. (2000). Environmental management systems and SMEs. *Greener Manag Int, 2000*, 55–69.
10. Iraldo, F., Testa, F., & Frey, M. (2009). Is an environmental management system able to influence environmental and competitive performance? The case of the eco-management and audit scheme (EMAS) in the European Union. *Journal of Cleaner Production, 17*, 1444–1452.
11. Fithritama, A., Rondeau, E., Bombardier, V., & Georges, J-P. (2015). Modeling fuzzy rules for managing power consumption of Ethernet switch. In *Software, Telecommun. Comput. Networks (SoftCOM), 2015 23rd Int. Conf. IEEE* (pp. 42–47).
12. Hankel, A., Oud, L., Saan, M., & Lago, P. (2014). A maturity model for green ICT: The case of the SURF green ICT maturity model. In *EnviroInfo* (pp. 33–40). Germany: BIS Verlag Oldenburg.
13. IRI. (2014). Sustainability maturity model from Industrial Research Institute (IRI). 2015.
14. Swiss Informatics Society. (2015). Fachgruppe Green IT Action Plan model. 2016.
15. UK Government. (2013). UK HM Government Green ICT Maturity Model. 2015.
16. Molla, A., Cooper, V., & Pittayachawan, S. (2011). The green IT readiness (G-readiness) of organizations: An exploratory analysis of a construct and instrument. *Communications of the Association for Information Systems, 29*, 67–96.
17. de Monsabert, S., Odeh, K., & Meszaros, J. (2012). SustainaBits: A framework and rating system for sustainable IT. In *Green Comput. Conf. (IGCC), 2012 Int* (pp. 1–9).
18. Davis, R. (1986). Knowledge-based systems. *Science, 231*, 957–963. doi: 231/4741/957 [pii].
19. Beta, Q. (2008). Volume I Basic Programming Guide.
20. Friedman-Hill, E. (2003). *JESS in action*. Greenwich: Manning.
21. Bali, M. (2009). *Drools JBoss rules 5.0 developer's guide*. Birmingham: Packt Publishing Ltd.
22. OpenRules Inc. (2016). *OpenRules Rule Engine. 2016.*
23. Giarratano, J. C., & Riley, G. (1998). *Expert systems*. Boston: PWS Publishing Co..
24. Sun, R. (1995). Robust reasoning: Integrating rule-based and similarity-based reasoning. *Artificial Intelligence, 75*, 241–295. doi: http://dx.doi.org/10.1016/0004-3702(94)00028-Y.
25. Hossain, M. S., Khalid, M. S., Akter, S., & Dey, S. (2014). A Belif Rule Based expert system to diagnose influenza. In *Strateg. Technol. (IFOST), 2014 9th Int. forum* (pp. 113–116).
26. Yang, J.-B., Liu, J., Wang, J., et al. (2006). Belief rule-base inference methodology using the evidential reasoning approach-RIMER. *IEEE Transactions on Systems, Man, and Cybernetics Part A: Systems and Humans, 36*, 266–285. doi:10.1109/TSMCA.2005.851270.
27. Chen, Y.-W., Yang, J.-B., Xu, D.-L., et al. (2011). Inference analysis and adaptive training for belief rule based systems. *Expert Systems with Applications, 38*, 12845–12860.
28. Wang, Y.-M., Yang, J.-B., & Xu, D.-L. (2006). Environmental impact assessment using the evidential reasoning approach. *European Journal of Operational Research, 174*, 1885–1913. doi: http://dx.doi.org/10.1016/j.ejor.2004.09.059.

29. Chen, S., Liu, J., Wang, H., & Augusto, J. C. (2015). A group decision making model for partially ordered preference under uncertainty. *Inf Fusion, 25*, 32–41.
30. Yang, J.-B., & Xu, D.-L. (2002). On the evidential reasoning algorithm for multiple attribute decision analysis under uncertainty. *IEEE Transactions on Systems, Man, and Cybernetics Part A: Systems and Humans, 32*, 289–304. doi:10.1109/TSMCA.2002.802746.
31. European Commission JRC. (2015). Best environmental management practice in the telecommunications and ICT services sector.
32. Zadeh, L. A. (1965). Fuzzy sets. *Information and Control, 8*, 338–353.
33. Zadeh, L. A. (1997). Toward a theory of fuzzy information granulation and its centrality in human reasoning and fuzzy logic. *Fuzzy Sets and Systems, 90*, 111–127.
34. Bazarhanova, A. (2016). A Belif Rule Based Environmental Responsibility assessment system for small and medium-sized enterprises, An unpublished Masters Degree Dissertation, Leeds Beckett University. URL: https://www.doria.fi/bitstream/handle/10024/124773/Thesis%20 Bazarhanova.pdf?sequence=2

Chapter 9
Measurement and Classification of Smart Systems Data Traffic Over 5G Mobile Networks

Mohammed Dighriri, Gyu Myoung Lee, and Thar Baker

9.1 Introduction

In the world of mobile networks, Machine-to-Machine (M2M) communication is a term that reflects the automatic operation of smart devices that are connected to each other, with or without human intervention [2]. Along with the tremendous increase in mobile traffic, it is also expected that M2M traffic will rise quickly due to the growing use of M2M devices (e.g. smartphones, traffic control and blood pressure sensors) in numerous applications. The application areas of M2M include smart office, smart traffic monitoring, smart alerting systems, smart healthcare systems and logistics systems [6, 7]. Furthermore, M2M communication provides ubiquitous connectivity between M2M devices. These interconnected devices (e.g. laptops, smart sensors, computers) can perform several automatic operations in various M2M applications. Unlike traditional Human-to-Human (H2H) communication (e.g. voice call, video chat, email), M2M data traffic has distinctive features, such as a large number of M2M devices, group based communication between devices, dynamic mobility scenarios, small sized data transmission, bursty transmission and extra low power consumption.

Mobile network data traffic is predicted to increase more than 24-fold between 2010 and 2015, and more than 500-fold between 2010 and 2020 [8]. Mobile networks are expected to face challenges due to future Machine-to-Machine (M2M) data traffic with various quality of service (QoS) requirements, such as a provision of radio resources to a large number of smarts devices, prioritization and inter-device communication [13]. The existing cellular systems might reach capacity in

---

M. Dighriri (✉) • G.M. Lee • T. Baker
Department of Computer Science, Liverpool John Moores University, Liverpool L3 3AF, UK
e-mail: M.H.Dighriri@2015.ljmu.ac.uk; G.M.Lee@ljmu.ac.uk; T.baker@ljmu.ac.uk

M. Dastbaz et al. (eds.), *Technology for Smart Futures*,
DOI 10.1007/978-3-319-60137-3_9

the future due to significantly increasing M2M traffic, resulting in performance degradation of regular mobile traffic [13]. M2M devices transmit small and large sized data with different QoS requirements. For example, smart healthcare system devices convey big sized data but are delay sensitive [18]. The physical resource block (PRB) is the smallest radio resource unit allocated to a single device for data transmission in 5G mobile networks. In smart systems, there are different devices transmitting various sized data and sometimes the capacity of the PRB is not fully used. This results in significant system performance degradation.

This research chapter proposes a novel data traffic aggregation model for 5G radio resources based on efficiently utilizing the smallest unit of PRB by aggregating the data of several M2M devices. In addition, a new 5G network slicing model relies on smart systems in the smart city case study, where network slices will differentiate smart systems data traffic in terms of QoS requirements (e.g. smartphones, smart traffic monitoring, smart healthcare systems). The Optimized Network Engineering Tool (OPNET) will be used to assess the performance of the proposed models in terms of QoS for data traffic-slices isolation. The simulated 5G data traffic classes contain file transfer protocol (FTP), voice over IP (VoIP) and video users. The scenarios, categorized into three slices of M2M communication based on the data traffic, include popular, sensitive and heavy data traffic. The results will show the proposed models M2M data traffic impact on QoS of 5G data traffic. The network end-to-end performance will be tested by aggregating data of several M2M devices in each slice enhanced by different protocols, such as FTP, VOIP and video, in terms of average cell throughput, average upload response time, average packet end-to-end delay and radio resource utilization [12]. The following sections study the literature in more detail and highlight the limitations of the existing data traffic models.

9.2 Background

9.2.1 Cellular Networks Evolution

It has taken less than 25 years for the number of mobile subscribers to rise from few to several billion users. This tremendous growth reveals the strong desire of the users to stay reliably connected all over the world without missing a fraction of a second even if they are moving at high speed. Cellular communication plays a significant role in improving human life standards by providing various services including voice calls, short message services (SMS), web browsing, multimedia calls, etc. According to [22], the idea of mobile communication was formulated at AT&T's Bell Labs in the 1970s and has now emerged as the higher generation cellular networks. The overview of the evolution of cellular technologies is demonstrated in Fig. 9.1. The evolution of cellular technologies from the first generation (1G) to the fifth generation (5G) communication standards is given in [22, 28].

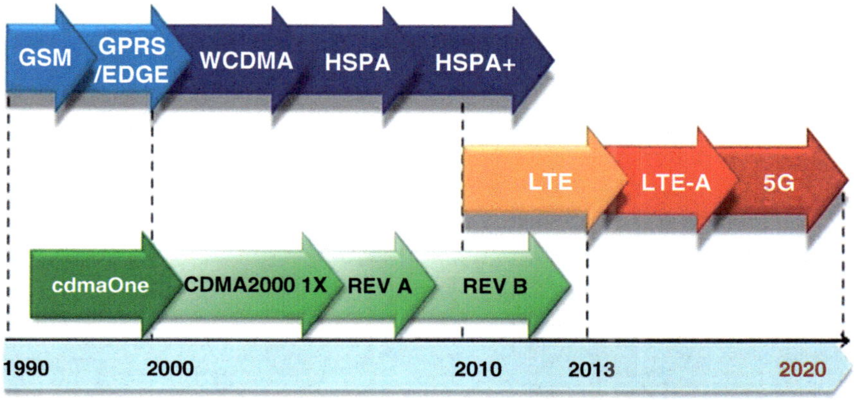

Fig. 9.1 Cellular networks evolution

9.3 5G Protocols Architecture

The 5G mobile network interfaces between UE, E-UTRAN, EPC and SGWs are linked with protocol stacks that are used by the network component to communicate signalling messages and data [1]. Consequently, the 5G–protocol stack can be separated into the following groups that are defined as:

- *Signalling protocols*

  Signalling protocols are utilized to exchange signalling data between numerous devices within the network.

- *User plane protocols*

  These protocols enhance routing of user messages between UEs and S-GWs.

- *Transport protocols*

  Transport protocols convey data and signalling messages between several networked devices on the air (Uu) interface, and the MME control the UE high-level functionalities. Nevertheless, there is no straight connection route between UE and the MME. The connection path between UE and the MME is established using E-UTRAN eNB that supports the level of hardware complexity in the network. To decrease this complexity, the Uu interface is additionally separated into two protocol levels; access stratum (AS) and non access stratum (NAS). The MME high-level signalling lies at the NAS level, however, it is transported within the network using AS protocols. The control and user plane termination protocols are maintained by the eNB. A high-level outline of the protocol stack is portrayed in Fig. 9.2. The user plane protocols contain the packet data convergence protocol (PDCP), the medium access control (MAC) and the physical

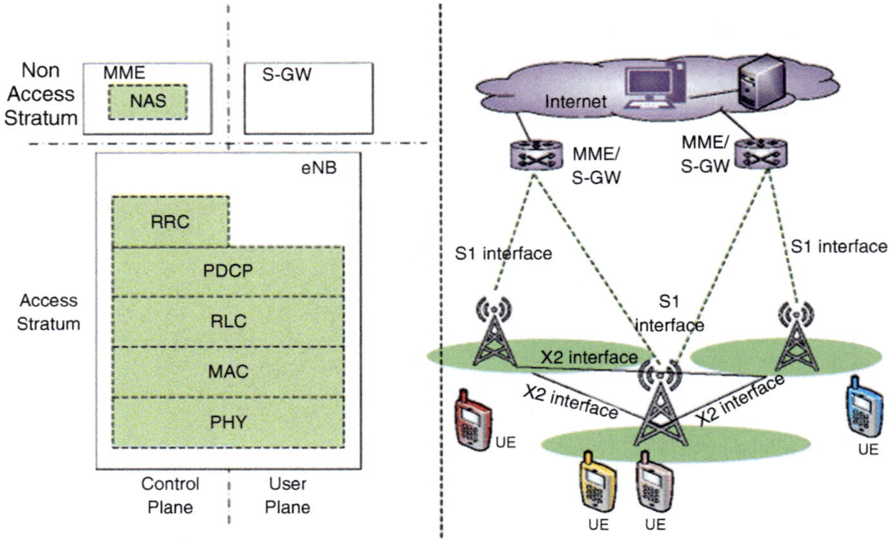

Fig. 9.2 5G protocols stake

(PHY) layer protocols. In addition, the control plane includes the radio resource control (RRC) protocols. According to [31], the main functionalities of these protocols are as shown below:

- *Access Stratum (AS)*
  The AS protocols create UE able to access the abilities and services of the communication networks. The numerous functionalities of the radio AS protocols include dynamic allocation of radio resources, controlling bearers, radio admission control, traffic management, scheduling techniques, throughput and bit error rate.
- *Non Access Stratum (NAS)*
  The NAS establishes radio links between the network and the UE. Furthermore, the functionalities of the NAS protocols also include registration, authentication and local registration management.
- *Radio Resource Control (RRC)*
  The key functionalities of RRC include distribution of system information, an establishment of RRC connections, mobility functions, security management, QoS management and direct connection between UE and NAS.
- *Packet Data Convergence Protocol (PDCP):*
  The PDCP layer re-transmits the session data units (SDUs), performs header compression, ciphering and detection of duplicate data.

- *Radio Link Control (RLC)*
  In RLC there are a number of functionalities including the segmentation of data packets according to the available size of the transport block, adjustment of errors through automatic repeat request (ARQ) and resegmenting for retransmission. Other functions contain a concatenation of the same bearer SDUs, packet transfer and error discovery in the protocols.
- *Medium Access Control (MAC)*
  The main functionalities of MAC protocols are RLC SDUs multiplexing/demultiplexing, error corrections using hybrid ARQ (HARQ), scheduling, local channel prioritization and padding.
- *Physical (PHY) Layer*
  A transmission time interval (TTI) of a 1 ms period is used by the PHY layer to share a channel with the upper layer of the protocols stack. The frequency and time variation in cellular are exploited over OFDM and SC-FDMA techniques, e.g. the physical layer sub-carrier spacing in LTE is 15 kHz.

9.4 5G Enabling Technologies

5G Release 8 specified the next generation network requirements and components. Those main objectives include LTE and SAE for the specification of EPC, E-UTRAN and E-UTRA. UE and E-UTRAN communication is accomplished using IP, which is delivered by the EPS. In 5G, air interface and radio access networks are modified while the architecture of EPC is kept almost the same. The EPS is the basis for LTE, LTE-A and 5G networks. The main 5G features include carrier aggregation (CA), enhanced multiple input multiple output (MIMO) technology, coordinated multipoint (CoMP) and relay node (RN). We will give more details about each technology in the future, including CA, MIMO techniques and CoMP due to the use of RN for the aggregation of M2M traffic [1].

9.5 Infrastructure Based RNs

The RNs are categorized into fixed and mobile RNs depending upon the infrastructure. RNs are used in distinct scenarios to improve data rates, coverage and to facilitate UEs indoor and outdoor movements. Also, UEs experience satisfactory coverage through mounted RNs, such as at the top of a bus or train. Further classifications of the infrastructure based RNs are given below [21]:

- Fixed RNs are mainly used to advance the coverage for those UEs which are not close to the regular donor eNB (DeNB) or a base station usually exists at the

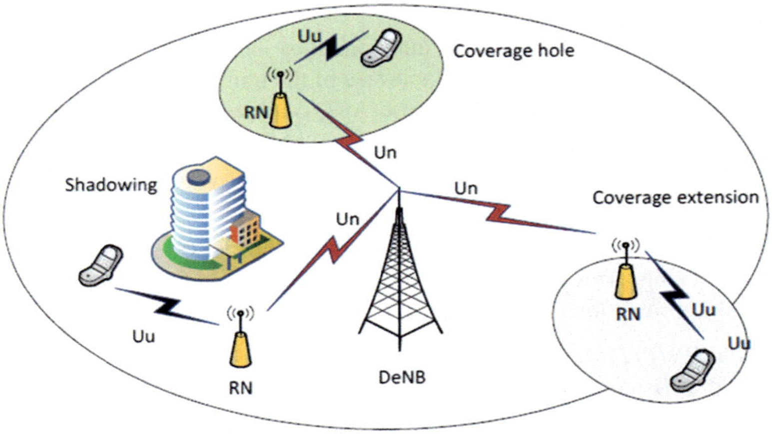

Fig. 9.3 Fixed RN

corner of the cells. Furthermore, the coverage holes due to shadowing are also improved. Fixed RNs are able to extend the cell coverage for the users outside the coverage of the regular base stations. The functionalities of fixed RNs are shown in Fig. 9.3. The fixed RNs contain comparatively small antennas to the ones at the base stations, and are normally positioned at the top of a building, tower, pole, etc.

According to [25], 3GPP considered mobile RNs to provide satisfactory services to users in fast moving trains. However, recent literature has shown that the mobile RNs can also professionally improve the services in public vehicles, e.g. buses and trams. The purpose of mobile RNs is to offer coverage within a moving environment. The mobile RNs are positioned on the vehicle and create a communication path between the mobile UEs and the base station. The RNs communicate with the base station through the mobile relay link (backhaul), whereas they use the access link with the mobile UEs. Due to the vehicle restrictions and other safety measures, antenna size of the mobile RNs is kept small. The functionalities of mobile RNs are shown in Fig. 9.4.

9.6 M2M Communication

The 5G mobile network considers the requirements of M2M communication, whereas the concept of cellular-based M2M communication offers lower costs and increased coverage. Currently, cellular standards (e.g. GSM/GPRS) are used to support M2M communication in some applications. It is expected that the network operators will upgrade the existing cellular standards to support M2M communication. The 5G mobile network is considered the promising cellular technology to provision a huge number of future M2M devices with high capacity, data rates, support for mobility and QoS requirements [6].

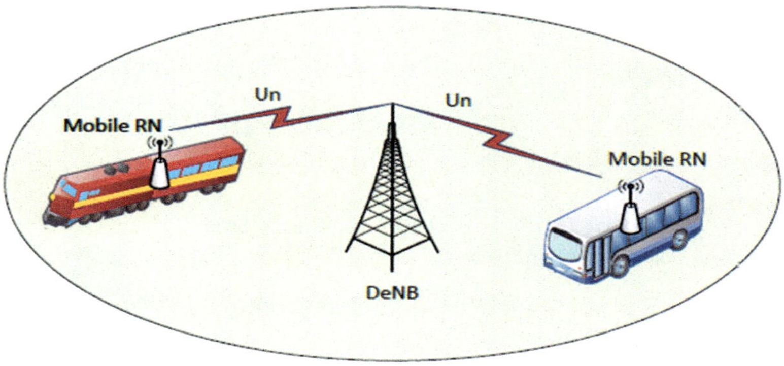

Fig. 9.4 Mobile RN

9.7 5G Network Slicing

5G as a new generation of the mobile network is being actively discussed in the world of technology; network slicing surely is one of the most deliberated technologies nowadays. Mobile network operators (e.g. China Mobile and SK Telecom) and merchants (e.g. Nokia and Ericsson) know it as a model network architecture for the coming 5G period [3]. This novel technology allows operators to slice one physical network among numerous, virtual, end-to-end (E2E) networks, each as a rationally isolated counting device, and access transport and core networks like separating a HDD into C and D drives devoted to diverse services with different features and QoS requirements. Every network slice is committed resources, e.g. resources within network functions virtualization (NFV), software defined networking (SDN), cloud computing, network bandwidth, QoS, as seen in Fig. 9.5 [14].

9.8 State of the Art

This section presents a survey on the studies related to the data traffic aggregations and M2M applications data traffic models as well as the advantages and drawbacks associated with these models, particularly focusing on the data traffic QoS requirements.

9.8.1 Data Traffic Aggregation Models

Data aggregation models have continued as a research topic mainly in mobile networks. It is normally used to decrease the number of sensor transmissions, and thus to recover the network lifetime. In recent years, extraordinary research efforts have been conducted on the topic of data traffic aggregation. For example, in [5], the

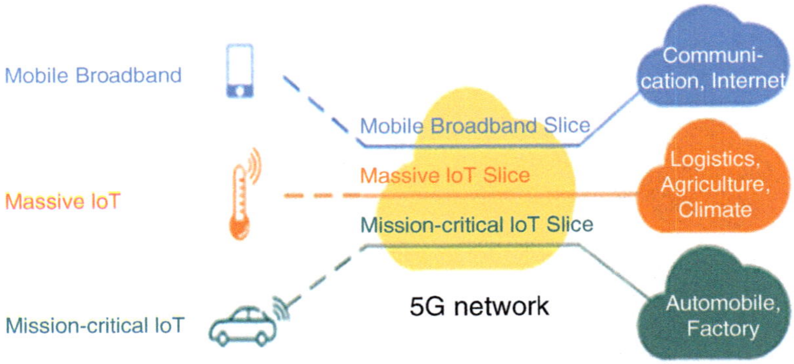

Fig. 9.5 5G network slicing

authors discovered the advantages of data-centric routing algorithms and proposed a routing algorithm relying on the tree and query model. The authors of [23, 24] suggested aggregation tree algorithms to discover the aggregation node in order to develop network lifetime. Moreover, the authors in [26] improved the network bandwidth by choosing multiple appropriate aggregation nodes over clustering algorithms that led to a decrease in the number of transmitted packets to the gateway. In order to decrease energy consumption, bandwidth effectiveness and improve system throughput, data aggregation accumulates and collects data packets from multiple nodes before transmitting to the next hop. For example, an energy-aware aggregation algorithm is suggested in [10] in which the authors deliberated through experimental analysis that data aggregation increases the network lifetime by 80%.

Furthermore, the authors of [10] showed the disadvantages of data aggregation, e.g. improved transmission time consumption. Resultantly, the authors in [18] revealed two timing control methods that contain periodic simple and periodic per-hop. In the former case, packets are aggregated until the number of buffered packets reaches the extremely defined threshold. While in the latter case, the aggregation node retrieves packets until the expiry of the timer. In addition, the authors in [9] examined the impact of buffer bloat on delay and system throughput in IEEE 802.11n mobile networks. Their simulation figures presented that large buffers could degrade the fairness in multi-hop networks based on a parking lot. Also, the authors in [27] recommended a queue management scheme called WQM for mobile networks that adapts the size of the buffer depending on measured link features and network load.

9.8.2 M2M Data Traffic Models

M2M communication has played a significant role in interconnecting physical systems with the Internet world. The concept of cellular-based M2M communication is flourishing. Numerous researchers have investigated new and advanced methods to upgrade the existing cellular standards to provision M2M communication. The authors of [18] suggested 5G cellular technologies as the most promising for future M2M communication. The authors of [6] delivered an overview of cellular-based M2M communication and highlighted the growing M2M traffic. Traffic modelling is the scheming of stochastic processes, which represent the physical parameters of the measured traffic. The initial detailed characterization of M2M was done in [11] where the authors gathered traffic data for 7 days from one of the main TIER-1 cellular service providers in the USA. The authors determined that there is a clear difference between the features of M2M and traditional human-type-communication (HTC). Their experimental results revealed that the volume of M2M traffic in the uplink is much larger than in the downlink. Moreover, M2M devices generate bursty traffic compared to traditional mobile devices.

In [10], the authors modelled the M2M traffic as a simple Poisson process due to numerous machines assigned to one server. The authors of [5] stated that the data streams in various M2M applications follow different statistical patterns which are difficult to capture. According to [7], M2M traffic is classified in the source and aggregated traffic. Some aggregated traffic models for M2M are identified in [10], i.e. defining M2M data traffic as one stream from numerous devices. Contrarily, to model and understand the behaviour of the M2M traffic more accurately, source traffic modelling is required, i.e. modelling each M2M device on its own. However, in M2M source traffic modelling there are many challenges that need to be overcome. For example, it is very hard to model the behaviour of traffic being produced from a massive number of devices in parallel with a strong spatial and temporal correlation between devices [16]. In [10], a coupled Markov modulated Poisson processes (CMPP) model mostly aimed at the source traffic in M2M networks was presented.

9.9 Limitations of Existing Models

As per the literature review for data traffic aggregation models, limited research efforts have been done to study the impacts of data aggregation in the case of the mobile network for M2M communication. For example, the authors in [10] assess the performance of data aggregation in relation to energy consumption in order to grow the lifetime of capillary M2M networks; although, the authors neglected the emerging mobile M2M applications in their work. Furthermore, the authors in [13] suggested a system of bundling M2M data packets at the macro station, also called donor eNBs (DeNBs), to decrease the risks of congestion in backbone networks.

Therefore, this research presents a novel data aggregation model over 5G mobile network slices. For this purpose, based on the wireless layer a 3 inband RN is used to enhance coverage and aggregate uplink M2M data traffic to each slice separately.

9.10 Problem Statement

In mobile networks with LTE and 5G, massive access (e.g. H2H, M2M, personal devices) can lead to serious system challenges in terms of radio access network (RAN) congestion and overload. Since radio resources are a rare, essential

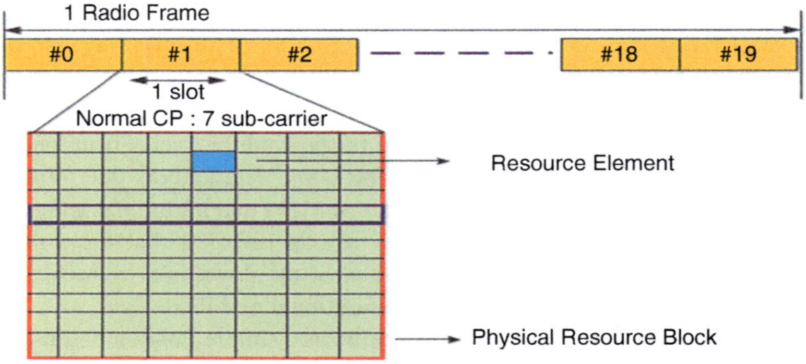

Fig. 9.6 Physical resource block (PRB)

component, efficient utilization is required. The novel communication technologies (e.g. long term evolution (LTE), long term evolution advanced (LTE-A) and 5G) make use of multiple carriers schemes to offer better data rates and ensure high QoS. The smallest resource unit allocable in the 5G system to an M2M device is the PRB as illustrated in Fig. 9.6. Under favourable channel conditions, PRB is capable of transmitting several kilobytes of data. These multiple carrier schemes are capable of transmitting a large amount of data. However, in the case of M2M communication, both narrowband and broadband applications have to be considered to enhance QoS requirements. Especially, these applications have different data traffic size, which need specifications for QoS, such as real-time, accuracy and priority. If one PRB is allocated to a single M2M device for data transmission of just a few bytes, then it might cause severe wastage of radio resources. The different types of data traffic should also be considered in the 5G slices approach as illustrated in Fig. 9.7. Therefore, utilizing the full radio resources and classifying data traffic should be brilliant solutions for data traffic explosion and the fairness of services in the near future.

9.11 Proposal Work

M2M communication is so various that the M2M messages range from bit level bandwidth consumption (e.g. smartphones, light sensing, metering) to byte level bandwidth consumption applications (e.g. smart healthcare systems). The M2M applications with bit level (i.e. M2M narrowband communication devices) transmit small sized data in regular time intervals. While in broadband M2M applications, e.g. smart healthcare systems, thousands of bytes are conveyed in order to monitor regularly. According to the cellular network standardization, a PRB is the smallest resource, which is given to every device for data transmission over the network. 1 PRB is capable of transmitting several hundreds of bytes. Although, allocating 1

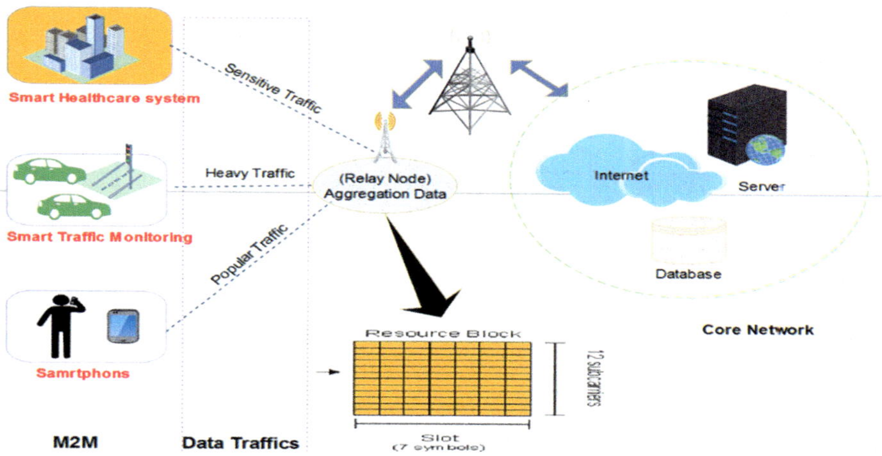

Fig. 9.7 M2M data traffic on 5G mobile network

PRB to an individual M2M device operating in the narrowband applications significantly influences the network performance due to inefficient utilization of the capacity of a PRB, i.e. fewer number of bits per PRB, as indicated in [22]. The M2M devices usually comprise individual bearer buffers, which are used for segmentation and re-assembly purposes. The PRBs are allocated separately to the M2M devices by the base station to send data collected by the M2M devices to the backend M2M servers through Uu interface in the uplink. Increasing the number of M2M devices results in higher demand for resources for transmission of M2M data. In the M2M data aggregation technique, the PRBs are shared by multiple devices according to the capacity of PRBs. In this way, more and more bits per PRB are sent to the network, which significantly enhances the network performance in terms of spectral efficiency and cell throughput. In order to transmit more bits per PRB, M2M data traffic aggregation and slicing models are proposed, which can be implemented for RN in the 5G mobile network. To send and receive M2M data through RN, the M2M devices and the RN communicate through the Uu interface. The Un interface is used to start the communication path between RN and the base station. In the implementation, an M2M device is considered as an aggregation device RN at which data from several M2M devices is aggregated into the data of a single device.

9.12 Data Traffic Aggregation Model

The proposed model relies on aggregating data from several M2M devices at the PDCP layer of the RN. The PDCP layer performs header compression, retransmission, delivery of PDCP session data units (SDUs), duplicate detection, etc. In the proposed model, the PDCP layer is used for the aggregation of the M2M data in the

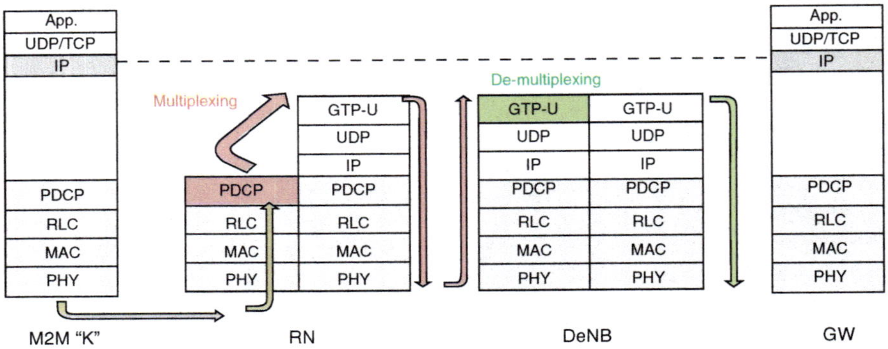

Fig. 9.8 M2M data packets flow diagram

uplink in order to aggregate data with a minimum number of additional headers, as shown in Fig. 9.8.

The individual data packets from several M2M devices approach the PHY layer of the aggregation device with various intact headers, such as MAC, RLC and PDCP. The headers are removed as the received data is transported to the upper layers. On arriving at the PDCP, all the headers are removed and only the payloads from the individual devices are available, which are then aggregated.

One single aggregation buffer *B* at the RN is considered to aggregate M2M data traffic. This buffer aggregates data from different M2M devices ensuring QoS for both the 5G and M2M data traffic. In this implementation, RN is used for M2M devices and base station for 5G data traffic. In order to reach the maximum performance improvements in spectral efficiency, packet propagation delay and cell throughput, we consider scenarios in which all the M2M devices communicate with the base station through the RN. The M2M data aggregation algorithm is shown in Fig. 9.9 and described as follows:

- Data from *K* M2M devices is considered for aggregation.
- The essential parameter for M2M data aggregation is the maximum delay time *Tmax* for the packet at the RN.

The maximum delay time *Tmax* is an essential parameter for M2M data and is calculated according to the various traffic classes of the M2M devices. M2M data have different priorities according to their applications. For example, data packets received from the M2M device deployed in the smart healthcare system scenario for the measurement of temperature or pulse rate of the patient have higher priority over the packets from M2M devices which are deployed in smartphones. The data packets from a device having the highest priority face the smallest delay. Therefore, we initiate the *Tmax* value as the inter-send time of the M2M device data with the highest priority. For example, in the simulation setup for distinct M2M applications, the inter-send time of the M2M traffic models is 1 s, which is the maximum time a packet is delayed at the RN. Thus, the value of the *Tmax* is initiated as 1 s, which means that the data packets received from the distinct M2M devices are delayed for 1 s at the RN.

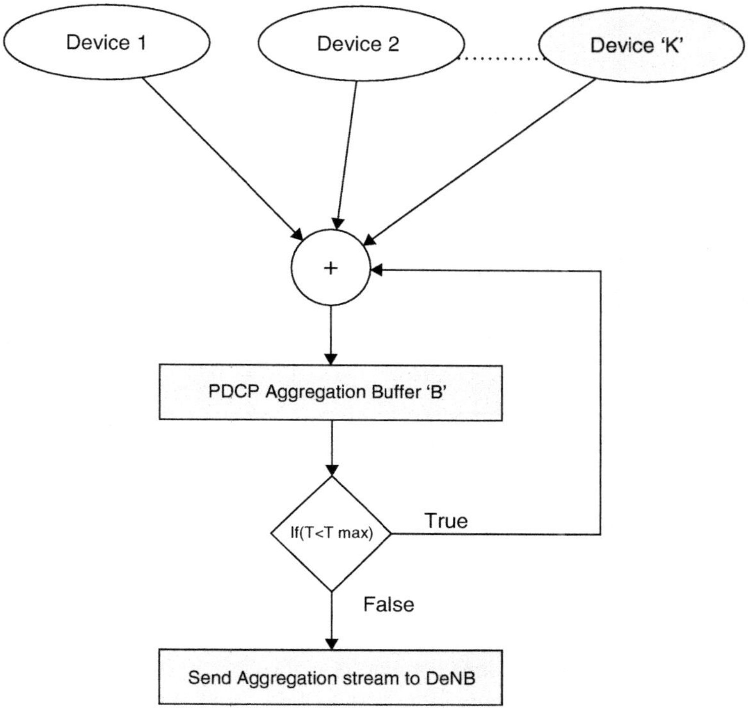

Fig. 9.9 M2M data aggregation algorithm

The value of *Tmax* is adaptive, i.e. the algorithm updates the value of *Tmax* if RN receives packets from a device with higher priority than the priorities of all the other devices in the queue of the RN. The data from all the M2M devices is buffered at the RN. The individual IP headers of all the M2M devices are kept intact. The data packets are buffered until time delay approaches *Tmax*. In order to compare the performance of the data aggregation model in narrowband and broadband M2M application scenarios, the aggregation scale for M2M devices is kept at 1 (unaggregated), 5, 10, 15 and 20 in both cases. The aggregation scale represents the number of devices which are aggregated. For example, in a scenario with 180 M2M devices, the aggregation scale of 5, 10, 15 and 20 means that the data from the group of 5, 10, 15 and 20 devices is aggregated at the RN respectively.

The aggregated data is sent to the base station through the *Un* interface where the data is demultiplexed. The individual IP streams are then sent to the respective application server by the base station.

The M2M packet flow from the M2M devices to the aGW through the RN is described in Fig. 9.13. *K* number of M2M devices transmit data packets to the RN, which are collected at the PHY layer of the RN. The packets are transported to the PDCP layer of the RN on the uplink. The IP packets are packed according to their quality control identifier (QCI) values in the aggregation buffer. The aggregation buffer collects packets from several M2M devices. The data packets are placed in

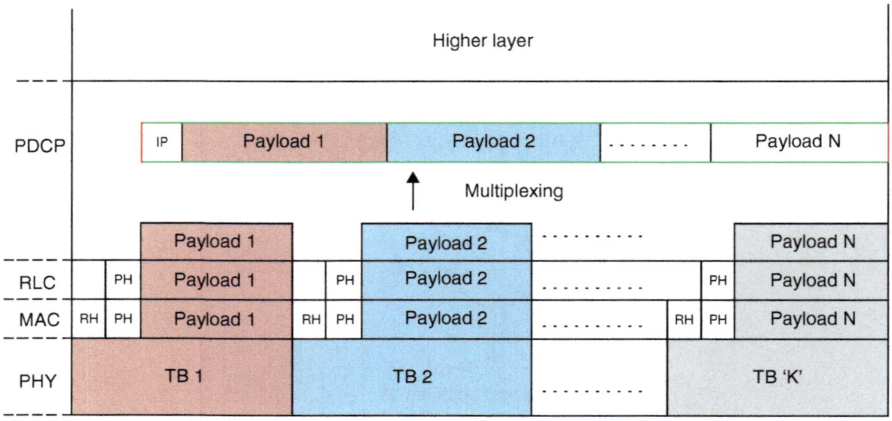

Fig. 9.10 M2M IP packets multiplexing at PDCP

the aggregation buffer according to the packet arrival from the different devices. The detailed structure of the aggregated data stream is depicted in Fig. 9.10 where only the layer two protocols are presented to illustrate the aggregation of the M2M data. The RN PHY layer receives the data packets in the form of distinct transport block size (TBS). The TBS transmitted by the M2M devices at the RN are shown from 1 to K. The data packets arrive at the RLC through the MAC layer. The RLC headers are removed and the remaining protocol data unit (PDU) is transported to the PDCP. The received PDUs at the PDCP layer comprise the individual IP headers of each M2M device and pack into a single PDCP buffer.

9.13 Data Traffic Slicing Model

The application layers in the 5G mobile networks are the main terminal to offer exceptional QoS over a variety of networks for M2M devices. The proposed 5G network-slicing framework will be based on data traffic aggregation and multiplexing models as we mentioned above, which is focused on QoS requirements for each service (application) layer. Therefore, we will clarify the main 5G network architecture layers, which are physical/MAC layers, network layers, open transport protocol (OTA) layers and service layers as shown in Fig. 9.11.

9.14 Applications (Services) Layer

The 5G wireless network terminal is to offer exceptional QoS through a diversity of networks. Currently, mobile internet users do not have the opportunity to utilize QoS history to choose the suitable mobile network link for a provided service. In the

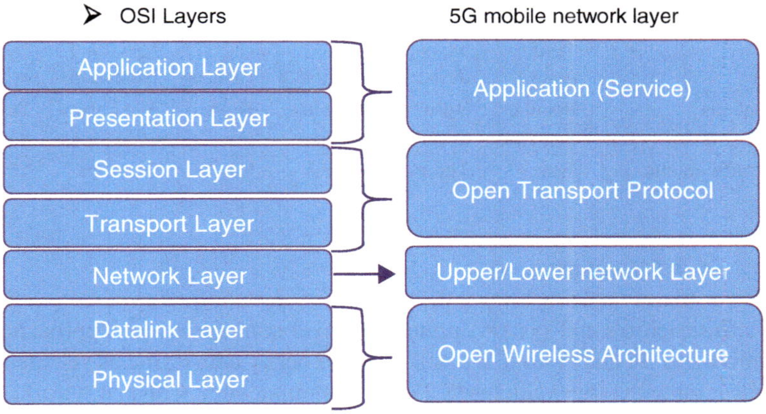

Fig. 9.11 5G architecture layers

future, the 5G phone will deliver an opportunity for QoS analysis and storage of measured data traffic in the mobile network terminal. There are diverse QoS parameters (e.g. bandwidth, delay, congestions, jitter and reliability) that will be stored in a database in the 5G mobile running as system processes that eventually will automatically offer the best appropriate wireless connection based on needed QoS. Therefore, we will consider different types of data traffic (e.g. sensitive, popular and heavy flow traffics) as service slices models, as shown in Table 9.1 [17].

- *Smartphones*
  Smartphones and tablets are recent technologies that represent popular data traffic. Although smartphones are expected to continue as the key personal device and develop further in terms of performance and ability, growth is driven by wearable or sensor devices, which are expected to reach millions by 2020. The content type of mobile streaming in these devices is video. The total of the flow packets is regularly several megabytes or even tens of megabytes, in many packets, usually with continual transmission, and the priority is generally low because the video needs broad bandwidth and is likely blocked in congestion [3].

- *Smart Healthcare System*
  The smart healthcare system as sensitive data traffic is a promising model, which has currently achieved extensive attention in research and industry. A sensor body-area-network (BAN) is generally positioned around the patient to gather information about numerous health parameters, e.g. blood pressure, pulse rate and temperature. Moreover, the patients are also monitored repeatedly by placing smart M2M sensors on the body of the patient when they are outside the hospital or home. To handle critical situations, alarms are triggered to send messages to the related physicians for urgent treatment [7]. In the smart healthcare system scenario, in order to monitor the patients frequently outside the medical centres (i.e. hospitals) they are equipped with smart M2M devices that monitor various health parameters.

Table 9.1 5G services requirements

Services	Traffic types	No of devices	Priority	Latency	Min bandwidth	Mobility
Smartphones	Popular	Billions	10 ms	High	100 kbps	Yes
Smart healthcare system	Sensitive	Thousands	1 ms	Low	100Mbps	Yes
Smart traffic monitoring	Heavy	Thousands	5 ms	Low	25 Mbps	Yes

- *Smart Traffic Monitoring*
  Smart traffic monitoring allows the passage of alerts information between vehicle infrastructures and system applications over communication approaches and technologies. In this system, we will consider heavy data traffic as vehicles communication with other vehicles (V2V) or communication with smart traffic monitoring servers, vehicle to infrastructure (V2I). This system application includes collision avoidance and safety, parking time, Internet connectivity, transportation time, fuel consumption, video monitoring, etc. [4]. In the case of emergency, the information from devices positioned to monitor emergency situations is transmitted to other networked vehicles within the communication range. To prevent any more accidents, the communication between the server and vehicles to detect emergency messages and deliver alerting messages should be very fast. Since the response time of the warning messages is very small, the collision avoidance services request for a high level of QoS and low latency can be supported by the 5G cellular networks. According to [4], the alerting messages are small in size and must only be sent in critical circumstances for effective use of the communication network bandwidth. Traffic and infrastructure management play an important role in controlling the issue of traffic congestion.

9.15 Data Traffic Slices Model

Our 5G slicing framework will focus on classifying and measuring QoS requirements and data traffic of M2M applications, e.g. smartphones, smart healthcare systems and smart traffic monitoring (Fig. 9.12). The results of M2M applications (services) have data traffic characteristics in the 5G network slicing framework that rely on the content type of data, amounts type of flow data, priority of data transmission and data transmission mode. Content type contains voice and video streaming; amounts type consists of different size, large size refers to a number of packets that is more than 1 K bytes, small size refers to a number of packets less than 1 K bytes. Transmission methods are periodic transmission, continuous transmission, burst transmission and time-response transmission; priority of transmitting consists of low, medium and high. Depending on the M2M applications slicing, our research classifies them into three main slices based on QoS and data traffic types.

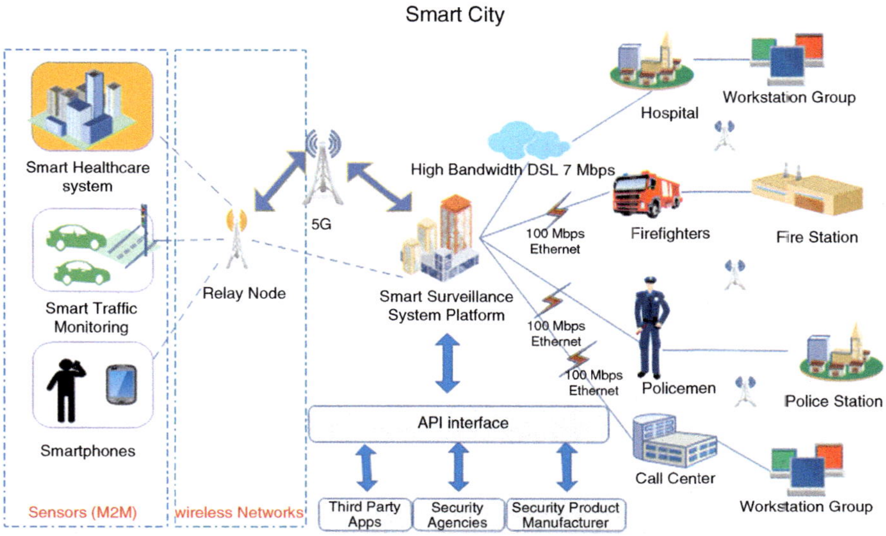

Fig. 9.12 Smart systems in smart city use case

9.16 Data Traffic Slices Algorithm

In the data traffic slices model, we will consider associating our previous data traffic aggregation model, which enhanced QoS by efficient utilization of the 5G radio resources for M2M, and the principle idea of the priority queueing (PQ) mechanism. PQ normally confirms the fastest service of high priority data at each point where it is applied [19]. It provides firm priority to the traffic, which is very essential. The location of each packet in one of four queues (high, medium, normal or low) is achieved depending on the allocated priority of each packet [19]. The possible disadvantage of this scheduling mechanism is that the lower level traffic cannot be served for a long time while the higher priority is present [19]. Consequently, the lower class will have a starving issue that leads to a major discard of the packets as shown in Fig. 9.13.

Therefore, these smart devices have various priorities queuing based on the priority of the packets, the highest priority is transferred on the output port first and then the packets with lower priority and so on as illustrated in the data traffic slices algorithm shown in Fig. 9.14 [30]. Therefore, we design our smart systems environment in three levels of priority; high *(slice1),* medium *(slice2)* and low *(slice3),* which rely on the data traffic types as follows:

- Smart healthcare system as sensitive data with high priority (1 ms)
- Smart traffic monitoring as heavy data with medium priority (5 ms)
- Smartphones as popular data with low priority (10 ms)

  As shown in Table 9.1, these data traffic will work in a form of slicing over the 5G mobile network in the uplink path between RN and base station based on user plane interface as shown in Fig. 9.15.

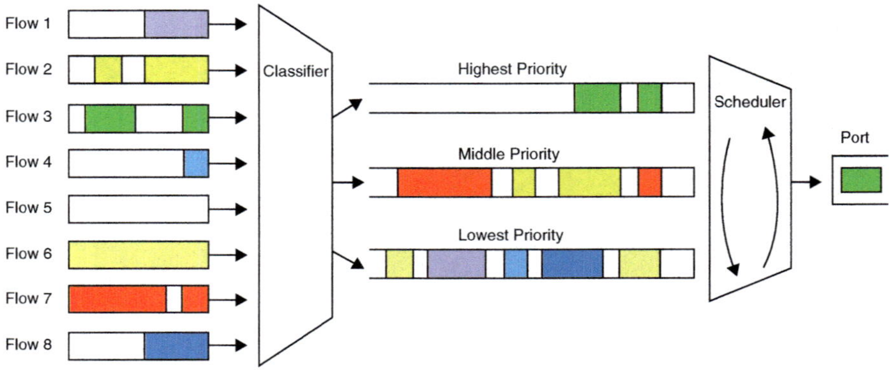

Fig. 9.13 Priority queuing mechanism

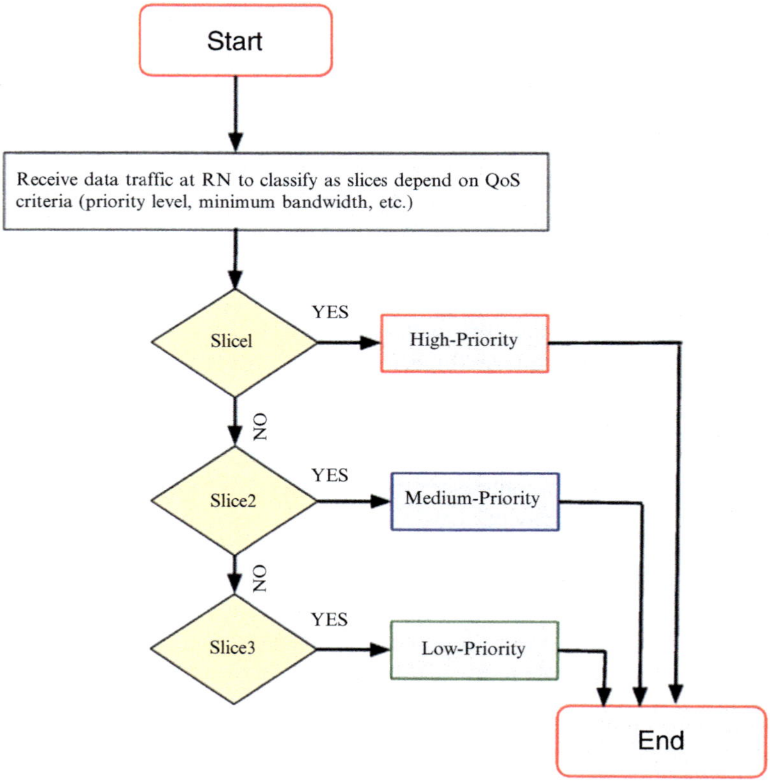

Fig. 9.14 Data traffic slices algorithm

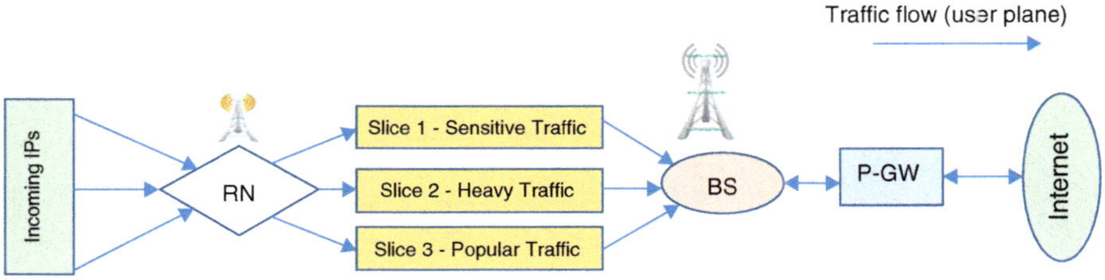

Fig. 9.15 Data traffic slices proposed model

9.17 Measurement QoS in Slices

QoS can be defined as the capability of the network to provision decent services in order to be accepted by quality of experience (QoE). Moreover, QoS measures the level of user satisfaction and network performance. Smart devices, such as smartphones in our case study, have different protocols, e.g. HTTP, FTP and e-mail are not sensitive to delay of transferred information to evaluate QoS in significant issues. Whereas, other applications such as smart healthcare systems have a number of protocols, e.g. video and voice, that are more sensitive to loss, delay and jitter of the information. Consequently, QoS of our smart systems have several protocols, FTP, VoIP and video users require an essential attention to ensure these IPs packets are not delayed or lost while being transmitted over the network [15]. VoIP QoS is measured according to ITU recommendations that rely on diverse parameters (e.g. delay, jitter and packet loss), these parameters can be altered and controlled within the suitable range to enhanced VoIP, FTP and video QoS requirements. Parameters affecting QoS are briefly shown in the following sections [15].

9.17.1 Latency

A delay sensitive application such as a smart healthcare system has voice, file and video that cannot sustain too much delay. Latency is the average time it takes for a packet to convey from its source to its destination. For a patient who has sensors to control and track his/her condition, sensor sources such as voice and video should reach the end destination in a few milliseconds without delay. Moreover, the delay should remain as low as possible but too much traffic on the uplink side of the 5G mobile network could cause congestion; or if there is a call, a voice packet could get stuck behind a bunch of data packets, e.g. an email attachment, and the voice packet will be delayed to the point that the quality of the call is compromised [20]. The maximum amount of latency that a voice call can tolerate one way is 150 ms (0.15 s), but in our case study it is consider to be 10 ms (0.1 s) for the sensitive application system.

Equation (9.1) illustrates the calculation of delay where *Average delay (D)* is expressed as the sum of all delays (*di*), divided by the total number of all measurements (*N*) [29].

$$D = \sum_{N}^{i=1} di / N \tag{9.1}$$

9.17.2 Jitter

Jitter (variation of delay): In order for voice to be understandable, voice packets should arrive at regular intervals. Jitter defines the degree of fluctuation in packet access, which could be affected by too much traffic on the line [29]. Voice packets can sustain only about 75 ms (0.075 s), but 40 ms (0.040 s) is preferred, of jitter delay [15].

Equation (9.2) illustrates the calculation of jitter (*j*). Both average delay and jitter are measured in seconds. Clearly, if all (*di*) delay values are equal, then $D = di$ and $J = 0$ (i.e. there is no jitter) [15].

$$J = \sqrt{\frac{1}{N-1} \sum_{n}^{i=0} \left(di - D\right)^2} \tag{9.2}$$

9.17.3 Packet Loss

Packet loss is the term used to refer to the packets that do not arrive at the purposed destination when a device such as a switch, link or router is overloaded and cannot accept any incoming data at a given moment [30]. Packets will be dropped during periods of network congestion. For example, voice traffic can sustain less than a 3% loss of packets (1% is optimum) before callers feel gaps in conversation [29]. Equation (9.3) illustrates the calculation of packet loss ratio defined as a ratio of the number of lost packets to the total number of transmitted packets, where *N* equals the total number of packets transmitted during a specific time period, and N_L equals the number of packets lost during the same time period [15].

$$\text{Loss packets ratio} = \left(\frac{N_L}{N}\right) \times 100\% \tag{9.3}$$

9.17.4 Simulation Environment

The simulation tool implemented for this research proposal is OPNET version 17.5 OPNET is an object-orientated simulation tool for creation network modelling and QoS analysis of the simulation of network communication, network devices and protocols. OPNET Modeller has an enormous number of models for network basics, and it has many different real-life network configuration competences. These create real-life network environment simulations in OPNET extremely close to reality and offer full phases of a study. Therefore, we will consider smart systems application in this research consists of different priorities based on data traffic types in each slice, such as slice one with high priority, slice two with medium priority and slice three with low priority. Also, we will prepare simulation applications sources VoIP, FTP and video to measure QoS such as latency, jitter and packet loss for each slice [30].

9.18 Research Expected Contributions

This research chapter offers contributions to an efficient radio resource exploitation scheme, which relies on the M2M data traffic aggregation in 5G cellular networks. It provides for M2M devices in which the data packets from several M2M devices will aggregate and transmit as a single device data. In the implementation, an M2M device is used to aggregate data of various M2M devices (i.e. functions as an aggregation node). That will enhance the performance in terms of cell throughput and end-to-end delay of 5G data traffic for different scenarios. The M2M applications in this research concentrate on smart systems based on a smart city case study to support and assist the operations of diverse systems, e.g. smart traffic monitoring, smart healthcare systems and smartphones. These smart systems are used in the 5G slicing data traffic model to classify and measure QoS requirements (e.g. priority and packet loss), which will consider data traffic types with different priorities in diverse cases (e.g. sensitive, popular and heavy traffic).

Overall, these proposed models and algorithms will offer better solutions for 5G as the next generation mobile network:

- Allow network operators to offer fair QoS requirements for different applications and users.
- Abilities to enhance massive connectivity and massive capacity.
- Provision for a progressively various set of service applications and users, all with extremely differing QoS requirements for business and lifestyle.
- Efficient and flexible use of all available non-contiguous spectrum for wildly different network deployment scenarios.
- Latency from 1 ms to a few seconds, to match end user requirements.

  The simulation model will consider presenting results of the M2M data aggregation model to minimize the impact of M2M devices on 5G data traffic in the case of network slicing. The results of the data aggregation model will be collected for both the M2M narrowband and broadband scenarios.

9.19 Conclusion

This chapter proposed two models and algorithms. We proposed a M2M data traffic aggregation model and algorithm in fixed RNs for the uplink in 5G cellular networks. It improves the radio resource utilization for M2M commutations in 5G networks. It offers a maximum multiplexing gain in the PDCP layer for data packets from several M2M devices and considers diverse priorities to solve packets E2E delay. In the implementation, the M2M device is used to aggregate data of various M2M devices, i.e. functions as an aggregation node. That will enhance the performance in terms of cell throughput and E2E delay of 5G data traffic for different scenarios. Further, this research proposed a data traffic slicing model and algorithm for classifying and measuring the QoS requirement. It is based on a priority queuing mechanism and measuring QoS in each slice, which helped the network operators and researchers to differentiate the QoS requirements of the diverse systems, e.g. smart traffic monitoring, smart healthcare systems and smartphones.

In future works, we will reveal the simulation results and analyse the proposed data traffic slicing model in different data traffic scenarios (e.g. sensitive, popular and heavy traffics) in diverse classes including FTP, VoIP and video users. The proposed models can offer opportunities for future researchers in terms of resolving data traffic explosion and fairness of services area.

References

1. Abdalla, I., & Venkatesan, S. (2012). Remote subscription management of M2M terminals in 4G cellular wireless networks. In *Proceedings – conference on local computer networks, LCN* (pp. 877–885).
2. Ajah, S., et al. (2015). Machine-to-machine communications energy efficiencies: The implications of different M2M communications specifications. *International Journal of Wireless and Mobile Computing, 8*(1), 15–26. Available at: http://dx.doi.org/10.1504/IJWMC.2015.066752.
3. Annunziato, A. (2015). 5G vision: NGMN - 5G initiative. In *IEEE Vehicular Technology Conference*. doi:10.1109/VTCSpring.2015.7145586.
4. Booysen, M. J., Zeadally, S., & van Rooyen, G.-J. (2011). Survey of media access control protocols for vehicular ad hoc networks. *IET Communications, 5*(11), 1619–1631. Available at: http://ieeexplore.ieee.org.sci-hub.org/xpl/articleDetails.jsp?tp=&arnumber=5969652&queryText=survey+security+VANETs+multihop.
5. Casale, G., Zhang, E. Z., & Smirni, E. (2010). Trace data characterization and fitting for Markov modeling. *Performance Evaluation, 67*(2), 61–79.
6. Chen, K. C., & Lien, S. Y. (2014). Machine-to-machine communications: Technologies and challenges. *Ad Hoc Networks, 18*, 3–23.
7. Chen, M., Wan, J., & Li, F. (2012). Machine-to-machine communications: Architectures, standards and applications. *KSII Transactions on Internet and Information Systems, 6*(2), 480–497.
8. Chin, W. H., Fan, Z., & Haines, R. (2014). Emerging technologies and research challenges for 5G wireless networks. *IEEE Wireless Communications, 21*(2), 106–112. Available at: http://ieeexplore.ieee.org/lpdocs/epic03/wrapper.htm?arnumber=6812298.
9. Conti, J. P. (2006). The internet of things. *Communications Engineer, 4*(6), 20.
10. Ghavimi, F., & Chen, H. H. (2015). M2M communications in 3GPP LTE/LTE-A networks: Architectures, service requirements, challenges, and applications. *IEEE Communication Surveys and Tutorials, 17*(2), 525–549.
11. Gohil, A., Modi, H., & Patel, S. K. (2013). 5G technology of mobile communication: A survey. In *2013 International Conference on Intelligent Systems and Signal Processing, ISSP 2013* (pp. 288–292). doi:10.1109/ISSP.2013.6526920.

12. Hasan, M., Hossain, E., & Niyato, D. (2013). Random access for machine-to-machine communication in LTE-advanced networks: Issues and approaches. *IEEE Communications Magazine, 51*(6), 86–93.
13. Iftikhar, M., et al. (2011). Service level agreements (SLAs) parameter negotiation between heterogeneous 4G wireless network operators. *Pervasive and Mobile Computing, 7*(5), 525–544.
14. Iwamura, M. (2015). NGMN view on 5G architecture. In *IEEE Vehicular Technology Conference*. doi:10.1109/VTCSpring.2015.7145953.
15. Jiang, D., Wang, H., Malkamaki, E., & Tuomaala, E. (2007). Principle and performance of semi-persistent scheduling for VoIP in LTE system. In *2007 international conference on Wireless Communications, Networking and Mobile Computing, WiCOM 2007* (pp. 2861–2864). doi:10.1109/WICOM.2007.710.
16. Jian, X., et al. (2013). Traffic modeling for machine type communication and its overload control. *Tongxin Xuebao/Journal on Communications, 34*(9), 123–131.
17. Lee, C. S., Lee, G. M., & Rhee, W. S. (2014). Smart ubiquitous networks for future telecommunication environments. *Computer Standards and Interfaces, 36*(2), 412–422.
18. Lien, S. Y., & Chen, K. C. (2011). Massive access management for QoS guarantees in 3GPP machine-to-machine communications. *IEEE Communications Letters, 15*(3), 311–313.
19. Miaji, Y., & Hassan, S. (2010). Comparative simulation of scheduling mechanism in packet switching network. In *Proceedings – 2nd international conference on Network Applications, Protocols and Services, NETAPPS 2010* (pp. 141–147). doi:10.1109/NETAPPS.2010.54.
20. Nisar, K., Said, A. M., & Hasbullah, H. (2010). Enhanced performance of packet transmission using system model over VoIP network. In *Proceedings 2010 International Symposium on Information Technology – Engineering Technology, ITSim'10* (pp. 1005–1008). doi 10.1109/ITSIM.2010.556159.
21. Niyato, D., et al. (2009). Relay-centric radio resource management and network planning in IEEE 802.16 j mobile multihop relay networks. *Communications IEEE, 8*(12), 6115–6125. Available at: http://ieeexplore.ieee.org/lpdocs/epic03/wrapper.htm?arnumber=5351729.
22. Pashtan, A. (2006). Wireless terrestrial communications: Cellular telephony. *Telecommunication Systems, 1*, 1–12.
23. Shafiq, M.Z., et al. (2013). A first look at cellular network performance during crowded events. *Proceedings of the ACM SIGMETRICS International Conference on Measurement and Modeling of Computer Systems, 41*(1), 17–28. Available at: http://dl.acm.org/citation.cfm?doid=2494232.2465754.
24. Sriram, K., & Whitt, W. (1986). Characterizing superposition arrival processes in packet multiplexers for voice and data. *IEEE Journal on Selected Areas in Communications, 4*(6), 833–846. Available at: http://ieeexplore.ieee.org/lpdocs/epic03/wrapper.htm?arnumber=1146402.
25. Sui, Y., et al. (2013). Moving cells: A promising solution to boost performance for vehicular users. *IEEE Communications Magazine, 51*(6), 62–68.
26. Utkovski, Z., Ilik, D., & Kocarev, L. (2013). Wireless Communication Systems (ISWCS 2013), Proceedings of the tenth international symposium on, Wireless Communication Systems (ISWCS 2013). Proceedings of the tenth international symposium, pp. 1–5.
27. Weyrich, M., Schmidt, J. P., & Ebert, C. (2014). Machine-to-machine communication. *IEEE Software, 31*(4), 19–23.
28. Yang, H. (2005). A road to future broadband wireless access: MIMO-OFDM-based air interface. *IEEE Communications Magazine, 43*(1), 53–60.
29. Zhang, Y., Li, Z., Mei, S., Xiao, L., & Wang, M. (2009). A new approach for accelerating IPSec communication. In *1st international conference on Multimedia Information Networking and Security, MINES 2009* (pp. 482–485). doi:10.1109/MINES.2009.151.
30. Zirong, G., & Huaxin, Z. (2009). Simulation and analysis of weighted fair queueing algorithms in OPNET. In *Proceedings – 2009 international conference on Computer Modeling and Simulation, ICCMS 2009* (pp. 114–118). doi:10.1109/ICCMS.2009.51.
31. Sesia, S. (2009). The LTE network architecture, LTE — The UMTS long term evolution: From theory to practice (pp. 23–50). Wiley. doi:10.1109/IPPS.1995.395978.

Chapter 10
Monitoring the Energy Consumed by a Network Infrastructure to Detect and Isolate Faults in Communication Architecture

Dimitar Minovski, Eric Rondeau, and Jean-Philippe Georges

10.1 Introduction

The energy-efficient network infrastructures have recently become a hot topic in the business world, as the concept of Green IT strives to reduce the overall operational costs, but in the same time also to eliminate the inefficiencies from the enterprises' IT systems. The network infrastructures are already massively deployed and even projected to have exponential growth [1] due to the evolution of the Internet, user demands, and trending topics such as Internet of things [16]. Thus, as a shared resource, they have to be constantly available, which exacerbates the sustainability issues. Researchers have already proposed different network-wide energy management schemes targeting various areas such as datacenters [13], mobile networks [19], and WANs [8]. However, it is quite challenging to tackle the energy efficiency issues within the households and from small to large enterprises. One obstacle in making enterprise networks more energy efficient is the range of devices from multiple vendors deployed on the network. Also, it is difficult to operate with enterprise networks because of their unpredictable growth, frequently changed topology and architecture, regarding the energy consumption.

Current research and developed network management systems (NMS) are not fully automated when executing energy efficiency policies. The introduction of

D. Minovski (✉)
Université de Lorraine, Vandœuvre-lès-Nancy, France
e-mail: dimitar.minovski1@etu.univ-lorraine.fr

E. Rondeau • J.-P. Georges
Université de Lorraine, CRAN, UMR 7039, Campus Sciences, BP 70239, Vandœuvre-lès-Nancy Cedex 54506, France

CNRS, CRAN, UMR 7039, Vandœuvre-lès-Nancy, France

M. Dastbaz et al. (eds.), *Technology for Smart Futures*,
DOI 10.1007/978-3-319-60137-3_10

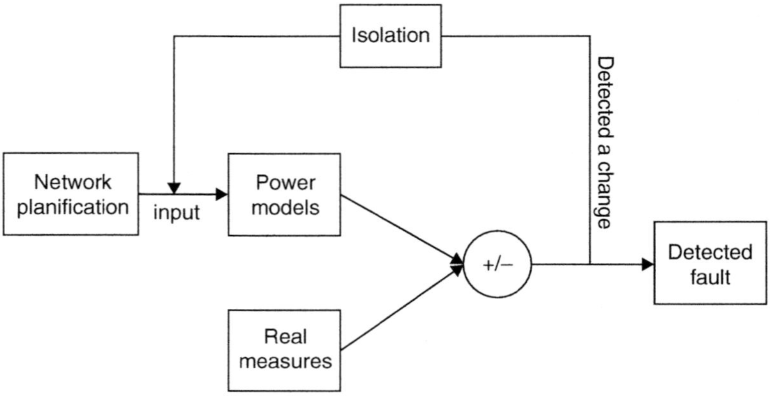

Fig. 10.1 The general approach of the paper

SDN paradigm brought new opportunities for managing networks through abstraction of lower-level functionality. This means that a centralized SDN controller with only one transaction is able to reconfigure a group of devices. However, the monitoring process of existing NMS to report for changes in the state of the devices is based on point-to-point (p2p) communication with every device on the network. This creates a great deal of traffic and puts additional burden on the network, which raises a sustainable issue in complex networks. Due to improvements in the field of Green IT and smart meters, it is feasible to build a NMS that recons only on power data fetched from the power distribution unit. A pattern for augmenting the values extracted from the power consumption of the network could provide useful information about different network states, for instance detecting changes in the topology. The idea is to use fault detection and isolation (FDI) approach to monitor the network state based on the energy usage. Fig. 10.1 shows that we need two information, a model representing the expected behavior of the devices and the real-time measurements to analyze deviation between the two processes. A deviation corresponds to fault detection in the network, which is in a form of misconfiguration or improper use of the equipment. This means that monitoring the energy consumption could be used not only for Green IT purposes, for raising awareness and reducing the electricity costs, but as well for a classical ICT monitoring system. Also, as the concept of smart grid is making use of digital networks to improve the transportation of energy, the work presented in this paper could be explained as the reverse process – how the use of energy could improve the data transport.

The remaining part of this paper is organized as follows: Sect. 2 presents the related work, while the Sect. 3 presents the objective of the paper. The system is described in Sect. 4, which includes the architecture for network management, the design of the experiments, and the implementation of the developed application. In Sect. 5 the obtained results are presented and discussed, while the Sect. 6 concludes the paper.

10.2 Related Work

In recent years a new approach for network management emerged under the name of software-defined networking (SDN) that allows the network operators to manage the network services through abstraction of high-level functionality. A study by [12] proposes an event-driven network control framework based on SDN paradigm and OpenFlow protocol to manage a complex campus network. The focus is on enabling frequent changes to network conditions and states, providing support for network configuration in high-level language, and providing better visibility and troubleshooting. Having a global knowledge of the network state, the developed control framework introduces a centralized approach for network configuration, opposed to distributed management. Meaning that the network operators will not have to configure all the devices individually, but instead let the software make network-wide traffic forwarding decision from a logically single location. Moreover, the network operators provide high-level network policies which are translated by the controller in a centralized manner into a set of forwarding rules, which are used to enforce the policy on the underlying network equipment, by using OpenFlow, as depicted in Fig. 10.2. They offer a set of control domains which can be used by the network operators to define conditions by assigning a suitable packet forwarding actions which corresponds to each condition. Even though the proposed solution reduces the workload of network configuration and management due to the SDN paradigm, the study mainly focuses on the algorithm for translating the policies into a set of reconfigurations of the devices. The system is based on event sources that dynamically collect the current state of the devices, which are inputs to the controller for forming the policies. The event sources monitor the network state and report the changes to the controller, such as bandwidth usage of every end-host device. The monitoring is based on the SNMP and OpenFlow for pulling data with p2p

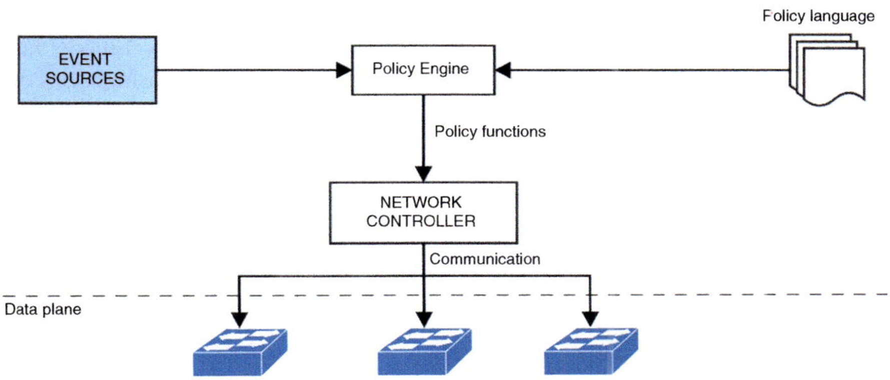

Fig. 10.2 Classical SDN architecture

communication which is difficult to manage especially during an expansion of the network, when adding new devices to the network, or during changes to the physical network topology which alters IP address modifications.

The strategy presented in [3] for studying the effects of unknown induced delays in network architecture suggests the use of concepts such as fault detection and isolation (FDI) and fault tolerance control (FTC). The study defines a threshold for the expected delay on the basis of the network characteristic and network calculus theory. A faulty situation is then generated and compared to the defined threshold in order to successfully detect which elements are causing the delays and deals with them in a controlled manner. Similarly, [18] proposes the use of power profiles for each device on the network to determine their expected energy usage under different circumstances. However, the power profiles are considered in special case when the real-time energy measurements are not accessible. They suit as a backup figure to proceed with the FDI's calculations to determine a faulty situation and produce energy-efficient policies for the network.

The study by [5] proposes models that give a global overview of the impact that the network and the ICT equipment is having on the environment, developing the following equation:

$$E = E_{\mathrm{m}} + E_{\mathrm{u}} + E_{\mathrm{d}}$$

$$E_{\mathrm{m}} + \int_{t=0}^{end\,\text{of lifecycle}} Pu(t)\,dt + E_{\mathrm{d}} \tag{10.1}$$

where E_{m} is the energy required for manufacturing and transportation of the equipment, E_{u} is a factor related to the energy consumed during the usage of the equipment, and the energy required to dismantle the equipment is E_{d}. For the second part of (1), Pu is related to the power consumption by the network architecture during its use phase.

Two strategic approaches exist for modeling the energy consumed during the usage phase of the devices. The first involves a high-level modeling [7] of the whole architecture, and the second approach is more precise, meaning that it provides power models for each of the devices part of the network. A high-level model has less interactions with the devices and therefore fails to give accurate estimations on how much energy a network architecture consumes at a particular point of time. On the other hand, the use of energy consumption models developed for a particular device would require constant p2p interactions and cause certain amount of additional traffic in the network. However, by giving a more precise figure on the expected behavior of the devices, the system would be more responsive to minor changes, faults, and anomalies, which is the purpose of the experiments part of this paper. Therefore, the power models for switch [10, 17], router [4], PC [2], and access point [6] are added to the power model registry, as shown in Fig. 10.5, explained in the next sections.

10.3 Objective

The goal is to develop sustainable software application suitable for network monitoring systems, which operates only with the retrieved energy consumption data from the ICT devices part of the network infrastructure. The developed application communicates with the power grid illustrated in Fig. 10.3 and, by following the FDI approach shown in Fig. 10.1, is able to detect different states that are occurring on the ICT devices. The network of the power grid includes the Raritan power distribution unit (PDU), which provides the real-time measures for the energy consumption through SNMP periodic pulling. With this, the PDU grid forms the power monitoring process. The detection of the states is achieved by the developed application, which analyzes the retrieved raw power data to distinguish possible faulty situations and anomalies in the usage phase of the ICT devices. The output of the application or the detected new state suits as an input parameter to the SDN controller. This means that the developed application is envisioned to replace the classical event sources, part of Fig. 10.2, which are using point-to-point communication to report for changes in the state of the devices. To achieve this, the developed software requires two information to be considered: (i) power data from *real-time measurement* procedure and (ii) values for the expected power consumption of each device extracted from the *power models*.

(i) The process of *real-time measurement* requires probes and standardized protocols to access the needed metrics which are altogether coupled in a monitoring system. In practice, Raritan power distribution unit (PDU) is the monitored equipment which has a predefined energy management information base (MIB) as a hierarchical structure with properties that is pulled out periodically. The monitoring system itself is then able to collect, analyze and modify the information stored in the MIB through SNMP. It is important to note that the PDU in this case acts as a separate network from the existing network architecture. This means that the monitoring process has a separate channel to the PDU and

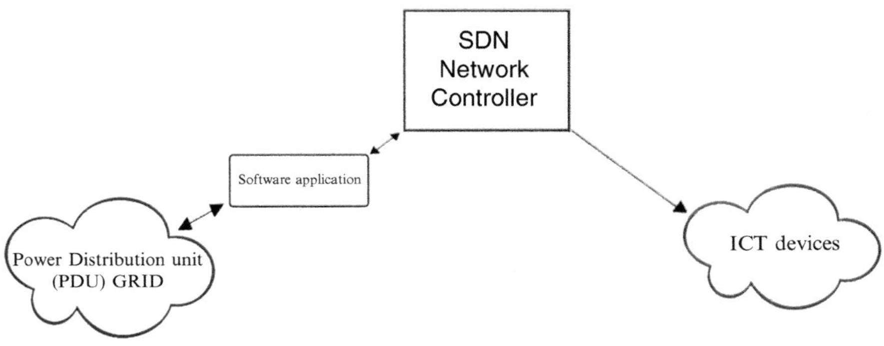

Fig. 10.3 Objective of the developed system

the packets does not cross the existing network architecture, which releases the burden on the network.

(ii) From the other side, as discussed in the section II, the selected *power models* represent the expected behavior of the equipment. Therefore, currently there is an open field to combine the real-time measures, which are translated to the real behavior of a device, with the developed models. A deviation between the models and the monitored power data is used to detect anomalies and to anticipate fault according to a trend analysis.

The output of the developed system is a continuous detection of changes in the network infrastructure, for instance, detecting new device on the network, changes in the Spanning Tree Protocol, detection in changing the operating state of a switch port, etc. Finally, the objective is the developed monitoring system based on the power grid as shown in Fig. 10.3, which will be a replacement for the existing monitoring system [12] which does a p2p communication with every device part of the network.

10.3.1 Building a Knowledge Base

The developed application that incorporates the monitoring system and the power models has to form thorough foreknowledge in order to successfully anticipate and interpret the readings for the energy consumption. Mostly because the power data is just a raw value and does not contain rich information. As suggested by [11], some types of computing systems within the enterprise can exhibit large variations on the power data even when comparing two instances of the same device model. More precisely, two Dell Optiplex 760 PCs were observed to have over 40% discrepancy in their average power draw. To remedy this, the application has to augment new ways to isolate the problem with the inconsistent energy consumption of the devices, to locate the fault or the anomaly, and to classify the problem. One way to deal with this is to predict the fault that may occur during the usage of the equipment. It is important to test the responsiveness of the application by generating different faults and misconfigurations that may occur during a typical workday in an enterprise network. Thus, closely observe the figures for energy consumption and build a knowledge base. A system with such knowledge comprised with smart algorithms is capable of analyzing the deviations and benchmark the faulty situations. From there a pattern could emerge to later detect the problem of aberrant values regarding the energy consumption of the devices during the real-life usage phase of the equipment.

List 1 represents the states which are chosen as most common during a typical enterprise workday, tested on various heterogeneous combinations of network architectures, prior to building the knowledge base. To identify where the faults are applied, a classification based on the objective is depicted in Fig. 10.4.

List 1: States to be detected on the network

Testing the operational vs. the sleeping mode on the devices such as switches, routers, access points, LCD monitors, PCs, and laptop computers

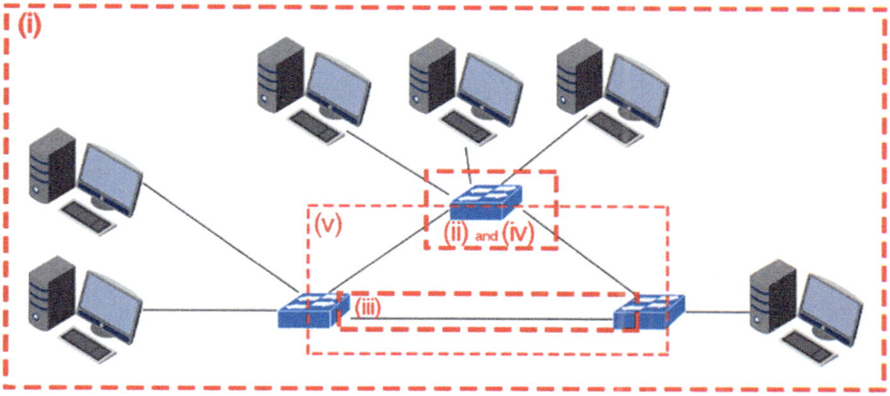

Fig. 10.4 The classification of the faults

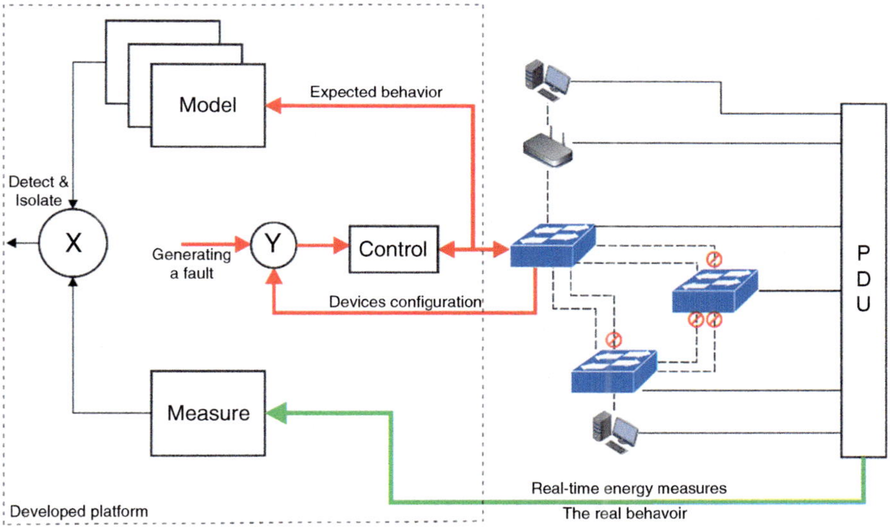

Fig. 10.5 System behavior

Turning off/on a switch port
Performing link adaptation on the switch's ports
Evaluating the effect on energy-efficient Ethernet (EEE) and Cisco EnergyWise
Testing a Spanning Tree Protocol (STP) reevaluation on the network

One possible case study of network architecture is depicted in Fig. 10.5, where the generated faults are represented as the *Y* input parameter. At disposal for the

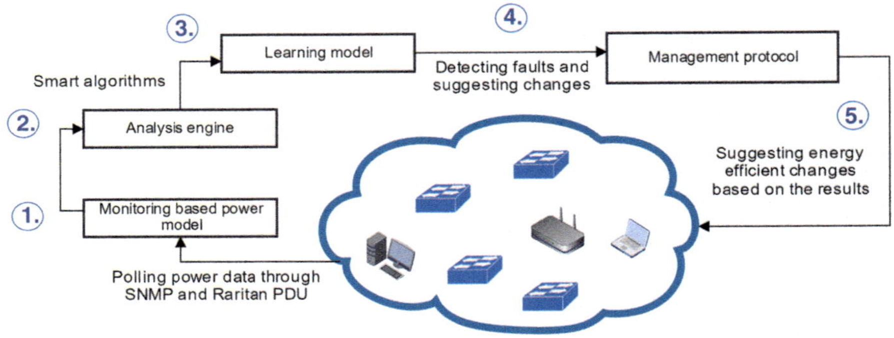

Fig. 10.6 The experiment's logic

experiments, there were eight Cisco switches from the series 2960 and 3560, eight Cisco 1941 Routers, ten Dell Precision T1700 stations, two HP Pavilion laptops, six Raritan PX2 PDUs, and two Raritan PX3 PDUs.

10.3.2 The Logic Behind the Monitoring System

The states illustrated in Fig. 10.6 represent the logic throughout the execution process. Having developed a standalone application able to monitor the energy consumption as a first state allows further analysis process on the retrieved and stored data. The analysis engine as a second state is tightly coupled with the learning model of the system by developed smart algorithms which are able to detect and identify the changes in the power data. Big part of the analysis process plays the inclusion of the power models, discussed in the related work, which are defining the power profiles of the devices and are calculating their expected behavior. The output of the models is then validated with the retrieved real-time measurements for possible deviations. Raritan PDU offers the possibility to monitor its overall consumption as well as the individual power consumption per socket. This means that if a deviation is detected in the comparison process of the overall consumption, the analysis engine will further seek for the device causing the energy disparity. A recent history of energy consumption is retrieved from a database for the detected device, and the results are logically passed to the third state, where they are further dismantled toward understanding and interpreting the power data values. The emphasis of the third state, the learning process, is given to the simulation of the faults that are most frequent during the regular usage phase of the ICT equipment. The established knowledge base is used as a baseline on which the learning model is constructed. This means that series of experiments have to be conducted for a range of different network states and faults in the interest of acquiring the desired foreknowledge to benchmark the anticipated states. The benchmarking process is also the key concept for the isolation part as a last state of the logic depicted in Fig. 10.6.

Ever since the commencement period of the learning process, the system is constantly updating its knowledge base depending on the behavior of the devices, and by following the figures for the energy consumption, it is able to discern the aberrant states of the devices. For instance, a possible overloading on the devices, detecting sleep/hibernate states, physical or logical changes in the configuration of the devices, etc. To have seamless operations and accurate output, the learning process also needs to update the static values in the power models based on the detected new state of a particular device. Namely, as seen in Fig. 10.1, the equations in the power models depend on static parameters from the network planification, which are gathered during one-time measurement. This means that the learning model has to recompute the input parameters in the power models after a detection of a new network state. For instance, the equation in the power model of a switch has to be updated with a new value if a switch port transitioned from offline to online state and vice versa. The outcome of the learning process regarding the benchmarking part concludes the fourth state of the logic behind the experiment, the engine for power management. An accurate benchmarking provides the possibility to identify the changes that occurred on the network infrastructure and suggests a recuperation or a reconfiguration of the network.

10.4 System Description

This section provides more technical explanation of the developed application. The architecture of the system has been previously described in [15].

10.4.1 Implementation

The application, which includes the power models and the monitoring module, is implemented in Java language using the SNMP library and MySql database. The energy monitoring task is developed in a multi-threaded fashion to periodically pull the energy values from the Raritan PDU, by importing Java libraries for SNMP to access the functions for requesting the right Raritan's MiB OIDs. There is an ability to optimize the frequency of the probes sent to the Raritan PDU throughout a developed GUI; thus the accuracy of the measures is practically defined by the operator. During the test phase of the application, probes were requested every second, and the responses are stored locally with the interest to enable further analysis on the data. The response from each probe is a group of values in the following format: $I_{(A)}$ – *Electric current significant with* 10^{-3} *amps (A) base*; $V_{(V)}$ – *Voltage*; *PF* – *Power factor*. To get the current energy consumption (*W*), the following equation has to be performed:

$$W = I_{(A)} \times V_{(V)} \times PF \tag{10.2}$$

Moreover, the probe response does not only contain data for the total energy consumed by the whole PDU, but also it reports the consumption of each socket individually. This means that the application with additional analysis can successfully build its own schemas and detect what kind of device is connected to a certain socket. Giving this ability, the process flow can successfully discover which device specifically is not consuming the energy as it is expected. Algorithm 1 describes the process flow of the application.

Algorithm 1: The process flow

```
input: PM, the power models; PR, the probe response
output: F_fault, detected fault, I_isolation, isolating the fault
R is the Raritan PDU
E denotes the attached equipment
KB represents the knowledge base in a form of database

foreach X in PR do
   W_total ← X(I_(A)) * X(V) * X(PF)
   Store(W_total)
   if W_total > PM_total then
      foreach E in R do
         W_individual ← E(I_(A)) * E(V) * E(PF)
         Store(W_individual)
      if W_individual > PM_individual then
         F_fault ← Detection(E, KB, History(W_individual)
         I_isolation ← Isolation(F_fault)
         Store(F_fault, I_isolation)
return F_fault, I_isolation
```

The support of power models (PM) in the application creates the premises to set the expected threshold regarding the energy spent by each detected device. The discussed power models are represented with static parameters gathered by one-time measurement at the initialization phase of the application or when a new device is plugged to the PDU socket.

Having real-time availability probes enables the application to perform parallel comparison on the retrieved data for individual sockets and, as a whole, with the data provided by the power models. In case of aberrant values, the application tries to locate the device with the unexpected behavior, load the latest stored power data history of the device, and perform analysis for fault detection. The FDI concept is based on calculations directed by the predefined fault foreknowledge, achieved by the experiments for generating the faults in order to benchmark the consumption that a particular occurring state is causing. The knowledge base (KB) is also integrated in the algorithm in a form of a database.

The Detection() function is the core of the algorithm due to the fact that it couples the analysis and the learning process, described in the previous section.

It begins by locating the device that is observed to have disparity in the energy consumption. Also, a recent history of the spotted device is passed to the function, which includes two separate components. First is the recent energy consumption history of the device, and second is the recent history of the noted running states and configuration of the device. This process maps the previous running configuration and understands the current states of the devices that might be causing the energy disparity. Moreover, the knowledge base is as well loaded which can interpret the difference in the expected and the real energy measures into a new network state. The function combines the abovementioned parameters and the calculations result with detection of the changed state. For instance, a switch is detected to have powered off a port, changed the link speed, or perhaps reevaluated the STP.

The Isolation() function as an input parameter receives the detected new state and performs further analyses to examine the implications, especially on the energy consumption. The analysis determines the actions that have to be taken upon detection of new state. For instance, if a switch port is detected to transition from online to offline mode, this function recomputes the input parameters of the static power models to make them having accurate calculations of the expected energy consumption of the switch. This means that the examination of the implications results with discovery whether the detected new state is a misconfiguration of the network, a faulty situation, or just changes in the network state that have to be noted. As an output, the Isolation() function updates the database of the spotted device recent history and the knowledge base and reports to the network operator for the changes.

The first version of the GUI is shown in Fig. 10.7. The network operator is able to see live chart of the energy consumption of the whole network, as well as select specific PDU and monitor the consumption. Based on the detected states, the network operator could map the running configuration of the whole network, as well as separately for each PDU. When a specific PDU is selected, the operator could monitor the individual sockets, see the detected device, and read the changes in the state. There is a possibility to download a report with a complete history of the changes on the network. The obtained results are discussed in the next section.

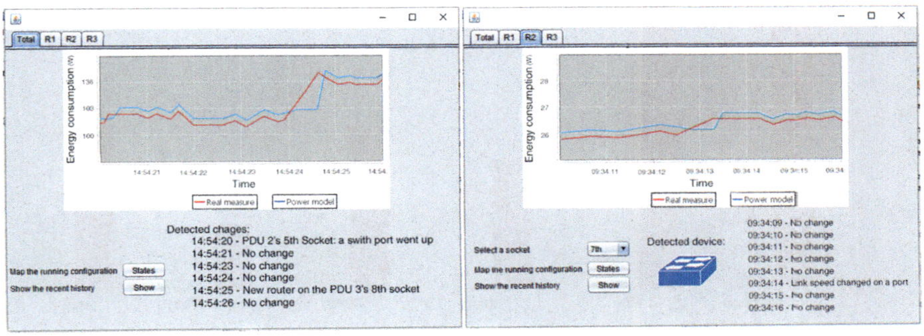

Fig. 10.7 Developed GUI

10.5 Results and Discussion

This section refers to the List 1, presenting the obtained results during the test phase of the developed application. Each of the experiments was tested on different network architectures and monitored for 10 min continuously, as well as under different amount of load generated to flow through the network.

(i) The possibility to automatically detect the working hours of an enterprise allows the application dynamically to report on the equipment that was left in an operational state. For instance, if a wireless access point is observed to consume constant level of energy during non-working hours, the application will indicate that the device is powered, but does not transmit any data, meaning that it should transition into a sleep mode. The cost of powering the hardware components of each device dominates the overall power profile, and therefore the opportunity for energy savings is the highest. The difference in the energy consumption between the operational and the sleeping mode is illustrated in Fig. 10.8. Typically, there is a burst of energy consumption when a device goes to operational state, either from sleep/hibernate or offline mode. Those values are also used to benchmark the current state of the device.

(ii) Managing the active ports on a switch is one important and exquisite objective because it is the only way of providing network accessibility. Therefore, to have an overview on the status of the ports at one point of the time by observing the switch's values for the energy consumption is a delicate task. Mainly because of the introduction of EEE as a set of enhancements able to dynamically put a single link temporarily to sleep when not in use. The difference in

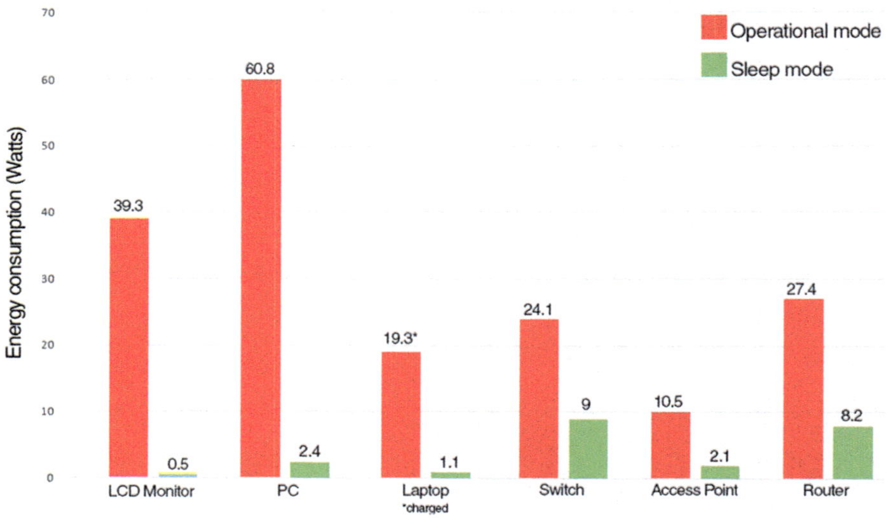

Fig. 10.8 Obtained results – experiment (i)

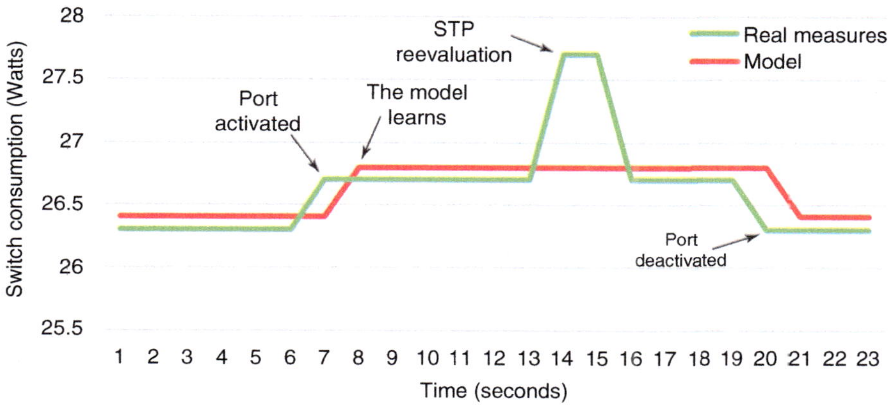

Fig. 10.9 Obtained results – experiments (ii) and (v)

the power consumption after a transition from active to inactive switch port is observed to be from 0.3 to 0.4 W, as illustrated in Fig. 10.9. This result allows the developed application to benchmark the state and have detailed matrix of the port statuses of each switch. Having this feature allows issuing reports for possible unnecessarily active ports, but also is part of the analysis for a possible misconfiguration of the network.

(iii) The utilization of the Ethernet links is, on average, extremely low [9]. This suggests that there is an ample opportunity for energy savings by operating Ethernet links at a low data rate for most of the time with no perceivable performance impact to the user. To keep track of each switch's port assigned speed, by following the energy consumption, means an opportunity for adjustments according to the needs. But it also helps the application to correlate the assigned link speed with the congestion, or the load on the switch, which is as well affecting the overall power consumption. During the observations, there is no notable change in the consumption when comparing 10 Mb/s to 100 Mb/s speed link, but the difference when making a transition from 1Gb/s to 100 Mb/s is from 0.2 to 0.4 W per port. Thus, the developed application by benchmarking this value is forming a schema of each port link speed that is assigned. The outcome is a suggestion when to reduce the speed of a link when a low utilization is perceived.

(iv) Energy-efficient Ethernet is not yet a feature supported by many network devices currently on the market; however it is important to observe its impact on the energy consumption. During an idle period of a switch with no traffic flowing through the ports, EEE will put all of the active ports in the low power idle (LPI) state and therefore achieve the same result as the second *(ii)* experiment of 0.3–0.4 W savings per port. Hence, the developed application has to be aware of the possibilities of EEE to dynamically, for a temporary time, put a port to idle mode.

(v) Spanning Tree Protocol (STP) is commonly used Layer 2 protocol that runs on switches and bridges, with purpose to ensure that there are no loops created on redundant paths in the network. By default STP dynamically manages the elimination of the loops with the process of electing a tree based on priorities with a particular switch as a root. This means that there is a possibility of a network reconfiguration without the assistance of the operator, which could lead to additional power demands if it is not well managed. STP, enabled between the switches in Fig. 10.6, does not save any energy with its ability to block certain port in order to discard the loops in the network. Namely, a port in a blocking state is consuming the same amount of 0.4 W as the other active ports. The only noticeable difference is the spike of energy consumption when STP is trying to reevaluate the tree. Namely, during a physical or a logical generated error, STP will elect a new root and construct the new tree, thus will add up to 1 W for few seconds to the overall consumption of the switch, depending on the architecture. This burst of energy, also shown in Fig. 10.9, is used to benchmark a change in the STP, with the interest of reporting it for possible misconfiguration of the network.

10.6 Conclusion and Future Work

This paper presents the achieved results from evaluating the proposed novel extension for energy-efficient network management application emulated in a heterogeneous network environment. The results show that a reliable application can be designed to monitor the changes in the state of the devices with minor impact on the network regarding the traffic. The proposed monitoring process is designated as a replacement for the existing p2p pulling process, part of the SDN typical systems. The SDN controller, and thus the network operator, has the ability to receive reports for the changed state of the network in an energy-efficient way by following the energy consumption of the network infrastructure. The approach of monitoring the power consumption directly from the PDU, instead of p2p communication with every device, brings own complexity as the network grows. For instance, in a scenario of multiple faults happening simultaneously, the FDI concept presented in this paper is able to cope with the occurring situation. This is achieved by the possibility of following each PDU's socket individually from a centralized application. The support for real-time availability calculation enables the operator to take service-centric decisions, rather than network centric as proposed currently in the literature.

As future work, we envision a deployment of the proposed application as part of a SDN controller. Moreover, the application should be scalable in terms of network size and adapt to any SDN software controller developed on different programming language, with diverse libraries. The anticipation is also to include other common network anomalies and to follow their impact on the energy consumption, for instance the packet loss rate and highly congested networks, which could possibly trigger a use of other techniques in order to gather all the necessary information for the analysis process, rather than just monitoring the energy values.

Acknowledgment This work is part of the Erasmus Mundus Master program in Pervasive Computing and Communication for Sustainable Development (PERCCOM) ([14]) of the European Union (www.perccom.eu). The authors thank all the partner institutions, sponsors, and researchers of the PERCCOM program.

References

1. Cisco. (2011). How the next evolution of the internet is changing everything. *White Paper*, April 2011.
2. Agarwal, Y., et al. (2009). *Somniloquy: Augmenting network interfaces to reduce PC energy usage*. NSDI. Vol. 9. Korea (South): 2014 the IEEE sponsered 16th International Conference on Advanced Communication Technology (ICACT) in Pyeongchang.
3. Aubrun, C., et al. (2008). *Network calculus based fault diagnosis decision-making for Networked Control Systems*. Emerging Technologies and Factory Automation, 2008. ETFA 2008. IEEE International Conference on. Korea (South): IEEE.
4. Ahn, J., & Park, H. (2014). *Measurement and modeling the power consumption of router interface*. Advanced Communication Technology (ICACT), 2014 16th International Conference on. Korea (South):IEEE.
5. Drouant, N., Rondeau, E., Georges, J. P., & Lepage, F. (2014). Designing green network architectures using the ten commandments for a mature ecosystem. *Computer Communications, 42*, 38–46.
6. Demir, M. Ö., Kurt, G. K., & Karaca, M. (2014). An energy consumption model for 802.11ac access points. *Software, Telecommunications and Computer Networks (SoftCOM), 2014 22nd International Conference on, Split,* (pp. 67–71).
7. Foll, L. S. (2008). *TIC et Energetique: Techniques d'estimation de consommation sur la hauteur, la structure et l'evolution de l'impact des TIC en France*. France: Institut national des telecommunications.
8. Gupta, M., & Singh, S. (2003). *Greening of the Internet. Proceedings of the 2003 conference on Applications, technologies, architectures, and protocols for computer communications*. Karlsruhe, Germany: ACM.
9. Gunaratne, C., Christensen, C., Nordman, B., & Suen, S. (2008). Reducing the energy consumption of ethernet with adaptive link rate (ALR). *IEEE Transactions on Computers, 57*(4), 448–461. April 2008.
10. Hossain, M. M., et al. (2015). *Modeling the power consumption of ethernet switch. International SEEDS Conference 2015*. Leeds UK: Sustainable Ecological Engineering Design for Society.
11. Kazandjieva, M., Heller, B., Gnawali, O., Levis, P., & Kozyraki, C. (2013). Measuring and analyzing the energy use of enterprise computing systems. *Sustainable Computing Informatics and Systems, 3*(3), 218–229.
12. Kim, H., & Feamster, N. (2013). Improving network management with software defined networking. *Communications Magazine IEEE, 51*(2), 114–119.
13. Koutitas, G., & Demestichas, P. (2010). Challenges for energy efficiency in local and regional data centers. *Journal of Green Engineering, 1*(1), 1–32.
14. Klimova, A., Rondeau, E., Andersson, K., Porras, J., Rybin, A., & Zaslavsky, A. (2016). An international Master's program in green ICT as a contribution to sustainable development. *Journal of Cleaner Production, 135*, 223.

15. Minovski, D., Rondeau, E. & Georges, J.P. (2016). Using ICT Energy consumption for monitoring ICT usage in an enterprise. *Future Internet of Things and Cloud (FiCloud), 2016 4th International Conference on, Vienna, 2016.*
16. Ovidiu, V., & Friess, P. (2013). *Internet of things: Converging technologies for smart environments and integrated ecosystems*. London: River Publishers.
17. Reviriego, P., Sivaraman, V., Zhao, Z., Maestro, J. A., Vishwanath, A., Sánchez-Macian, A., & Russell, C. An energy consumption model for energy efficient ethernet switches. *Journal Computer Communications archive, 50, 2014*, 110–118.
18. Riekstin, A. C., et al. (2016). Orchestration of energy efficiency capabilities in networks. *Journal of Network and Computer Applications, 59*(2016), 74–87.
19. Wang, X., et al. (2012). A survey of green mobile networks: Opportunities and challenges. *Mobile Networks and Applications, 17*(1), 4–20.

Chapter 11
Reducing Energy Consumption of Network Infrastructure Using Spectral Approach

Khan Mohammad Habibullah, Eric Rondeau, and Jean-Philippe Georges

11.1 Introduction

We are now living in the age of information and communication technology and the energy consumption, and carbon emission by computing and communications equipment is increasing rapidly. Mingay [24] stated that, at present, ICT is responsible for about 2% of global carbon emissions which is similar to the aviation industry. The enhancement of energy consumption has large negative economic and environmental impact. In recent years, we are struggling with climate change, and in coming years it is the greatest challenge to tackle it. For that reason, energy consumption reduction is very important because it is directly related to greenhouse gas emission which is responsible for climate change. Recent research [30] suggests that 15–30% volume of emission needs to be decreased in order to keep the global temperature increase below 2 °C.

The growth of Internet users has been astonishing over the last few years. From 1990 to 2010, the number of Internet users increased from 3 million to 2 billion, and by mid-2012, the user number had increased to 2.73 billion already. Regional growth rates between 2000 and 2012 have been exceeding 3600% in Africa and 2600% in the Middle East. As a result, the electricity demand of ICT is growing at much faster than overall electricity demand. Network-enabled device electricity demand is growing at a rate of 6% per year, and total ICT energy demand reached 1560 TWh in 2013 [17]. Because of this continuous growth of Internet users, the

K.M. Habibullah (✉)
Université de Lorraine, Vandoeuvre-lès-Nancy, France
e-mail: mohammad.habibullah2@etu.univlorraine.fr

E. Rondeau • J.-P. Georges
Université de Lorraine, CRAN, UMR 7039, Campus Sciences,
BP 70239, Vandoeuvre-lès-Nancy Cedex 54506, France

CNRS, CRAN, UMR 7039, Vandœuvre-lès-Nancy, France

M. Dastbaz et al. (eds.), *Technology for Smart Futures*,
DOI 10.1007/978-3-319-60137-3_11

spreading of broadband access, and the increasing number of online services being offered by telecoms and Internet service providers, the energy efficiency issue has also become one of the highest priority objectives for wired networks and service infrastructures. These continuously rising trends in network energy consumption depend on new services being offered, as well as on data traffic volume increase [25]. According to Zhang et al. [38], data traffic is increasing rapidly which follows Moore's law, by doubling every 18 months.

In recent years, some efficient steps have been taken to reduce the energy consumption of network infrastructure. This paper is a contribution to this mitigation in focusing only to reduce the energy consumption of network switches in a local area network (LAN). Mahadevan et al. [22] stated that a network switch consumes energy during its manufacturing phase, usage phase, and dismantling phase with the highest amount of energy consumed by network switch during usage phase.

The concern of this research work is only during the usage phase. The objective is to reduce the energy consumption that simple network architecture builds up with Ethernet switches and end devices. It is described in [17] that up to 80% of energy consumption is used for some devices just to maintain the network connection and more than 600 TWh of electricity was consumed on 2013 by such devices. It is possible to reduce up to 65% of energy consumption of such devices using the best available technologies and strategies. Hossain et al. [16] and Gunaratne et al. [14] found bandwidth and number of connection are mostly responsible for energy consumption of Ethernet switch in usage phase. Krommenacker et al. [19] and Rondeau et al. [27] showed proper network architecture designs, and cabling plan using clustering approach increases the efficiency of the network. An efficient network design reduces traffic propagation delay, jitter, and packet loss which also reduces energy consumption. Therefore, in this paper, optimization, reorganization, and clustering approach of network architecture have been proposed by analyzing bandwidth and the traffic load in point-to-point communication. The feasibility to hibernate or switch off devices which only remain alive to maintain network connection was also investigated, with the objective of reducing energy consumption.

The experiment was done with cisco switches, and strategies have been applied to optimize and reorganize the network topology. In the experiment, the power consumption was used to measure energy usage of network architecture. In the first phase of experiment, we designed network topology which consists of connected Ethernet switches which support energywise and desktop computers that were connected to each switch. Then we measured the power consumption of all switches using different bandwidth with same traffic load according to traffic matrix. In the second phase, we reorganized and optimized our network architecture using clustering method based on spectral approach, with the objective to reorganize network cabling in order to be able to sleep or hibernate unused components of network infrastructures for a certain amount of time during low-traffic periods.

Hibernation mode decreases a large amount of power consumption compared to the operation mode of an Ethernet switch, and switched off mode does not consume any power. In the third phase, we investigated the feasibility of putting various components connected with Ethernet switches or the switch itself to sleep or hibernate in considering the changes of communication activities. After applying the spectral approach to different cases of communication with the same network topology, it was possible to hibernate or switch off the network switch which saved a significant amount of energy.

11.2 Related Work

Numerous studies have explored both in the wired and wireless network to analyze the energy consumption pattern of network devices and to reduce energy consumption using different strategies.

11.2.1 Related Work in Network Energy Consumption Evaluation

Christensen et al. [8] have studied the life-cycle energy use of network devices and explained how network devices can impact on environment pollution. The life-cycle energy use of network devices has been studied by Nordman [26], but mostly in the context of home and office environments. Rivoire et al. [34] explained from a device manufacturer's point of view that network devices such as routers and switches are power proportional which means they consume energy proportional to their usage. Gupta et al. [15] showed how different parameters affect the energy consumption of network switch. After that, they defined a model which shows the relationship between parameters related to the Ethernet switch and energy consumption. Fithritama et al. [13] proposed a method based on fuzzy logic to identify the relationships between network parameters and their effect on power consumption when deploying new network equipment. Their proposed method is also applicable to control the desired level of energy consumption by tuning the network parameters. In the research by Reviriego et al. [33], the energy consumption of small Energy-Efficient Ethernet (EEE) switches is analyzed in several experiments. Based on the experiment result, the authors proposed a model for the energy consumption of Energy-Efficient Ethernet switch.

11.2.2 Related Work in Network Energy Consumption Optimization

Gupta et al. [15] explored the feasibility of power management schemes at network switches in the LAN. They examined the possibility to put various components on LAN switches to sleep for reducing energy consumption. Experiments to evaluate energy management studies for network switches were performed by Mahadevan et al. [22]. The authors found energy usage in the operational stage is dominating, and they parametrically examine various energy management techniques to reduce the operational energy footprint of network switches. Nedevschi et al. [25] presented the design and evaluation of two forms of power management schemes to reduce the energy consumption of network. They have shown that simple schemes for sleeping or rate adaptation have significant energy savings without noticeably increased packet loss and latency.

11.2.3 Related Work Based on Clustering Approach

Consideration of information flows is very important to design network architecture. The objective of network designer is to confine the strong cooperation and communication with subnetwork to avoid flooding and overloading the whole network. In order to minimize overloading, intragroup communication should be maximum and intergroup communication should be minimum. To achieve this network scalability objective, clustering method has been widely pursued by the research community. In this research work, clustering algorithm has been used to reduce communication and isolate groups for hibernating or switched off part of the network.

Many clustering algorithms have been proposed [1–5, 7, 20, 23] to fulfill various objectives based on wireless sensor network, but none of these algorithms aim to minimize energy consumption in network architecture. Most of these algorithms are heuristic in nature, and the criteria for cluster selection and node grouping are intra- and intercluster connectivity if the application is sensitive to data latency and the length of data routing paths. The objective of these algorithms is to generate the minimum number of clusters. White and Smyth [37] proposed the spectral algorithm to find communities in a graph and showed that spectral algorithm is effective and efficient at finding both good clustering and the appropriate number of clusters from a variety of real-world graph data sets. They observed spectral algorithm is faster for large sparse graphs. Von Luxburg [36] introduced and presented the most common spectral algorithm from scratch by different approaches. He also discussed the advantages and disadvantages of different spectral clustering algorithms, different graph Laplacians, and their basic properties. Krommenacker et al. [19] defined some criteria to reorganize the network architecture and explored algorithms, especially the spectral algorithm to define cabling plan of switched network for real-time applications. The spectral algorithm to design the cabling plan for industrial Ethernet architecture has been used by Rondeau et al. [27], which can reduce handling delays of messages inside the time cycle of applications.

11.3 System Architecture and Criteria to Reorganize Network Switches

This section describes the general network architecture that has been used in the research and the criteria to reorganize and design efficient architecture for switched Ethernet networks. Figure 11.1 shows the network architecture and the cabling plan used during the experiment. Three Cisco Ethernet Switch 2960-X (S1, S2, and S3) were used during the experiment, and three desktop personal computers (PCs) were connected to each switch.

In the section below, we have described the criteria to reorganize network switches in different actions.

11.3.1 Action 1

It illustrates that if PC 1 has frequent communication with large data size with PC 7, then it is better to connect either PC 1 in S3 or PC 7 in S1. Thus, it is very important to consider information flows during designing a network. In this case, the goal is to maintain communication among devices by avoiding overloading the whole network which will increase network efficiency and reduce energy consumption.

11.3.2 Action 2

The objective of action 2 is to gather the PCs strongly communicating among them, with the idea to group PCs without communication in other clusters in order to be able to hibernate or switch off these clusters.

11.3.3 Action 3

If the capacity of each port of all switches is 1 Gbps and total traffic load for the communication between S1 and S2 or S2 and S3 is more than 1Gbps, then more than one trunk port needs to be connected to support the communication. In this

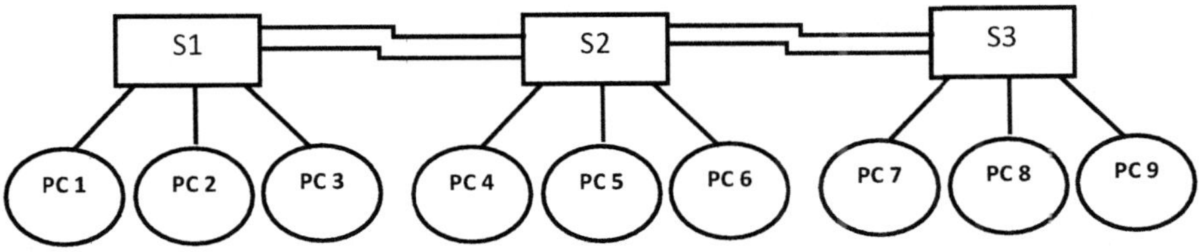

Fig. 11.1 Cabling plan and network architecture

situation, if inter-switch communication can be reduced, it is possible to eliminate redundant trunk between two switches. Thus, reducing the traffic between switches allows reducing bandwidth and, as a result, reduction of energy consumption.

Therefore, in general, the criteria to reorganize the communication among switches are to minimize intergroup dialogues in order to maximize the intragroup exchanges or to create groups without communication. Some other criteria should also take into account such as the number of ports and capacity of each port with the objective to use the smaller number of switches with a higher number of connected ports.

11.4 Methodology

The spectral algorithm is used to cluster the end device according to the usage of switch ports and to ensure the minimal interaction between two clusters. As an initial step of the approach, a graph G_d has been introduced and translated that into traffic matrix according to the communication among end devices.

This section describes clustering method based on the spectral approach which has been used to reorganize and optimize network topology.

11.4.1 Spectral Approach

The spectral algorithm is one of the most successful heuristics for partitioning graphs and matrices. This approach is used to solve scientific numerical problems such as solving parse linear systems explained by Pothen et al. [31], partitioning for domain decomposition [6], Ethernet architecture segmentation [27], and so on. In this research, spectral approach has been used to:

- Adjust the size of the clusters in regard to the capacity of the switches in terms of port number.
- Reorganize and minimize the cabling of the switch by breaking graphs into subgraphs.
- Investigate the feasibility of putting unused switches to hibernate or switched off according to usage.

Given an undirected weighted graph $G=(V,E)$, where G is a division of its vertices v into two disjoint subsets, V_1 and V_2. Let $E(V_1,V_2)$ is the set of edges (links for communication) e_{ij} with one endpoint in $V_1$and other in V_2. The cut size of the partition is the sum of the edge weights of E:

$$c = \sum_{eij \in E} w(eij)$$

where $v_i \in V_1$ and $v_j \in V_2$.

Here the objective is to find a disjoint subset of $V, P = \{v_i\}$ which minimizes c.

The adjacency matrix $A(G_d)$ of the graph G is $n \times n$ matrix where (ij)th entry is the weight of edge $e_{ij} : w_{ij}$ and 0 otherwise. For adjacency matrix, the diagonal entries are always 0.

As the spectral algorithm considers only undirected graph, thus it needs to work with

$$A(G) = A(G_d) + A(G_d)^t$$

11.4.2 Algorithm Description

Let D (G) is an undirected graph converted into $n \times n$ diagonal matrix such as

$$d_{ii} = \sum_j w(eij)$$

The Laplacian matrix $L(G)$of the graph G is $L(G) = D(G) - A(G)$.

According to [9–12], a good cut size is the second smallest eigenvalue λ2 and its associated eigenvector $u = (u_1, u_2, \ldots \ldots \ldots \ldots \ldots u_n)$ (Fiedler vector).

Now, sorting the vertices on the incremental way according to the values of the components of Fiedler vector gives the reorganization of the vertices.

The objective of the partitioning is to use exactly d devices (in this case network switches) and to minimize the communication through the switches if there are two levels of switches. A splitting value s should be found for spectral partitioning which divides the vertices of G into V_1 such that $u_i > s$ and V_2 such that $u_i < s$, and that is called a Fiedler cut. The choices for the splitting values s are:

- Bisection cut, where s is the median of $(u_1, u_2, \ldots \ldots \ldots \ldots \ldots u_n)$
- Sign cut, where s is equal to 0
- Ratio cut, where s is the value that gives the best cut ratio denoted ∅with$\varnothing(V_1, V_2) = |E(V_1, V_2)| / \min(|V_1|, |V_2|)$
- Gap cut, where s is the value in the largest gap in the sorted list of the Fielder vector components

To partition a graph into the power of two numbers of groups, recursive spectral bisection (RSB) algorithm explained by Simon [35] is applied. In this paper, optimized recursive bisection algorithm has been used for partitioning the graph.

11.4.3 Optimized Recursive Spectral Bisection Algorithm (RSB in [d/2])

Simon [35] proposed improvement on RSB algorithm for designing a switched Ethernet network cabling plan as follows:

1. Compute the Fiedler vector for the graph.
2. Sort the vertices according to the values of the components of the Fielder vector.
3. Calculate d for the group to be partitioned.
4. Assign $n[d/2]$ first vertices to one subgroup and the others to the second one.
   Cut of value ($n[d/2]$) = number of ports * [number of devices/2].
5. Apply steps 1–4 recursively to each subgroup until the size of each subgroup becomes$\leq n$.

11.5 Research Method and Experiment Details

11.5.1 Parameter Selection

Parameter selection is one of the most important issues for power management of Ethernet switch. Bandwidth and the number of connections are mostly responsible for energy consumption of Ethernet switch [16], and the packet size has no impact on energy consumption of the switch [21]. So in our experiment bandwidth, the traffic load on each switch port and traffic matrix according to the communication among end devices were used as the input parameter, and power consumption of switch was measured as the output parameter. Though traffic load has very little impact on energy usage, but it represents the service that the network must offer (QoS). In the experiment, traffic load was used to maintain the frequency of communication among the end devices and to draw traffic matrix depending on that communication.

11.5.2 Detailed Experiment

The experiment was done according to Fig. 11.1. Powerspy2 sensor was used to measure the total energy consumption of three switches which can send real-time energy consumption data via Bluetooth [16, 32]. Ostinato traffic generator was used to generate traffic into the switch ports [29]. Three sets of experiment were done and each set contains two experiments. During each set of experiment, for all switch ports, the same bandwidth was used which were 100 Mbps and 1 Gbps. Depending on the traffic load and frequency of communication, we have designed 9×9 traffic

Table 11.1 Traffic load and frequency of communication

	For large transmission (10)	For small transmission (1)
Frame size (byte)	1125	1125
Packet per second	10,000	1000
Total size of transmission per second (Mbps)	90	9

matrix and measured energy consumption of the network topology. According to the frequency of communication among end devices, two types of traffic loads have been used which are defined as large transmission and small transmission which are shown in traffic matrix as "10" and "1" in., respectively. More information about traffic loads and frequency of communication is shown in Table 11.1. Each experiment was run and monitored for 15 min.

11.6 Results and Discussions

This section presents the obtained result during the test phase. In traffic matrix, the red-marked PCs are connected with S1, blue-marked PCs are connected with S2, and green-marked PCs are connected with S3.

11.6.1 Case Study

For conducting the experiment, communication among the end devices during daytime or working hours and nighttime or nonworking hours was considered. The working hour was defined as $T_1 = 16$ h and the nonworking hour was defined as $T_2 = 8$ h. The communication among the end devices is more than nonworking hours which have been shown by traffic matrix1 and matrix2 accordingly in Fig. 11.2. Both of the matrices in Fig. 11.2 are unoptimized.

The architecture is composed of nine PCs connected to three Ethernet switches (respectively S1, S2, and S3) with the capacity of five ports each switch as explained in Fig. 11.1. Figure 11.2 shows that PCs in red (1, 2, and 3) are connected to switch S1, PCs in blue (4, 5, and 6) to switch S2, and PCs in green (7, 8, and 9) to switch S3.

In this case, most of the frequent communications are intra-switch communication, and, in optimized organization, most of the frequent communications are inter-switch communication among end devices shown in Fig. 11.2. The total amount of traffic load is same for both matrices.

The yearly power consumption for Fig. 11.2 can be calculated by the formula below:

$$P_{\text{total}} = \left(T_1 P_{\text{matrix}1} + T_2 P_{\text{matrix}2}\right)^{*} 365$$

Matrix 1

PC No	1	2	3	4	5	6	7	8	9
1				10				10	1
2					10				10
3						10	10		
4	10							10	
5		10				1		1	10
6		1	10				10		
7			10			10		1	
8	10			10			1		
9		10			10				

T_1

Matrix 2

PC No	1	2	3	4	5	6	7	8	9
1									
2					10				10
3									
4									
5		10							10
6									
7									
8									
9		10			10				

T_2

Fig. 11.2 Traffic matrix according to the communication among end devices during working and nonworking hours

Table 11.2 Power consumption for 100 Mbps bandwidth in the unoptimized organization

Time	Working hours (T_1)	Nonworking hours (T_2)	Total
Power consumption (KWh/year)	598.6	299.6	898.2

Table 11.3 Power consumption for 1 Gbps bandwidth in the unoptimized organization

Time	Working hours (T_1)	Nonworking hours (T_2)	Total
Power consumption (KWh/year)	620.8	308.9	929.7

Here P_{total} is the yearly power consumption; $T_1 P_{\text{matrix1}}$ is daily power consumption for Matrix 1; and $T_2 P_{\text{matrix2}}$ is daily power consumption for Matrix 2.

Table 11.2 and Table 11.3 show the experimental results of power consumed by the network architecture shown in Fig. 11.1 for 100 Mbps and 1 Gbps bandwidth. For all measurements, bandwidth for both trunk ports and the ports connected to end devices was same, and two trunk ports were connected between each switch to support communications.

There is no notable change in power consumption for traffic load, but there is a noticeable change in power consumption when there is a change in bandwidth. The power consumption increases 0.2–0.4 W per port if the link speed is changed from 100 Mbps to 1 Gbps, respectively. Reducing bandwidth from 1 Gbps to 100 Mbps could save yearly 33.3 KWh in the unoptimized organization where five ports in each switch were active. Another noticeable observation is an extra trunk link was needed in working hours for the unoptimized organization to support communication and to avoid packet loss for traffic overload between S1 and S2.

For clustering and optimizing the communication and cabling plan for the whole day, the weights of Matrix 1 and Matrix 2 have been calculated considering the time T_1 and T_2, respectively, and the spectral algorithm was implemented on that matrix. It has been assumed that Matrix 1 and Matrix 2 are derived from connected graph G1 and G2, respectively. The weight of G1 and G2 is defined by C which is shown in Fig. 11.3.

PC No	1	2	3	4	5	6	7	8	9
1				160				160	16
2					240				240
3						160	160		
4	160							160	
5		240				16		16	240
6		16	160				160		
7			160			160		16	
8	160			160			16		
9		240			240				

$C = T_1G1 + T_2G2$

PC No	1	2	3	4	5	6	7	8	9
1				320				320	16
2					480	16			480
3						320	320		
4	320							320	
5		480				16		16	480
6		16	320		16		320		
7			320			320		32	
8	320			320	16		32		
9	16	480			480				

A(C)

Fig. 11.3 Combined matrix and undirected adjacency matrix for the communication

PC No	1	2	3	4	5	6	7	8	9
1	656			−320				−320	−16
2		976			−480	−16			−480
3			640			−320	−320		
4	−320			640				−320	
5		−480			992	−16		−16	−480
6		−16	−320		−16	672	−320		
7			−320			−320	672	−32	
8	−320			−320	−16		−32	688	
9	−16	−480			−480				976

L(C)

Fig. 11.4 Laplacian matrix for graph C

As the spectral algorithm works with only undirected graph, so $A(C) = C + C'$ has been calculated shown in Fig. 11.3, and from there, Laplacian matrix $L(C)$ has been calculated which is shown in Fig. 11.4.

After computing the Fielder vector for the graph C and sorting the vertices according to the values of the components of the Fielder vector, sorted vertices shown in Fig. 11.5 were obtained.

Here nine end devices are connected to three network switches with the capacity of five ports where two ports are connected as trunk. The sorted vertices from Fig. 11.5 have been used as end devices. After applying improvement on RSB algorithm [27], the obtained result of Fig. 11.6 partitioned the end devices into three parts, and the group of three end devices is connected to each switch.

The optimized adjacency matrix was drawn based on working hours and non-working hours according to the serialization found in Fig. 11.7.

The power consumption by the optimized solution is measured which is shown in Table 11.4.

Table 11.5 shows the yearly total power consumption of optimized solution using the spectral algorithm. The result shows that optimized solution consumes less energy than unoptimized architecture. It was also possible to reduce extra trunk port between S1 and S2 during working hours because spectral algorithm produced

VFIEDLER	Sorted Vertices
0.45919	3
-0.12275	6
-0.34744	7
0.46906	2
-0.11657	5
-0.33986	9
-0.32144	8
0.43358	1
-0.11378	4

Fig. 11.5 Fiedler values and sorted vertices for the graph G

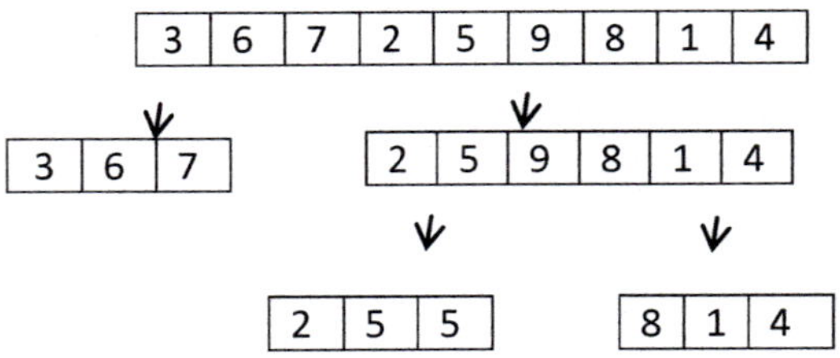

Fig. 11.6 Partitioning of end devices by using RSB in $n[d/2]$

PC No	3	6	7	2	5	9	8	1	4
3		10	10						
6	10		10	1					
7	10	10					1		
2					10	10			
5		1		10		10	1		
9				10	10				
8			1					10	10
1						1	10		10
4							10	10	

T_1
Optimized Matrix 1

PC No	3	6	7	2	5	9	8	1	4
3									
6									
7									
2					10	10			
5				10		10			
9				10	10				
8									
1									
4									

T_2
Optimized Matrix 2

Fig. 11.7 Optimized Matrix 1 and Matrix 2

Table 11.4 Power consumption for working hour and nonworking hours for optimized solution

Power consumption at working hours (T1) by optimized organization		
Bandwidth	*100 Mbps*	*1 Gbps*
Energy consumption (KWh/year)	596.3	613.8

Power consumption at nonworking hours (T2) by optimized organization (KWh/year)			
Bandwidth	**With trafic**	**Hibernate**	**Switched off**
100 Mbps	298.7	218.1	99.9
1 Gbps	306.9	219.6	100.7

Table 11.5 Total yearly power consumption for the optimized solution for different bandwidth

	Power consumption for 100 Mbps bandwidth (KWh/year)	Power consumption for 1 Gbps bandwidth (KWh/year)
Always active	895.0	920.7
With hibernation	814.4	833.4
With switched off	596.1	613.8

Table 11.6 Power saving by using optimized solution

	Power saving for 100 Mbps bandwidth (KWh/year)	Power saving for 1 Gbps bandwidth (KWh/year)
Always active	3.2	9.0
With hibernation	83.8	96.3
With switched off	202.1	315.9

clusters of PCs which have strong communications among those which reduce inter-switch communication shown in optimized Matrix 1. After applying spectral clustering on end devices and serialization of network switches, it was possible to hibernate or switch off two network switches (S1 and S3) during nonworking hours because it created two clusters of PCs without communication which are connected with S1 and S3.

Table 11.6 shows the power savings after using the spectral algorithm for different bandwidth. For 100 Mbps bandwidth, the optimized solution could save a yearly minimum of 3.2 KWh when no hibernate or switched off feature activated to maximum 202.1 KWh with switched off feature activated during nonworking hours. For 1 Gbps bandwidth, the optimized solution could save 9.0 KWh when no hibernate or switched off feature activated to maximum 315.9 KWh for switched off feature activated during nonworking hours.

For hibernation or switched off Ethernet switch, it should be considered the quality of service. The hibernation mode consumes about 20 W and switched off mode consumes nothing. For the Cisco switch model 2960-X, it takes 260 s to get ready after wake up from hibernation and 290 s from switched off mode. The time difference to be fully functional from hibernation and switched off mode is 30 s which can be critical for few cases. But for normal office works or local area network, it doesn't have that much impact. Frequent on-off can put extra load which can reduce lifetime, and some devices are vulnerable to this situation. On the other hand, always keeping on a device can also reduce lifetime; therefore, switched off can be a good choice too. Hibernation and wake-up time can be scheduled by command line which is not possible for switched off mode, but by using rack power distribution units (PDUs), it is also possible to switch off and on automatically. Therefore, the choice of hibernation or switched off Ethernet switch can be according to the necessity of the user.

The spectral approach is also applicable for optimizing bigger network architecture. So by using clustering method based on spectral approach, it is possible to save a large amount of energy according to traffic loads and pattern for the bigger picture and more complex architecture. The Ethernet switch is only a part of network architecture, and to put efficient and greater impact, other network components like router, Wi-Fi hotspot, etc. should be considered. It is also possible to apply the spectral algorithm for clustering those devices according to traffic and optimize the whole architecture. Moreover, the experiment was done with one type of Ethernet switch, so the result may vary for different type of switches. Rondeau et al. [28] mentioned the correlation between the power consumption and carbon emission. Therefore, by optimizing network architecture using the spectral algorithm, it is also possible to reduce carbon footprint caused by network equipment. As a big picture, with proper design and development of network architecture, it is possible to reduce and control global energy usage and carbon footprint caused by network devices.

11.7 Conclusion

The enhancement of energy consumption by network infrastructures has large negative impact on sustainability which should be controlled. This paper presents a novel way of energy-efficient network design and management system by using the clustering approach based on spectral algorithm to reduce energy consumption of network infrastructure. By using this approach, it is possible to hibernate or switch off part of network during low-traffic hours in order to save energy. Experimental result shows that significant amount of energy can be saved by reorganizing the network switches and clustering the end devices using the spectral algorithm. For future work, this approach can be applied to the bigger network architecture and big enterprise which has the potential to save a large amount of energy usage. Similar experiments can be done for other network devices such as routers to reduce more energy consumption.

Acknowledgment This work is part of the Erasmus Mundus Master program in Pervasive Computing and Communication for Sustainable Development (PERCCOM) of the European Union (www.perccom.eu). The authors thank all the partner institutions, sponsors, and researchers of the PERCCOM program [18].

References

1. Abbasi, A. A., & Younis, M. (2007). A survey on clustering algorithms for wireless sensor networks. *Computer Communications, 30*(14), 2826–2841.
2. Amis, A. D., & Prakash, R. (2000). Load-balancing clusters in wireless ad hoc networks. In *Application specific systems and software engineering technology, 2000. Proceedings of 3rd IEEE Symposium on* (pp. 25–32). Richardson, Texas: IEEE.
3. Baker, D. J., & Ephremides, A. (1981). The architectural organization of a mobile radio network via a distributed algorithm. *IEEE Transactions on Communications, 29*(11), 1694–1701.
4. Bandyopadhyay, S., & Coyle, E. J. (2003). An energy efficient hierarchical clustering algorithm for wireless sensor networks. In *INFOCOM 2003. Twenty-second annual joint conference of the IEEE computer and communications. IEEE societies* (Vol. 3, pp. 1713–1723). San Francisco, California: IEEE.
5. Basagni, S. (1999). Distributed clustering for ad hoc networks. In *Parallel architectures, algorithms, and networks, 1999.(I-SPAN'99) Proceedings. Fourth International Symposium on* (pp. 310–315). Perth/Fremantle, Australia: IEEE.
6. Chan, T. F., & Smith, B. F. (1994). Domain decomposition and multigrid algorithms for elliptic problems on unstructured meshes. *Electronic Transactions on Numerical Analysis, 2*, 171–182.
7. Chiasserini, C. F., Chlamtac, I., Monti, P., & Nucci, A. (2002). Energy efficient design of wireless ad hoc networks. In *Networking 2002: Networking technologies, services, and protocols; performance of Computer and communication networks; mobile and wireless communications* (pp. 376–386). Berlin/Heidelberg: Springer.
8. Christensen, K., Nordman, B., & Brown, R. (2004). Power management in networked devices. *Computer, 37*(8), 91–93.
9. Donath, W. E., & Hoffman, A. J. (1973). Lower bounds for the partitioning of graphs. *IBM Journal of Research and Development, 17*(5), 420–425.
10. Fiedler, M. (1973). Algebraic connectivity of graphs. *Czechoslovak Mathematical Journal, 23*(2), 298–305.
11. Fiedler, M. (1975a). A property of eigenvectors of nonnegative symmetric matrices and its application to graph theory. *Czechoslovak Mathematical Journal, 25*(4), 619–633.
12. Fiedler, M. (1975b). Eigenvectors of acyclic matrices. *Czechoslovak Mathematical Journal, 25*(4), 607–618.
13. Fithritama, A., Rondeau, E., Bombardier, V., & Georges, J.P. (2015). Modeling fuzzy rules for managing power consumption of ethernet switch. In *Software, telecommunications and computer networks (SoftCOM), 2015 23rd International Conference on* (pp. 42–47). Island, Brac: IEEE.
14. Gunaratne, C., Christensen, K., Nordman, B., & Suen, S. (2008). Reducing the energy consumption of Ethernet with adaptive link rate (ALR). *IEEE Transactions on Computers, 57*(4), 448–461.
15. Gupta, M., Grover, S., & Singh, S. (2004). A feasibility study for power management in LAN switches. In *Network Protocols, 2004. ICNP 2004. Proceedings of the 12th IEEE International Conference on* (pp. 361–371). Berlin, Germany: IEEE.
16. Hossain, M.M., Rondeau, E., Georges, J.P., & Bastogne, T. (2015). Modeling the power consumption of Ethernet switch. In *International SEEDS conference 2015: Sustainable ecological engineering design for society*. Leeds UK.
17. Iea.org. (2014). *Publication:-More data, less energy: Making network standby more efficient in billions of connected devices.* [online] Available at: http://www.iea.org/publications/freepublications/publication/more-data-less-energy.html
18. Klimova, A., Rondeau, E., Andersson, K., Porras, J., Rybin, A., & Zaslavsky, A. (2016). An international master's program in green ICT as a contribution to sustainable development. *Journal of Cleaner Production, 135*, 223–239.

19. Krommenacker, N., Rondeau, E., & Divoux, T. (2001). Study of algorithms to define the cabling plan of switched Ethernet for real-time applications. In *Emerging technologies and factory automation, 2001. Proceedings. 2001 8th IEEE International Conference on* (pp. 223–230). Antibes-Juan Les Pins, France: IEEE.
20. Lin, C. R., & Gerla, M. (1997). Adaptive clustering for mobile wireless networks. *Selected Areas in Communications, IEEE Journal on, 15*(7), 1265–1275.
21. Mahadevan, P., Sharma, P., Banerjee, S., & Ranganathan, P. (2009). A power benchmarking framework for network devices. In *Networking 2009* (pp. 795–808). Berlin/Heidelberg: Springer.
22. Mahadevan, P., Shah, A., & Bash, C. (2010). Reducing lifecycle energy use of network switches. In *Sustainable systems and technology (ISSST), 2010 IEEE International Symposium on* (pp. 1–6). Crystal City, Arlington, VA, USA: IEEE.
23. Mellier, R., & Myoupo, J.F. (2006). A weighted clustering algorithm for mobile ad hoc networks with non unique weights. In *Wireless and mobile communications, 2006. ICWMC'06. International Conference on* (pp. 39–39). Bucharest, Romania: IEEE.
24. Mingay, S. (2008). IT's role in a low carbon economy. In *Presentation at Gartner Symposium/IT Expo, Cannes, France*.
25. Nedevschi, S., Popa, L., Iannaccone, G., Ratnasamy, S., & Wetherall, D. (2008). Reducing network energy consumption via sleeping and rate-adaptation. In *NSDI* (Vol. 8, pp. 323–336).
26. Nordman, B. (2008). Networks, energy, and energy efficiency. In Cisco Green Research Symposium. UK.
27. Rondeau, E., Divoux, T. and Adoud, H. (2001). Study and method of ethernet architecture segmentation for Industrial applications. In *4th IFAC Conference on Fieldbus systems and their applications (FET'2001). Nancy, France* (pp. 165–172).
28. Rondeau, E., Lepage, F., Georges, J. P., & Morel, G. (2015). Measurements and sustainability, chapter 3. In M. Dastbaz, C. Pattinson, & B. Akhgar (Eds.), *Green information technology: A sustainable approach* (1st ed.). Burlington, Massachusetts, USA: Elsevier.
29. Ostinato Network Traffic Generator and Analyzar User Guide. [online] Available at: https://github.com/pstavirs/ostinato/wiki/UserGuide
30. Pamlin, D., & Szomolányi, K. (2006). Saving the climate@ the speed of light. First roadmap for reduced CO2 emissions in the EU and beyond. *European Telecommunications Network Operators' Association and WWF*.
31. Pothen, A., Simon, H. D., & Wang, L. (1992). *Spectral nested dissection*. Pennsylvania State University, Department of Computer Science.
32. PowerSpy2 user manual. (2015). [Online] Available at:http://www.alciom.com/images/stories/downloads/rl1236004%20powerspy2%20user%20manual%20v0101.pdf
33. Reviriego, P., Sivaraman, V., Zhao, Z., Maestro, J. A., Vishwanath, A., Sánchez-Macian, A., & Russell, C. (2012). An energy consumption model for energy efficient ethernet switches. In *High performance computing and simulation (HPCS), 2012 International Conference on* (pp. 98–104). IEEE.
34. Rivoire, S., Shah, M., Ranganathan, P., Kozyrakis, C., & Meza, J. (2007). Modeling and metrology challenges for enterprise power management. *IEEE Computer, December*.
35. Simon, H. D. (1991). Partitioning of unstructured problems for parallel processing. *Computing Systems in Engineering, 2*(2–3), 135–148.
36. Von Luxburg, U. (2007). A tutorial on spectral clustering. *Statistics and Computing, 17*(4), 395–416.
37. White, S., & Smyth, P. (2005). A spectral clustering approach to finding communities in graph. In *SDM* (Vol. 5, pp. 76–84).
38. Zhang, G. Q., Zhang, G. Q., Yang, Q. F., Cheng, S. Q., & Zhou, T. (2008). Evolution of the internet and its cores. *New Journal of Physics, 10*(12), 123027.

Chapter 12
Virtual Vulnerability: Safeguarding Children in Digital Environments

Emma Bond and Vanessa Rawlings

12.1 Introduction

While there has been considerable academic, political and economic attention given to the development of smart futures, little attention has been paid to the impact of smart technologies on childhood. The chapters in this section focus on the impact of emerging technologies in society, and our intention here is to consider some of the challenges which arise in safeguarding children in digital environments. Children, previously ignored in sociological analysis, have increasingly become the centre of social science attention with the growth of childhood studies (see [35]) and framed, arguably often inappropriately, as digital natives [67]. Many of these analyses remain, as Holloway and Valentine argued in [31] either 'cybertopian' or 'cybercritical', yet adopting such polarised standpoints is unhelpful in understanding the challenges that smart technologies bring. This chapter sets out to consider some of the emerging trends, developments and challenges associated with the design and applications of future technologies and smart environments in safeguarding children online.

Following on from the Byron Review [10], understanding online risk has attracted considerable academic and media attention in the past few years. This chapter outlines some of the challenges faced in keeping children safe in digital environments and how the rapidly changing landscape of risk complicates the development of effective safeguarding strategies. It draws upon the growing body of research including the recent Ofcom studies from the UK and the extensive findings of the EU Kids online project across Europe. The EU Kids online project is a multinational research network led by Professor Sonia Livingstone and Professor

E. Bond (✉) • V. Rawlings
University of Suffolk, Ipswich, UK
e-mail: e.bond@uos.ac.uk

M. Dastbaz et al. (eds.), *Technology for Smart Futures*,
DOI 10.1007/978-3-319-60137-3_12

Leslie Haddon, which has provided an extensive and detailed understanding of European children's online opportunities, risks and safety. The project uses several methods to map children's and parents' experiences of the Internet, to promote an informed dialogue with national and European policy stakeholders. Online risk is now closely associated with what children are using the Internet for, how they are accessing it and what social media, websites and apps they are using. The ever changing technical landscape is widely recognised. Children are growing up in a world of hyperconnectivity that is very different to adults' own experiences of childhood which has implications for both parents and professionals in keeping children safe online.

Children aged 8–11 years use the Internet for entertainment and 'to have fun', with games dominating the online activities of both girls and boys, played via apps and websites that were often discovered via friends or family members [59]. The Ofcom [59] study found that choice of what to play tended to be quite spontaneous, influenced mainly by whichever device they had access to at the time. Most recently, Ofcom [61] in the UK found that 37% of 3–4 s, 54% of 5–7 s and 73% of 8–11 s watched YouTube, with the younger age group consuming traditional 'TV like' content but moving quickly to 'vloggers, music videos, game tutorials and joke or prank videos'. Equally touchscreen technology, which makes devices like tablets and phones very easy to use and navigate, enables children to go online from a very young age and often with parents having little awareness of what their children are doing and seeing online [45]. Notably, more traditional literacy skills are no longer required to access the Internet, with online gaming being especially popular even with very young children.

The UK Council for Child Internet Safety [78] suggest that while new technologies inspire children to be creative, communicate and learn, and although the Internet is a great resource, it is important that children and young people are protected from the risks they may encounter. 'Children's use of social media is constantly evolving, and that brings both new opportunities and risks' [61:3]. Recent research suggests that the likely ages that children encounter known risks is getting younger as more children are online at a younger age, have access to more technology and more mobile technologies in the form of tablets and smart phones [8, 60, 61]. This poses considerable challenges for educators, parents, professionals and policy makers as the importance of having an accurate understanding of the risks that children face is becoming increasingly apparent. The nature of risks must include that they are likely to encounter both currently and in the next few years as society moves towards a 'smart future'.

While this chapter considers risk in relation to children's interactions online, children's vulnerability and the challenges that currently dominate both policy and public discourse, it is important to acknowledge that academic research, policy and practice guidelines have shifted away from only emphasising the risks children face to also explicitly celebrate the positive opportunities afforded by online environments, social media and virtual worlds. It is also well established within safeguarding discourses that to deny children access to the Internet and online information has a detrimental effect on their opportunities to socialise, learn and interact, as Hope [32:350] observes:

> Through the lens of governmentality, the discursive construction of e-kids can be seen as a normalising, controlling process. Polarised, undifferentiated simulacrum of children are constructed, representing them as either 'socially naïve' and 'at risk', or potentially 'dangerous'. In both cases, it can be claimed that the state and its agents have a duty of care to intervene as protectors, while curbing the actions of miscreant students. Yet this unsophisticated envisaging of children, which draws upon notions of 'becoming', justifies the advocacy of crude, generic controls in e-safety policy, while failing to address what students might be losing in terms of their learning experiences.

Any consideration of risk, therefore, needs to be understood in balance with the very positive aspects of children engaging with digital worlds.

> Children's safety in connected media is vital, but it needs to be understood in the context of the spectrum of their digital rights, for example, in balance with children's rights of provision and participation in the Convention on the Rights of the Child. Agency is as crucial to positive, effective use of digital media as safety is. Without the agency needed to participate and exercise rights, children can neither take advantage of the opportunities digital media afford nor develop resiliency when facing risks. They must be encouraged to think critically and develop their own language, views, strategies, associations and interests as users of connected digital media [76:8].

The relationship between children's rights to protection and their rights to participation online is a complex one [8, 41]. Adopting a more rights-based approach to understanding children's experiences with digital media provides a more balanced view of their everyday encounters and how they can be safeguarded while making the most of the opportunities afforded by media technologies and digital environments [38].

12.2 Understanding Risk

Living in a risk society [3] with profound implications for self-identity [25], the individualisation of and management of risk has fuelled debates on modernity for decades. Both risk and childhood are social, cultural constructions [5, 7, 8] as such, they will change across time and space and what is understood as risky may not be seen as such to another group, or in another time. Educational, law enforcement, policy and media debate has focused primarily on addressing the potential impact of online content that may be harmful (Przybylski et al. 2014), yet in order to have an accurate understanding of online risk, we need to understand what school age children are actually doing online. Phippen's [64] study with 10- and 11-year-old children, for example, found that the vast majority of young people of this age already have active online social lives using a number of different technologies and platforms and that gaming plays a major role in many boys' lives, including online multiplayer environments and 18 certificate games. Most young people of this age have had to deal with some form of online abuse, and they turn to their peer group for help, while many have received unsolicited contact from strangers, either through social networks or mobile technology.

Similarly the Ofcom study [59: 14] concluded:

> Strategy game apps tended to have strong appeal with this age group. Games such as Angry Birds, Temple Run and Flappy Bird had multiple levels or lives that motivated them to keep playing. Multi-game websites such as MiniClip and Friv were also favoured options as they contain a large number of 'mini games' children can dip into for an instant, short gaming session. 'Virtual world' gaming websites with a social element were also popular among this age group in this research, with brands such as Club Penguin and Moshi Monsters cited as appealing, together with Minecraft, the current dominant online building blocks game, The 'craze' popularity of these games meant that children made a concerted effort to play them more and such titles spanned both app and website formats, helping to maximise appeal and uptake.

The Ofcom research clearly demonstrates that the diversity in which children connect to and interact online is rapidly increasing. This together with the rapid development of apps and online gaming platforms results in the diversification of online risk. Depending on their digital access, children are likely to face differing risks at different ages according to their activities, experiences and skills. Rudkin (in UKCCISS [77]) outlines online risk in relation to child development and proposes that, while 3–5-year-olds may be unaware of risks, 6–9-year-olds have developed a strong desire for immediate reward which triggers risk-taking behaviour. Just as children's understanding of risk changes with age, the risks they face online are also changing. Online bullying is now more common than face-to-face bullying [44]. According to the NSPCC [56], 7296 counselling sessions took place with children who talked to Childline[1] about online bullying and safety in 2014. However, Ey et al.'s [23] review of online bullying with primary school age children identified that at this age, there is very limited research on children's understanding of online bullying, the impacts of online bullying and bullying behaviours. Children can be involved in online bullying either as victim or perpetrator, and the bullying, which may begin at school (via traditional bullying), often continues and escalates outside of school via the Internet, mobile phone or tablet [50]. Vandebosch and Van Cleemput's study [79] with 2052 primary and secondary school children reveals that online bullying is not a marginal problem. They found that children who had been bullied via the Internet or mobile phone during the last three months were more dependent upon the Internet, felt less popular, took more Internet-related risks and were more likely to also be a bystander and perpetrator of Internet and mobile phone bullying. Yet peer-on-peer abuse – either through bullying or through sexting behaviour – is often overlooked in educational programmes, academic research and learning resources [65].

Access and exposure to pornography are linked to children's engagement in 'risky behaviours' [33:7]. Livingstone and Palmer [40] suggest that young people, mainly boys, are increasingly being affected by adult pornography. Viewing usually starts around the age of 11–12 years and can result in intimacy deficits, unrealistic

[1] https://childline.org.uk

expectations of their partners, inability to show empathy and higher likelihood of relationship breakdown. Horvath et al.'s study [33] found that there is some evidence that children and young people consider pornography easy to access and culturally prevalent. Contradictory findings exist, however, in relation to age of first exposure, with variations from 10 to 17 years old in research findings, yet there is a dearth of research on the complex relationship between pornography and children's expectations, attitudes and behaviours [33].

Given the varied use across many different platforms, diverse opportunities and a changing landscape of risk [6, 7, 8], the dangers children face online are complex, are multi-faceted and are often interrelated. Trying to identify which children may be more vulnerable online is, therefore, less than straightforward. Some children deemed vulnerable offline are more likely to be vulnerable online, but it is essential to contextualise when, why and how children may be at risk online. The 'four Cs' (see [39]) for potential risks facing children (viz., contact, content, conduct and commercialism) come into play at different stages of a child's development. Therefore, it is important to remember that 'vulnerability is not a static issue but one that needs contextualising within the emotional, psychological and physical developmental stages of childhood' [40:4].

Classification of online risks [39]

	Content: child as recipient	Contact: child as participant	Conduct: child as actor
Commercial	Advertising, spam, sponsorship	Tracking/harvesting personal information	Gambling, illegal downloads, hacking
Aggressive	Violent/gruesome/hateful content	Being bullied, harassed or stalked	Bullying or harassing another
Sexual	Pornographic/harmful sexual content	Meeting strangers, being groomed	Creating/uploading pornographic material
Values	Racist, biased information/ advice (e.g. drugs)	Self-harm, unwelcome persuasion	Providing advice, e.g. suicide/pro-anorexia

In developing effective responses and effective intervention strategies, it is important to recognise that children can be both victims and offenders in relation to the construction of online risk, for example, in cases of online bullying or the sharing of violent or sexual content. This is an important consideration as both child victims and child offenders require support and safeguarding. However, the binary categorisation of victim/offender may also be unhelpful in understanding children's online experiences, in that it masks the wide spectrum of children's characteristics, competencies, vulnerabilities, contexts, family circumstances and the wider socio-economic, political and cultural environment. The Byron Review [10] discussed vulnerability online but did not define what is actually meant by vulnerability [40], and in practice concepts of harm and 'vulnerability' are rarely defined and tend to be implicitly understood [48]. Livingstone and Palmer [40] highlight the importance of understanding vulnerability online in the broader context of children's lives with a focus on building protective environments and resilience. However, they suggest that there are many unresolved questions still to address which include the implications for children with disabilities; the impact of cultural context on behaviour, risk, aspiration, vulnerability, and young abusers; the impact of pornography – both solicited and unsolicited – resilience factors leading to sexual abuse and exploitation; and what works in reducing harm/risk.

In relation to online vulnerability, there are, according to Davidson and Martellozzo [19], three different categories of children: resilient children who form the majority of children and are the least likely to interact online and have a low risk of meeting offline, disinhibited children who are willing to interact online but who are unlikely to meet up offline and vulnerable children who, although the minority group, are willing to interact online, are seeking relationships and are at high risk of meeting offline. The factors that lead to online sexual abuse are, however, complex and intertwined. The EU Kids online study identifies how children whose parents lack education or Internet experience tend to lack digital safety skills, leaving them more vulnerable to online risk. The analysis by Göran Svedin [27] indicates that teenage girls with sexual abuse history and depressive feelings, who have poor relationships with their parents and where there is weak parental monitoring are more at risk of Internet-related sexual abuse. The most important and, also most examined, risk factor is young people's risk-taking behaviour online. It is apparent the that more young people are open to online sexual activities (especially flirting and having sexual conversations with strangers), the more probable it is that they may become victims of sexual harassment, solicitation or grooming [27:49]. Whittle et al. [86] also found that behaviours specific to vulnerability to online grooming include engaging in risk-taking behaviour online, high levels of Internet access and lack of parental involvement in the child's Internet use.

Also of interest here is Carrick-Davies' [11] work with vulnerable, excluded young people which questioned what can make these children vulnerable online. He found that they had low self-confidence and so their identity was that of 'outsiders'. Fluid learning environments, as well as gaps in their education and induction also contributed to them being very vulnerable online. Moreover, they are more likely to have experienced abusive relationships or environments. These children are also more likely to be influenced by alcohol, drugs and gang culture, and they are considered to be both risk takers and at risk. Plus, they have more unsupervised time, fewer structures and boundaries and a lack of supportive adults in their lives. Thus, there is a real 'need for further interventions to foster and develop these young people's strategies to avoid online risk as part of more general initiatives to develop digital literacy' [16:72]. Children with psychological difficulties also appear to encounter more risk online and be more upset by it [21]. Children with disabilities tend to have more digital skills but encounter more risk online and often lack peer support, and minority children also have more digital skills and encounter more risk, but their parents are keen to help them [42]. Yet it is important to remember that the absence of vulnerability is not fixed, but is a variable that can change very quickly [11]. As such 'vulnerable children and young people are not a self-contained or static group. Any child/young person may be vulnerable at some time depending on any one, or a combination of, the risks or challenging life events they face and their resilience' [18:9]. Therefore, an understanding of online vulnerability is particularly important in overcoming the challenges of safeguarding children in digital environments. Children who are understood to be more at risk than others can specifically benefit from improvements in self-protection [1]. However, although some studies have identified specific characteristics associated with online vulnerability, the relationship between online and offline vulnerability is a complex one. While children who are more vulnerable offline are also likely to be more vulnerable online, some children may appear highly resilient offline but can be highly vulnerable online and vice versa. Furthermore, 'what might be labelled as an online safety issue for a 5-year-old child could be perceived as unproblematic for a 15-year-old' [32:345]. However, it would be dangerous to stereotype and presume that because a child does not appear to 'fit' with a risky profile that they are safe. Many children who have been victims of online grooming, for example, do not have a 'troubled' background nor did they appear vulnerable previously but had strong friendships, good relationships with their parents and had a high level of educational attainment [40]. This suggests that 'online risks can manifest in unexpected ways' [21:9]. There is, to date, very little research in relation to sexual abuse online among children up to 9 years old relating to online activities leading to offline abuse (Ainsaar and Soo 2011:5), and further research is required to enhance understanding of effective service responses to reduce the number of children suffering harm as a result of online activities [51, 62].

While there is some evidence to suggest that some children may be more vulnerable than others in relation to certain risks, Livingstone and Palmer [40] found that vulnerability to grooming, the impacts of adult pornography and displaying risky and/or harmful behaviours online appear to be less about being a vulnerable child offline and more about the child's stage of development – namely pubescent – starting around 11–12 years of age. They observe from practice findings that for many of these child victims, there are few common indicators of vulnerability. The impact of these abusive behaviours on child victims, however, appears to be universal, and being made the subjects of abusive images may affect children of any age, sex and ethnicity. Yet little is known about the demographics of these children. It is also interesting to note from the Internet Watch Foundation's 2015 annual report [34] that there is a large increase in the number of reported illegal child sexual abuse images and videos. 68,092 reports were positively identified as containing illegal child sexual abuse imagery and taken down. This is a 417% increase in online confirmed reports over 2 years and a 118% increase in illegal child abuse imagery over the previous year. The report also looks at trends emerging from the 2015 data. It found that 69% of victims were identified as aged 10 or under, 1788 victims were assessed as aged two or under, and 34 percent of images were category A – which involved the rape or sexual torture of children.

12.3 Factors Influencing Vulnerability

Research to date reveals a number of factors related to online vulnerability. In relation to age, for example, the Net Children Go Mobile data shows that it is the youngest children who report higher rates of getting upset from being bullied [57]. Younger children are also less resilient online and are more likely to go offline and experience more harm [32, 43, 51, 84]. In relation to gender, girls appear to be more susceptible to the harmful effects of sexual risks, but they are, however, more talkative than boys. No matter what type of risk is faced, vulnerability is less acceptable among boys; girls were more likely to experience persistent cyberbullying than boys, but boys are more likely to have made friends with someone they don't know online and to meet in real life, mainly through online games [18, 43, 49, 51, 84]. Family structure also appears in the literature as a factor of interest in that some studies have found that the majority of victims of Internet-related sex crimes live in single-parent or reconstituted families, but this may be linked to weaker ties between the child and caregiver and less monitoring of the child's (online) activities [4, 15, 24, 27, 42, 71, 72]. Additionally, children who experience family difficulties and are brought up in 'chaotic' family/home environments, looked-after children and young carers are more vulnerable. Although different from family structure but related to the family context is socioeconomic status (SES), research has found that young people from households with higher SES have been more exposed to unwanted sexual material online than those from lower income families but that children receiving free school meals are more likely to experience online bullying [18, 27, 43, 51]. SES also plays a significant role in children's online resilience in that a lower SES is related to lower parental Internet use, and children from lower class families are not necessarily more vulnerable online, as they seem to be willing to solve the problem, especially when confronted with online bullying [44, 45].

According to Göran Svedin [27] and Livingstone et al. [41], the level of parental education appears as a protective factor though children with a more educated parent (or some other household member) are less likely to be victims of sexual solicitation. Also related to resilience is self-efficacy which, according to Livingstone et al. [43] and Vandoninck et al. [82], is important especially in relation to online bullying as self-efficacious children are more engaged in solving the problem. The diversity of children's lived experiences is also essential here in understanding the vulnerability matrix as sexuality also influences online experiences and children identifying as LGBTQI are more likely to experience online bullying and become victims of online sexual abuse and grooming [40, 52, 75]. Ethnicity is also a fundamental factor as research suggests that being of black or Asian ethnicity was identified as a risk factor for receiving a request for a sexual picture, and Gypsy-Roma, Traveller of Irish Heritage, European, East European, Chinese and those children of mixed ethnicity were more likely to be persistently cyberbullied, compared to white young people [18, 49, 51]. Children with learning disabilities, autism spectrum disorders (ASD) and attention deficit hyperactivity disorder (ADHD) also are more likely to encounter risk and harm online, and research suggests that children with a disability are also exposed to more sexual abuse offline than other groups [26, 42, 51, 66]. Similarly Livingstone et al. [42] and Munro [51] found that children with emotional or behavioural difficulties are more vulnerable to harm online. As the likelihood of receiving online sexual solicitations increases substantially if a child has been physically or sexually abused in their lifetime, previous experience of abuse – either domestic abuse or sexual abuse – is also a vulnerability marker [46, 49, 55]. Finally children who experience 'exclusion of access' in that they experience 'system neglect' and are unable to access services that are universally available to other children, or have chaotic lives, for example, children in pupil referral units or belonging to marginalised groups such as travellers, asylum seekers, trafficked children and migrant communities are especially vulnerable as there is an absence of supportive adults in their lives and they will have experienced staggered or fractured access to learning environments potentially missing online safety education [11, 16, 42].

While individual risk factors, such as those outlined above, do not simply lead to vulnerability, accumulation of these or in combination is considered to increase a young person's vulnerability towards online grooming [85] and online risk. Whittle et al. [85] found from interviews with eight young people who had been groomed online that six had separated parents (often acrimoniously) and/or came from a reconstituted family, and they described fights at home and difficult family relations, including issues with stepfamilies (Whittle et al. [85]:1188). They identify three scenarios which led to the young people becoming vulnerable:

1. Multiple long-term risk factors; young people who have increasing risk factors in day-to-day life, with few protective factors, and so take increasing risks online. These young people would be considered vulnerable offline.
2. Trigger events; young people who have some risk factors but are initially protected, until a trigger event or events result in the loss of those protective factors. These young people will be considered vulnerable offline, but only at a certain point in time.
3. Online behavioural risks; young people who have few risk factors and many protective factors but engage in risk-taking behaviour despite warnings. These young people will not be considered vulnerable offline.

> Young people at risk of harm online may not have any previous vulnerabilities that are often associated with being victims of sexual abuse and exploitation, such as being in care, from families facing adversities or having a history of sexual abuse. This has implications for identification, as they are less likely to be known to the authorities. The currently accepted indicators of possible sexual exploitation, such as going missing or school absence, may not be displayed, and the first parents may know that their child has been a victim of sexual exploitation is when the police contact the family. Certain groups, such as young people with learning difficulties, those with mental health problems and lesbian, gay, bisexual, transgender and questioning young people, appear to be particularly vulnerable to online harm. This is in part due to seeking social interaction online that they are not able to achieve offline and in part due to not fully understanding the consequences of sharing personal information, sending images or arranging to meet strangers met online [62:7].

Specifically focused on online grooming, The Robert Project review [27:49] identifies the reasons why some children are more vulnerable than others and concludes that 'it is difficult to compare the results of studies across countries and cultures when different questions are asked of different age groups'. Despite this, some of the observations from the available literature seem rather consistent. For example, although the probability that a child receives unwanted sexual contacts through the Internet varies between 6% (Mitchell et al. 2011) and 59% in relation to girls [9], girls are more exposed to risks of receiving an offer for sexual activities than boys. There are many different kinds of exposures to sexuality that a child can experience through the Internet, from rude sexual language to being encouraged to act sexually in front of a webcam or sending sexually explicit photos or suggestions to meet offline. Exposure to pornography on the Internet can be described as a normative experience, even if some children, especially girls, can find it both embarrassing and disgusting. Furthermore, children with a risk background tend to be both at risk for sexual solicitation on the Internet and exhibit sexually aggressive behaviour on the Internet.

12.4 Responding Effectively to the Challenges

The NFER [53] review identified 19 documents relating to safeguarding of children accessing the Internet. They found most evidence was available on the extent to which safety tools and safeguarding approaches are being used by schools, parents, Internet service providers and other organisations/stakeholders, but that there was less evidence on how aware parents, teachers and children are about ways of safeguarding from online risks, including specific safety tools. Although some studies have been undertaken (as discussed below), most focus on online bullying. Generally, according to the PSHE association [68], 'there is a lack of evidence around effective practice in online safety education specifically, although authors note that the same principles are often transferrable across different health and risk behaviours'. Davidson et al. [20] considered the UK and selected European and international initiatives in protecting children online and found that measures to protect children included school-based programmes aiming to educate children, parents and teachers about the dangers posed by sex offenders in cyberspace. One example of such an initiative is Schilder et al.'s [69] study on the impact on an online safety intervention, which found that the children in the intervention group were more likely to be aware of online risks directly after the intervention and that this effect was still noticeable four months later. Another intervention-based study by Schultze-Krumbholz et al. [70] that adopted multigroup structural equation modelling showed a significant effect of a short intervention on cognitive empathy and significant effects of the long intervention on affective empathy and cyberbullying reduction. They also found that groups without any intervention showed an increase in cyberbullying behaviours and that affective empathy decreased across the study period.

Although some schools have developed interventions and educational awareness about online bullying, more evidence-based intervention programmes are needed [80]. Couvillon and Ilieva [14] suggest that from their review, there was a clear consensus from the literature that cyberbullying prevention efforts need to be comprehensive and that an effective school-wide model needs to involve three principal groups of stakeholders: teachers, students and parents. Yet, the actions suggested across the literature are multiple, are widespread and differ in their scope or type of intervention. Additionally, Lam and Frydenberg's [36] study found that a programme teaching coping skills (Best of Coping skills program, BOC) and a programme teaching cybersafety (Cyber Savvy Teens program, CST) can optimise children's capacity to cope online. Participants in both intervention groups demonstrated improvements in their overall mental health and in making better online choices post-programme, and while the BOC program was found to be a better programme for improving general coping than CST alone, the CST programme was better at tackling cyber-specific issues. More recently Papatraianou et al. [63] propose that adopting a more ecological approach can be helpful in responding to online bullying and also in understanding resilience in that they propose that both the risk factors and the protective factors need to be considered.

However, one of the key components of responding to the challenges of safeguarding children in digital environments is the early identification of risk, but there is a fundamental difference between the discovery of abuse and the disclosure of abuse as children rarely disclose online abuse, and it is more likely to be discovered [40]. This makes it even more important to provide time and space to talk about online activities and for children to be listened to. The evidence suggests that often children are upset by something online, but they may not disclose it. Lilley and Ball's ([37]:6) study for the NSPCC concludes 'almost a quarter (23%) of 11 and 12 year olds who have a profile on a social networking site say that they have been upset by something on it over the last year'. These incidences vary, but examples include trolling, online stalking and being asked to send a sexual message. While most children quickly recover from their experience, around one fifth felt upset or scared for weeks or months after the incident occurred, and a fifth of these reported that they felt upset every day or almost every day. Yet many professionals do not ask young people about their online lives nor open up opportunities to talk about their online experiences.

> Safeguarding the cyborg child requires social workers to assess a range of different communities – not just the locality in which children live, go to school/college, but also their involvement in fluid and ever-changing online networks. Where serious concerns emerge, assessment methods might also trigger checks that require additional specialist technical expertise. Thus, as a way of taking social work safeguarding practice further, we do not ask what are the risks and harms posed by the internet, but rather when and how might a child be vulnerable in the everyday space of 'networked publics'? [47:611].

However, schools remain reluctant to talk to children about sexual bullying, to address the issue in their policies and responses to disclosures of online abuse in children's relationships depends largely on the teacher dealing with the issue [74:online]. Similarly Palmer [62] argues:

> Currently, there is no national multi-disciplinary guidance for best practice when online abuse is discovered. The standard of the response families receive is dependent on where they may be living and the expertise of the professionals in their area. Practice evidence to date informs us that the discovery of the online abuse of a child needs to be treated as seriously as any other kind of abuse. Careful, but timely, planning needs to occur on the part of the professionals at the outset and how we intervene in internet cases needs to be reconsidered. Immediate safety measures need to be assessed at point of discovery, including the decision regarding mobile phone possession, access by young people to online platforms

> and assessing the parents or carers abilities to safeguard their children's future online activities. There appears to be a poor understanding of the impacts of online abuse on the victims. The effect of the differential nature of the grooming process online, in particular the speed at which it occurs, and the seeming complicity by young people, appears to be little understood by the majority of the children's workforce. This has an impact on the nature of the referrals made to projects and the types of interventions requested.

'Research surrounding the vulnerabilities of young people to online grooming and abuse is extremely rare and more research is required, particularly from the perspective of those who have experienced the effects of abuse themselves. Young people across all vulnerability scenarios can be better protected through consistent, collaborative approaches by parents, carers and other adults in their lives' [85:1194]. Agencies in both statutory and nonstatutory organisations are required to collaborate effectively to safeguard children [83] in the UK. All organisations, including charitable organisations, must comply with the government interagency statutory guidance *Working Together to Safeguard Children* [28]. Furthermore, schools need to ensure that they also comply with the Department of Education's [29] revisions to the statutory guidance set out in *Keeping Children Safe in Education* that came into effect on 5th September 2016.

Furthermore, the Criminal Justice Joint Inspection [17] also emphasises that multi-agency responses are of paramount importance. Their inspection was informed by a body of research about assessment and interventions, in addition to studies exploring the range and effectiveness of responses to children and young people exhibiting sexually harmful behaviour. They found that although the research is largely inconclusive regarding the effectiveness of specific assessment tools or types of intervention, consensus exists in the literature about the need for a multidisciplinary approach to meeting the needs of the child or young person. Simpson [72] argues that 'social work practitioners need to acknowledge that the challenges of unregulated contact are here to stay and that as part of the complexities that arise, there is a need for them to have a greater knowledge and understanding of mobile technologies and social media'. May-Chahal et al. [47:803] propose that an on/offline distinction is unhelpful for ordinary social work practice and that, if it is 'replaced by an integrated ontological position (cyborg childhood) then social work has a great deal to offer in safeguarding children in an interconnected on/offline ecology'.

In effectively responding to risk, O'Neill and McLaughlin [58] draw on the findings from the EU Kids Online study to suggest:

- In recognition of the children who have been bothered by something on the Internet in the past year (12% of all children), schools and parents should reinforce the importance of reporting abuse while also encouraging children and young people to speak to an adult when they come across upsetting content.
- The most common way in which children come across sexual images online is through images that pop up accidentally (7% of all children; 12% of 15–16-year-olds). In order to avoid such accidental exposure to any unwanted content online, safety awareness messages need to give greater emphasis to the filter and safety settings of browsers and websites (including search engines and video hosting sites), informing parents and children about how to block such content.

- The easy availability of pornography online causes much public debate and anxiety with respect to children's use of the Internet. The finding that the Internet is now the most common way for children to see sexual images (14%), marginally more than on television, films or videos (12%), may fuel further concern in this regard. The principal implication arising is that safety messaging should be measured in approach, avoiding implications of harm and seeking to empower parents and children to talk about the subject of sexual images online.
- Of the 6% who have been bullied online, cyberbullying is fairly upsetting or very upsetting for over half (54%), more so for younger children for whom the effect was longer lasting and for children from lower SES homes. In other words, bullying is rarely trivial, and more vulnerable children need targeted supports to enable them to cope more effectively.
- In dealing with cyberbullying, Internet safety awareness should include responses and coping strategies targeted at children of different ages, enabling them to cope with situations that may arise in online communication and social networking.
- Awareness centres and educational authorities should provide teachers with resources, enabling them to be alert to, and be able to respond to, incidents of cyberbullying.
- Effective support can potentially ameliorate harm and minimise further abuse/risk encounters online.

> A minority of parents are providing guidance for children about internet dangers but few teachers are. This leaves many young children vulnerable. To leave even a small percentage of children susceptible to internet dangers is not socially responsible. Although protecting children while at school is important, education about internet protective behaviours is necessary to protect them elsewhere. It is essential that educational institutions work in partnership with families to teach children protective strategies they can use wherever they access the internet [22:63].

Currently in the UK, Australia and the UK children are being criminalised by the very laws that were introduced to protect them. *Sexting in schools and colleges: responding to incidents and safeguarding young people,* produced in consultation with the National Police Chief's Council, UKCCIS [78], provides comprehensive guidance to schools and law enforcement agencies. It is important to remember that 'creating and sharing sexual photos and videos of under-18s is illegal and therefore causes the greatest complexity for schools and other agencies when responding' [78:5]. This guidance provides important information for schools and professionals in managing sexting incidents, to avoid the unnecessary criminalisation of young people yet ensure that children and young people are effectively safeguarded in cases of online grooming and producing and viewing of indecent sexual images. However, responding to online abuse is no longer confined to specialist interventions and Children's Workforce (CWP) professionals. All services to children need to be able to recognise and respond appropriately to children who have been harmed or abused online and to effectively support them and their families. Key to effectively responding to risk is identifying that risk early. Taking time to understand what children are doing online and what their experiences are is essential. As Corish [13] points out ‘one thing is certain however; intelligence, knowing is the key. How can a school or parent make the right decisions for the safety and well-being of the children in their care if they don’t know what is going on?’ Understanding the behaviours associated with those technologies allows a school to shape its education programme to build the necessary resilience in children and young people for them to cope with and enjoy life in an online world.

In the UK, service response to the online abuse of children is ad hoc and uncoordinated and lacks a nationwide strategy [62]. Speaking at the Child Internet Safety Summit in London in November 2016, Keith Niven, Head of Safeguarding at CEOP, emphasised the importance of four key elements in effectively responding to risk: parental engagement, children’s awareness of parental engagement, educational awareness and reporting mechanisms. Much of the advice on interventions to prevent and reduce the online risk concern educating children and their parents in online safety and technological tools, such as blocking or reporting bullying behaviour [15]. While there has been considerable investment in designing and developing technological tools for blocking and reporting abusive behaviour and filters are being applied to prevent children accessing sexual or violent content online, the safeguarding challenges associated with smart environments will only continue to multiply with the diversification of technological applications. This volume considers the emerging trends, developments and challenges associated with the design and applications of future technologies and smart environments. We would argue that there is sufficient evidence to suggest that as we move rapidly towards our ‘smart futures’ with the Internet of things, it is worth exercising some caution in that these are our children’s futures too – smart they may be, but how can we ensure that they are safe futures too?

References

1. Ainsaar, M., & Lööf, L (Eds.) (2011). *Online behaviour related to child sexual abuse. Literature report: Robert*. Available from http://www.childcentre.info/robert/public/Online_behaviour_related_to_sexual_abuse.pdf. Accessed on 05 Sept 2016.
2. Anti-Bullying Alliance (online). *Cyberbullying and children and young people with SEN and disabilities: the views of young people – SEN and disability: developing effective anti-bullying practice*. Available from http://www.anti-bullyingalliance.org.uk/media/7443/disabled-young-peoples-views-on-cyberbullying-report.pdf
3. Beck, U. (1992). *Risk society*. London: Sage.
4. Blum-Ross, A. (2015). What foster and adoptive parents need to know about digital media – Part 2: The risks. *Parenting for a digital future* (29 Apr 2015). Blog Entry.
5. Bond, E. (2010). Managing mobile relationships: Children’s perceptions of the impact of the mobile phone on relationships in their everyday lives. *Childhood, 17*(4), 514–529.
6. Bond, E. (2010). The mobile phone = bike shed? Children, sex and mobile phones. *New Media & Society, 13*(4), 587–604.
7. Bond, E. (2013). Mobile phones, risk and responsibility: Understanding children’s perceptions. *CyberPsychology: Journal of Psychosocial Research on Cyberspace, 7*(1), article 3. Available from http://www.cyberpsychology.eu/view.php?cisloclanku=2013011602. Accessed on 19 Sept 2016.

8. Bond, E. (2014). *Childhood, mobile technologies and everyday experiences*. Basingstoke: Palgrave.
9. Brå. (2007). *The online sexual solicitation of children by adults in Sweden. English summary of Brå report 2007:11*. Stockholm: Swedish National Council for Crime Prevention. Available from https://www.bra.se/download/18.cba82f7130f475a2f1800024473/1371914733983/2007_11_online_sexual_solicitation_of_children.pdf
10. Byron, T. (2008). *Safer children in a digital world: The report of the Byron review*. Available from http://webarchive.nationalarchives.gov.uk/20101021152907/http://publications.education.gov.uk/eorderingdownload/dcsf-00334-2008.pdf
11. Carrick-Davies, S. (2012). *Munch, poke, ping: Vulnerable young people, social media and e-safety*. Available from http://www.carrick-davies.com/downloads/Munch_Poke_Ping_-_E--safety_and_Vulnerable_Young_People_-_Executive_Summary_-_updated_Nov_2012.pdf. Accessed 06 Sept 2016.
12. College of Policing. (2016). *Briefing note: Police action in response to youth produced sexual imagery ('Sexting') Version 1.0, Nov 2016*. Available from: http://www.college.police.uk/News/College-news/Documents/Police_action_in_response_to_sexting_-_briefing_%28003%29.pdf
13. Corish, K. (2014). *Making the connection: Engaging parents with online safety*. Available from http://www.kencorish.info/making-the-connection-engaging-parents-with-online-safety/. Accessed 11 Sept 2016.
14. Couvillon, M. A., & Ilieva, V. (2011). Recommended practices: A review of schoolwide preventative programs and strategies on cyberbullying. *Preventing School Failure. Alternative Education for Children and Youth, 55*(2), 96–101.
15. Cowie, H. (2011). Coping with the emotional impact of bullying and cyberbullying: How research can inform practice. *International Journal of Emotional Education, 3*(2). 50–56.
16. Cranmer, S. (2012). Listening to excluded young people's experiences of e-safety and risk. *Learning, Media and Technology, 38*(1), 72–85.
17. Criminal Justice Joint Inspection. (2013). *Examining multi-agency responses to children and young people who sexually offend*. Available from http://www.justiceinspectorates.gov.uk/prisons/wp-content/uploads/sites/4/2014/04/Children_and_Young_People_who_Sexually_Offend_Thematic_Report_310113_Final.pdf
18. Cross, E. J., Richardson, B., Douglas, T., & Vonkaenel-Flatt, J. (2009). *Virtual violence: Protecting children from cyberbullying*. London: Beatbullying.
19. Davidson, J., & Martellozzo, E. (2008). Protecting children in cyberspace. In G. Letherby, P. Birch, M. Cain, & K. Williams (Eds.), *Sex as crime*. Cullompton: Willan.
20. Davidson, J., Grove-Hills, J., Bifulco, A., Gottschalk, P., Caretti, V., Pham, T., & Webster, S. (2011). *Online abuse: Literature review and policy context*. European online grooming project. Available from http://childcentre.info/robert/extensions/robert/doc/99f4c1bbb0876c9838d493b8c406a121.pdf. Accessed on 15 Sept 2016.
21. Day, L. (2016). *Resilience for the digital world: Research into children and young people's social and emotional wellbeing online*. Young Minds and ECORYS. Available from http://www.youngminds.org.uk/assets/0002/5852/Resilience_for_the_Digital_World.pdf. Accessed on 18 Sept 2016.
22. Ey, L.-A., & Cupit, C. G. (2011). Exploring young children's understanding of risks associated with internet usage and their concepts of management strategies. *Journal of Early Childhood Research, 9*(1), 53–65.
23. Ey, L.-A., Taddeo, C., & Spears, B. (2015). Cyberbullying and primary-school aged children: The psychological literature and the challenge for sociology. *Societies, 5*(2), 492–514.
24. Fostering Network. (2011). *The heart of the matter: Supporting family contact for fostered children – Handbook on contact*. London: Fostering Network.
25. Giddens, A. (1991). *Modernity and self identity*. Cambridge: Polity Press.
26. Good, B., & Fang, L. (2015). Promoting smart and safe Internet use among children with neurodevelopmental disorders and their parents. *Clinical Social Work Journal, 43*(2), 179–188.

27. Göran Svedin, C. (2011). Research evidence into behavioural patterns which lead to becoming a victim of sexual abuse In M. Ainsaar, & L. Lööf. (Eds.). *Online behaviour related to child sexual abuse. Literature report: Robert*. 37–49. Available from http://www.childcentre.info/robert/public/Online_behaviour_related_to_sexual_abuse.pdf. Accessed on 05 Sept 2016.
28. HM Government. (2015). *Working together to safeguard children: A guide to inter-agency working to safeguard and promote the welfare of children*. Available from https://www.gov.uk/government/uploads/system/uploads/attachment_data/file/419595/Working_Together_to_Safeguard_Children.pdf. Accessed on 15 Sept 2016.
29. HM Government. (2016). *Keeping children safe in education: Statutory guidance for schools and colleges*. Available from https://www.gov.uk/government/uploads/system/uploads/attachment_data/file/550511/Keeping_children_safe_in_education.pdf. Accessed on 26 Sept 2016.
30. HMIC. (2015). *Online and on the edge: Real risks in a virtual world*. Available from http://www.justiceinspectorates.gov.uk/hmic/wp-content/uploads/online-and-on-the-edge.pdf. Accessed 26 Aug 2016.
31. Holloway, S., & Valentine, G. (2002). *Cyberkids: Youth identities and communities in an on-line world*. London: RoutledgeFalmer.
32. Hope, A. (2015). Schoolchildren, governmentality and national e-safety policy discourse. *Discourse: Studies in the Cultural Politics of Education, 36*(3), 343–353.
33. Horvath, M. A. H., Alys, L., Massey, K., Pina, A., Scally, M., & Adler, J. R. (2013). *"basically... Porn is everywhere": A rapid evidence assessment on the effect that access and exposure to pornography has on children and young people*. London: University of Middlesex and the Office of the Children's Commissioner. Available from https://www.childrenscommissioner.gov.uk/sites/default/files/publications/Basically_porn_is_everywhere.pdf.
34. Internet Watch Foundation. (2016). Annual report 2015. Cambridge. Available from https://www.iwf.org.uk/assets/media/annual-reports/IWF%202015%20Annual%20Report%20Final%20for%20web.pdf
35. James, A., Jenks, C., & Prout, A. (1998). *Theorizing childhood*. Cambridge: Polity Press.
36. Lam, C. W. C., & Frydenberg, E. (2012). Coping in the cyberworld: Program implementation and evaluation – A pilot project. *Australian Journal of Guidance and Counselling, 19*(2), 196–215.
37. Lilley, C., & Ball, R. (2013). Younger children and social networking sites: A blind spot. London: NSPCC. Available online from www.nspcc.org.uk/blindspot
38. Livingstone, S., & Haddon, L. (2009a). *EU kids online: Final report*. London: LSE. EU Kids Online. Available from http://www.lse.ac.uk/media@lse/research/EUKidsOnline/EU%20Kids%20I%20(2006-9)/EU%20Kids%20Online%20I%20Reports/EUKidsOnlineFinalReport.pdf. Accessed on 18 Sept 2016.
39. Livingstone, S., & Haddon, L. (2009b). Introduction. In S. Livingstone & L. Haddon (Eds.), *Kids online: Opportunities and risks for children* (pp. 1–6). Bristol: Policy Press.
40. Livingstone, S., & Palmer, T. (2012). *Identifying vulnerable children online and what strategies can help them*. London: UK Safer Internet Centre. Available from http://eprints.lse.ac.uk/44222/. Accessed on 02 Sept 2016.
41. Livingstone, S., Carr, J. and Byrne, J. (2015). *One in three: Internet governance and children's rights. Global commission of internet governance*. Available from https://www.cigionline.org/sites/default/files/no22_2.pdf. Accessed on 18 Sept 2016.
42. Livingstone, S., Davidson, J., Bryce, J., Millwood Hargrave, A., & Grove-Hills, J. (2012). *Children's online activities risks and safety: The UK evidence base*. London: UKCCIS. Available from http://www.saferinternet.org.uk/content/childnet/safterinternetcentre/downloads/Research_Highlights/UKCCIS_Report_2012.pdf
43. Livingstone, S., Haddon, L., Görzig, A., and Ólafsson, K. (2011). Risks and safety on the internet: The perspective of European children.Full findings and policy implications from the EU kids online survey of 9–16 year olds and their parents in 25 countries. London: LSE. EU Kids Online. Available from http://eprints.lse.ac.uk/33731/1/Risks%20and%20safety%20on%20the%20internet%28lsero%29.pdf

44. Livingstone, S., Haddon, L., Vincent, J., Mascheroni, G., & Ólafsson, K. (2014b). *Net children go mobile: The UK report*. London: London School of Economics and Political Science. Available from https://www.lse.ac.uk/media@lse/research/EUKidsOnline/EU%20Kids%20III/Reports/NCGMUKReportfinal.pdf
45. Livingstone, S., Marsh, J., Plowman, L., Ottovordemgentschenfelde, S., & Fletcher-Watson, B. (2014a). *Young children (0–8) and digital technology: A qualitative exploratory study – National report – UK*. Luxembourg: Joint Research Centre, European Commission.
46. Lööf, L. (2011). Introduction. In M. Ainsaar & L. Lööf (Eds.), *Online behaviour related to child sexual abuse literature report, Robert* (pp. 7–9.) Available from http://www.childcentre.info/robert/public/Online_behaviour_related_to_sexual_abuse.pdf. Accessed on 05 Sept 2016.
47. May-Chahal, C., Mason, C., Rashid, A., Walkerdine, J., Rayson, P., & Greenwood, P. (2012). Safeguarding cyborg childhoods: Incorporating the on/offline behaviour of children into everyday social work practices. *British Journal of Social Work, 44*, 596–614.
48. Millwood Hargrave, A., & Livingstone, S. (2009). *Harm and offence in media content: a review of the evidence*. Available from http://eprints.lse.ac.uk/49000/1/Livingstone_Harm_offence_media_2009.pdf. Accessed on 14 Sept 2016.
49. Mitchell, K. J., Finkelhor, D., & Wolak, J. (2007). Online requests for sexual pictures from youth: Risk factors and incident characteristics. *Journal of Adolescent Health, 41*(2), 196–203.
50. Monks, C. P., Robinson, S., & Worlidge, P. (2012). The emergence of cyberbullying: A survey of primary school pupils' perceptions and experiences. *School Psychology International, 33*(5), 477–491.
51. Munro, E. (2011). *The protection of children online: A brief scoping review to identify vulnerable groups*. Childhood Wellbeing Research Centre. Available from https://www.gov.uk/government/uploads/system/uploads/attachment_data/file/181476/CWRC-00085-2011.pdf. Accessed 06 Sept 2016.
52. NatCen. (2014). *Tackling homophobic, biphobic and transphobic bullying among school-age children and young people. Findings from a mixed methods study of teachers, other providers and pupils*. Available from http://natcen.ac.uk/media/563016/natcen-social-research-hbt-bullying-findings.pdf. Accessed on 16 Sept 2016.
53. NFER. (2010). *Children's online risks and safety a review of the available evidence*. Available from https://www.nfer.ac.uk/publications/COJ01/COJ01.pdf
54. Niven, K. (2016). *What are the police doing to help parents and children?* Presented at the 4th Annual Child Internet Safety Summit, 29th November 2016, London.
55. Noll, J. G., Shenk, C. E., Barnes, J. E., & Putnam, F. W. (2009). Childhood abuse, avatar choices, and other risk factors associated with internet-initiated victimization of adolescent girls. *Pediatrics, 123*(6), 1078–1083.
56. NSPCC. (2015). Always there when I need you. Available from https://www.nspcc.org.uk/globalassets/documents/annual-reports/childline-annual-review-always-there-2014-2015.pdf
57. O'Neill, B., & Dinh, T. (2015). Mobile technologies and the incidence of cyberbullying in seven European countries: Findings from net children go mobile. *Societies, 5*(2), 384–398.
58. O'Neill, B., & McLaughlin, S. (2011). *Report D7.1 recommendations on safety initiatives*. EU London: Kidsonline, LSE. Available from http://www.lse.ac.uk/media@lse/research/EUKidsOnline/EU%20Kids%20II%20(2009-11)/EUKidsOnlineIIReports/D71Policy_Implications.pdf
59. Ofcom. (2014). *Children and parents: Media use and attitudes report*. London: Ofcom. Available from http://stakeholders.ofcom.org.uk/binaries/research/media-literacy/media-use-attitudes-14/Childrens_2014_Report.pdf
60. Ofcom. (2015). *Half of UK homes turn to tablets – in just five years*. Available from http://media.ofcom.org.uk/news/2015/five-years-of-tablets/
61. Ofcom. (2016). *Children's media Lives – Year 2 findings*. Available from http://stakeholders.ofcom.org.uk/binaries/research/media-literacy/childrens-media-lives-year-2/children_media_lives_year2.pdf

62. Palmer, T. (2015). *Digital dangers: The impact of technology on the sexual abuse and exploitation of children and young people.* Barnardo's and the Marie Collins Foundation. Available online from http://www.barnardos.org.uk/onlineshop/pdf/digital_dangers_report.pdf. Accessed 24 Aug 2016.
63. Papatraianou, L. H., Levine, D., & West, D. (2014). Resilience in the face of cyberbullying: An ecological perspective on young people's experiences of online adversity. *Pastoral Care in Education, 32*(4), 264–283.
64. Phippen, A. (2012). *Sexting: An exploration of practices, attitudes and influences.* London: NSPCC. Available from https://www.nspcc.org.uk/globalassets/documents/research-reports/sexting-exploration-practices-attitudes-influences-report-2012.pdf. Accessed 12 Sept 2016.
65. Phippen, A. (2016). *Children's online behaviour and safety – Policy and rights challenges: A case study of child online safety in the UK 2010–2015.* Basingstoke: Palgrave.
66. Pinar, V., Yesim, U., & Vinnur, E. K. (2015). Relationship between symptoms of disruptive behavior disorders and unsafe internet usage in early adolescence. *Archives of Neuropsychiatry, 52*(3), 240–246.
67. Prensky, M. (2001). Digital natives, digital immigrants on the horizon (MCB University Press, 9: 5, 1–6 Available online from http://www.marcprensky.com/writing/Prensky%20-%20Digital%20Natives,%20Digital%20Immigrants%20-%20Part1.pdf
68. PSHE Association. (2016). *Key principles of effective prevention education.* Available online from https://www.pshe-association.org.uk/sites/default/files/PSHE%20Association%20report%20to%20CEOP%20April%202016%20FINAL.pdf
69. Schilder, J. D., Brusselaers, M. B. J., & Bogaerts, S. (2016). The effectiveness of an intervention to promote awareness and reduce online risk behavior in early adolescence. *Journal of Youth and Adolescence, 45*(2), 286–300.
70. Schultze-Krumbholz, A., Schultze, M., Zagorscak, P., Wölfer, R., & Scheithauer, H. (2015). Feeling cybervictims' pain – The effect of empathy training on cyberbullying. *Aggressive Behaviour, 42*(2), 147–156.
71. Simpson, J. E. (2015). Technology and the fostered child: A new social work response. *Sociological Studies of Children and Youth, 19*, 351–367.
72. Simpson, J. E. (2013). Managing unregulated contact in the age of new technology: Possible solutions. *Adoption & Fostering, 37*(4), 380–388.
73. Stanley, T., & Guru, S. (2015). Childhood radicalisation risk: An emerging practice issue. *Practice, 27*(5), 353–366.
74. STIR. (2015). *Safeguarding teenage intimate relationships (STIR) connecting online and offline contexts and risks.* Briefing report Available from http://stiritup.eu/wp-content/uploads/2015/06/STIR-Briefing-Paper-1-English-final-.pdf
75. Stonewall. (2012). *The school report: The experiences of gay young people in Britain's schools in 2012.* Stonewall. Available from https://www.stonewall.org.uk/sites/default/files/The_School_Report__2012_.pdf. Accessed 06 Sept 2012.
76. Third, A., Bellerose, D., Dawkins, U., Keltie, E., & Pihl, K. (2014). *Children's rights in the digital age: A download from children around the world.* Melbourne: Young and Well Cooperative Research Centre. Available from http://www.unicef.org/publications/files/Childrens_Rights_in_the_Digital_Age_A_Download_from_Children_Around_the_World_FINAL.pdf
77. UKCCIS. (2015). *Child safety online: A practical guide for providers of social media and interactive services.* Available from https://www.gov.uk/government/uploads/system/uploads/attachment_data/file/517335/UKCCIS_Child_Safety_Online-Mar2016.pdf. Accessed on 06 Sept 2016.
78. UKCISS. (2016). *Sexting in schools and colleges: Responding to incidents and safeguarding young people.* Available online from https://www.gov.uk/government/uploads/system/uploads/attachment_data/file/551575/6.2439_KG_NCA_Sexting_in_Schools_WEB__1_.PDF
79. Vandebosch, H., & Van Cleemput, K. (2009). Cyberbullying among youngsters: Profiles of bullies and victims. *New Media and Society, 11*(8), 1349–1371.

80. Vandebosch, H. (2014). Schools and cyberbullying: Problem perception, current actions and needs. *International Journal of Cyber Society and Education, 7*(1), 29–47.
81. Vandoninck, S., d'Haenens, L., & Roe, K. (2013). Online risks: Coping strategies of less resilient children and teenagers across Europe. *Journal of Children and Media, 7*(1), 60–78.
82. Vandoninck, S., d'Haenens, L., & Segers, K. (2012). Coping and resilience: Children's responses to online risks. In S. Livingstone, L. Haddon, & A. Görzig (Eds.), *Children, risk and safety on the internet: Research and policy challenges in comparative perspective* (pp. 203–216). Bristol: The Policy Press.
83. Watkin, A., Lindqvist, S., Black, J., & Watts, F. (2009). Report on the implementation and evaluation of an interprofessional learning programme for inter-agency child protection teams. *Child Abuse Review, 18*, 151–167.
84. Wespieser, K. (2015). *Young people and e-safety: The results of the 2015 London grid for learning e-safety survey*. Slough: NFER. Available from https://www.nfer.ac.uk/publications/LLGR01/LLGR01.pdf
85. Whittle, H. C., Hamilton-Giachritsis, C. E., & Beech, A. R. (2014). In their own words: Young peoples' vulnerabilities to being groomed and sexually abused online. *Psychology, 5*, 1185–1196.
86. Whittle, H. C., Hamilton-Giachritsis, C. E., Beech, A. R., & Collings, G. (2013). A review of young people's vulnerabilities to online grooming. *Aggression and Violent Behavior, 18*(1), 135–146.
87. Przybylski, A. K., Mishkin, A., Shotbolt, V., Linington, S. (2014). A shared responsibility building children's online resilience: An independent research paper commissioned and supported by Virgin Media and The Parent Zone. Available from https://parentzone.org.uk/sites/default/files/VM%20Resilience%20Report.pdf. Accessed on 18 Sep 2016.
88. Ainsaar, M. and Soo, K. (2011). Regional, thematic and methodological coverage of online activities related to child sexual violence literature – in Ainsaar, M. and Lööf, L. (Eds.) *Online behaviour related to child sexual abuse* (pp. 19–34). Literature report: Robert. Available from http://www.childcentre.info/robert/public/Online_behaviour_related_to_sexual_abuse.pdf. Accessed on 05 Sep 2016.
89. Mitchell, K. J., Finkelhor, D., & Wolak, J. (2011). Risk factors and impact of online sexual solicitation of youth. *Journal of the American Medical Association, 285*(23), 1–4.

Chapter 13
Evaluation of the Dynamic Cybersecurity Risk Using the Entropy Weight Method

T. Hamid, D. Al-Jumeily, and J. Mustafina

13.1 Introduction

According to current UK government research, two thirds of large UK businesses have been hit by a cyberattack, and 87% of the small companies and 93% of large houses experienced a cybersecurity breach in 2015. Eighty-one percent of the companies said even the senior management placed high priority on information security [1], they were mostly unable to translate this to a clear, effective cyber difference because the experts and the tools used in security system assessment only utilised certain values driven from statistics and probability ignoring enormous uncertainty values. Many significant research and developments have been carried out in the risk assessment theory domain; for example, Hicks et al. [2] formulate the risk assessment calculations as follows:

$$\Re = \text{Prob}(\text{Attack})^{*}\left(1-\text{Prob}\left(S_{\text{d}}\right)\right)^{*} C \qquad (13.1)$$

where Prob(Attack) is probability of malicious attack, $P(S_{\text{d}})$ is probability of security system defence in terms of preventing the attacker from compromising a vulnerability to reach the target and C is the impact of the attack, which R represents the risk.

T. Hamid (✉)
Cardiff Metropolitan University, Cardiff, UK
e-mail: thaier.hamid@outlook.com

D. Al-Jumeily
Liverpool John Moores University, Liverpool, UK
e-mail: d.aljumeily@ljmu.ac.uk

J. Mustafina
Kazan Federal University, Kazan, Russia
e-mail: jamila0111@hotmail.com

M. Dastbaz et al. (eds.), *Technology for Smart Futures*,
DOI 10.1007/978-3-319-60137-3_13

The security system defence effectiveness Prob(S_d) represents the probability of disturbance times. The majority of researchers use statistics and probabilities when they are calculating the risk. There are massive uncertainties in any calculations based on statistics and probabilities. For example, threats which are mainly based on probability are the most uncertain. Many researchers have used game theory, bounded intervals [3], to represent uncertainty in the risk assessment analysis. The main goal of this work is to introduce the information entropy theory of Shannon and use information entropy and entropy weight in security protections and assessments. We have considered the security systems as a collection of security entities (SEs) equipped with multiple types of protection; as security entities are having many uncertainties, risk entropy represents the middling step of uncertainties of security risk model. The system protection layers used in this work are applying numerical values for each layer driven from different statistics and the experts' references. Matlab was used to implement our algorithm and calculate the average entropy of different protection levels' values which represent the average level of uncertainties, and entropy weight used as an object index to represent the cost of protections.

This chapter has been structured as follows:

- Section 13.2 discusses the risk management and significant quantitative sciences by introducing the DVSS and the security protection levels.
- Section 13.3 covers the probability of security entities for all protection levels, the probability distributions of established protection on security entity and the risk entropy of security entity.
- Section 13.4 covers the entropy weight of the protection index, which represents the weight of the object index, points to the significant components of the index, explains the details of our algorithms to normalise the protective matrix and calculates the entropy and the entropy weight.
- Section 13.5 deals with cyber protection example, entropy weight example and the weight and paths to evaluate and calculate the total risk in the system and gives the system administrators a clear guidance on the vulnerable security entities, with a conclusion section to complete the chapter.

13.2 Information Entropy and Protection Effectiveness

American researcher Shannon in 1948 used information entropy to calculate the middling uncertainty of stream of data coming from a specific source [4]. As the information entropy represents the uncertain portion of probability dissemination, it is logical to say it is driven from certain probability dissemination $V = \{v_x\}$ of an event x with a set of variables V with n results:

V_x; where $\{x = 1, 2, 3 \ldots \mathrm{n}\}$ is the Shannon entropy method denoted by $H(v)$ and

Prob(vx) is the probability mass function of outcome Vx, in the sources of risk event, developed from "what can go wrong" scenarios to establish levels of risk weight. In our methodologies, we classified the protection levels of security entity to n stages giving to the value of protections O_x ($x = 1, 2, 3 \ldots .n$), and then we can represent the total protection as:

$$O_x = \{o_1,,,o_2,,,o_3,,,\ldots o_x\} \tag{13.2}$$

where [$x = 1, 2, 3, \ldots n$].

The risk is limited to be a mapping of the probability (likelihood) and the severity (impact) of the probable breaches on a scheme. The risk in IT systems could be exposed from the Internet, network, servers and local host.

A prototype uses the Dynamic Vulnerability Scoring System (DVSS) metric groups: intrinsic, time-based and ecological metrics developed to measure the risk rate and safeguard a system. CVSS v2 provides an individual static scoring for vulnerabilities; this strategy could make sense in the process of measuring an individual vulnerability score. Nevertheless, it is important to acknowledge that two or more apparently inoffensive vulnerabilities could be joined to create a critical impact in a multistep attack environment. The fixed scores are a problem, as the scores combining metrics do not take into account the changes in the parameters, when the attacker moves from one stage to another and the network architectures and configurations are affected, as these parameters will lead to ineffective scoring values of the vulnerabilities and will influence the accuracy of security decisions and polices. To solve this problem, the vulnerability scanners should combine with network resources and configurations; these resources should modify the parameters used to calculate the dynamic impact scores (DVSS). The method as shown in Fig. 13.1 can be summarised as follows:

1. Convert the Nessus scanning output to a spreadsheet XSL format (Nessus.xsl).
2. Create network architectures and configurations file Nconfig.xsl.
3. Establish the rules sheet (rules.xsl) which represents the influence of each network device on the reachability and other sub-scoring parameters used by CVSS v2.
4. Combine Nconfig.xsl with rules.xsl to produce a single network specification file (Network.xsl) which contains the necessary information of network architectures and configurations to be used in the DVSS calculation.

An open-source program is developed to convert Nessus scanning reports into Nessus. "Xsl" is used to be administered against network architectures and configurations, while Nconfig.xsl and rules.xsl are used to produce a single network specification file – Network.xsl. The DVSS takes the modified data from network.xsl to calculate a dynamic new score for each vulnerability and save it as Dynamic.xsl.

Different organisations are dealing with security impact severity in different manners; they are using own formulae and reporting mechanism. To calculate the intrinsic correction factor(Ω, a corrections.xsl has been used, which contains the vulnerabilities and the corresponding scoring from above organisations, to consider the impact factors on the scoring that other organisations believe are being ignored in CVSS base calculations to make the empirical dynamic impact scoring more accurate. The intrinsic metric score, which represents a threat, could change over

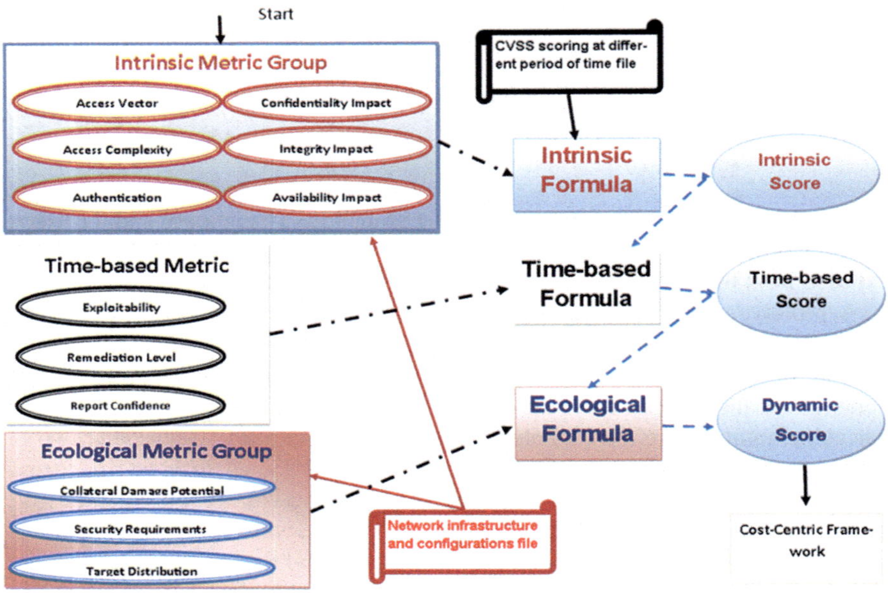

Fig. 13.1 The DVSS framework combines intrinsic, time-based and ecological metric groups

time according to the developments in exploiting methods, availability of mitigations and countermeasures for the vulnerability, exploit scripts and other information. The ecological metric score (ES) could modify the metric score of vulnerabilities, according to CDP, TD and security requirements. The primary ground for using the quantitative metric method is to prepare a unique cost-centric value for each vulnerability; the environmental score will represent the cost value for a vulnerability. The DVSS framework is built, using modified equations to simplify the calculations of the vulnerability scores and to benchmark against other models. The DVSS scores range from 0 to 100 with the higher score meaning worse security.

The following formulae will be used to calculate the operational level cost (OLC):

$$\text{Probability}_{\text{v}} = 1 - (\prod_{k=1}^{n} (1 - P_k)$$

$$OLC_V = \max(C) - \left(\left(\sum_{k=1}^{n-1} p_k{}^{*} C_k \right) / \left(\sum_{k=1}^{n} p_k / 100 \right) \right)$$

By estimating the effective costs using attack tree (AT) structure, the probability and the dynamic cost of the vulnerabilities are distributed across the nodes. For the purpose of framework analysis, a prototype network setup is being used, with two computers (this approach is applicable for much more complex enterprise level

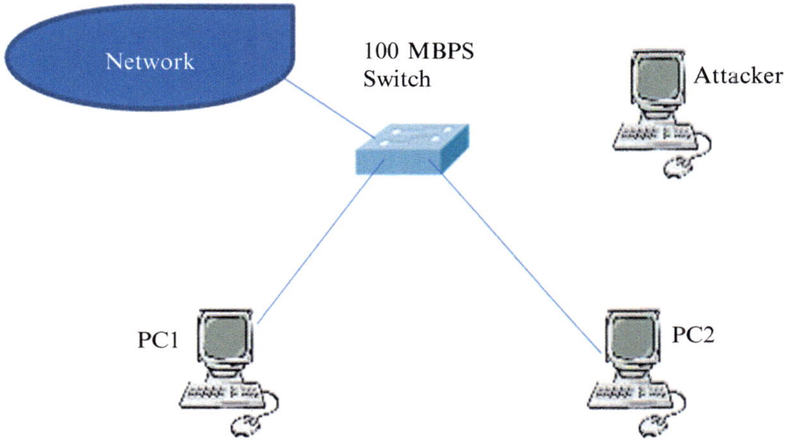

Fig. 13.2 Using Nessus scanner as a tool to discover known vulnerabilities on PC1 and PC2

network architectures as you will see in Sect. 13.5). PC1 (Windows XP platform with IP 192.168.153. 160) and PC2 (Linux platform with IP 192.168.153. 31) are connected with a 100 MBPS Ethernet switch; the Nessus 4.x scanner installed on PC1 and program has been used to convert the Nessus scanning output to a spreadsheet XSL format (Nessus.xsl) and network architectures and configurations file "Nconfig.xsl" and the rules sheet (rules.xsl) which represents the rules in binary to represent the influence of Ethernet switch configuration on the reachability and other sub-scoring parameters used by CVSS v2. Instead of specifying a state by network attributes, a dynamic cost-centric model is proposed; each state in the graph is stated using the features/vulnerabilities of an individual host. The dynamic cost-centric attack graph is a reference of this model by using the DVSS cost calculator and attacker privileges. The methodologies and all of the concepts mentioned are used to develop dynamic quantitative risk evaluation metrics and the existing results being used to build the dynamic cost-centric attack graphs [5]. Nessus 4.x scanner is being used as a tool to discover known vulnerabilities on PC1 and PC2 on the prototype network shown in Fig. 13.2. The Dynamic.xsl file is analysed adding the correction factor to the intrinsic score as the different security organisations are given the same vulnerability, different levels of severity [6]. The intrinsic correction factors are used to evaluate the intrinsic scoring; the corrections.xsl file has been used with a mathematical formula to calculate the correction factor, which might have negative values to adjust the intrinsic scoring.

In this model, attack trees (ATs) are represented using cost centric driven from DVSS as a new approach to building effective AT. Exploitation of more than one vulnerability in one asset is possible in terms of 'AND' and 'OR' refinement associations between vulnerabilities in different time slots. For calculating the effective cost, according to operational levels, only one vulnerability from an operational level group could be exploited at a given time. Based on this fact, the relationships between vulnerabilities in terms of calculating the effective cost for each operational

Table 13.1 Operational levels of PC1 host

	Vn1	Vn2	Vn5	Vn10	OLC	Value
Probability	0.31	0.25	0.20	0.24		
None privileges	57	47	38	44	CPC1,0	56.31
	Vn4	Vn6	Vn9	Vn11		
Probability	0.23	0.31	0.26	0.19		
User privileges	52	71	60	44	CPC1,1	70.47
	Vn3	Vn7	Vn8			
Probability	0.33	0.33	0.33			
Root/administrator privilege	89	90	87		CPC1,2	89.12

Table 13.2 Operational levels of PC2 host

	Vn1	Vn2	Vn6	Vn7	Vn11	Vn12		OLC	Value
Probability	0.16	0.21	0.23	0.18	0.17	0.06			
None privileges	50	64	69	54	51	18		CPC2,0	68.58
	Vn8	Vn10	Vn13	Vn14	Vn15	Vn16	Vn17		
Probability	0.13	0.21	0.15	0.18	0.18	0.10	0.06		
User privileges	48	77	54	65	66	38	23	CPC2,1	76.47
	Vn3	Vn4	Vn5	Vn9					
Probability	0.26	0.26	0.25	0.23					
Root/administrator privilege	92	91	90	82				CPC2,2	91.12

level will use only OR refinement to adjust the operational costs to be near to the higher cost vulnerability in the group. To calculate the OLC CPC 1,0, which means the average severity of PC1 vulnerabilities with no privileges, as shown in Table 13.1 for PC1 and Table 13.2 for PC2.

Both statistics and experts' feedback and opinions include high level of uncertainly.

We will use the information entropy and entropy weight in this book chapter to reflect the uncertain part of the risk assessments based on different vulnerabilities existing in Host-1 and Host-2.

13.3 Probability of Security Entities for All Protection Levels

Security entities' abilities of protection for all protection layers are represented as follows:

Prob(Ox) ($x = 1, 2, 3\ldots n$) refer to the probability of protection objects; for all protection layers we used Prob(x) where x is a specific type of protection:

$$\text{Prob}(x) = \left[\text{Prob}(o_1),,,\text{Prob}(o_2),,,\text{Prob}(o_3),,,,\ldots\text{Prob}(o_x)\right] \tag{13.3}$$

where Prob(o_1) + Prob(o_2) + Prob(o_3) + ….. + Prob(o_x) = 1

The probability distributions of established protection on security entity can be represented as follows:

$$\left[O,\text{Prob}\right]=\begin{pmatrix} o_1 & o_2 & o_x \\ \text{prob}_1 & \text{prob}_2 & \text{prob}_x \end{pmatrix}$$

where $x = (1, 2, 3 \ldots n)$.

$$\sum_{x=1}^{n}\text{Prob}_x = 1 \tag{13.4}$$

The value of a specific protection layer x that has a certain probability $\text{Prob}_x = \text{P}_x$ can be formulated as follows:

$$V\left(P_x\right)=C_x \ln\frac{1}{1-P_x} \tag{13.5}$$

where C_x represents the expert evaluation value related to certain protection. The risk entropy of security entity is as follows:

$$H\left(T\right)=\sum_{x-1}^{n}C_x P_x \ln\frac{1}{((1-P_x)} \tag{13.6}$$

where T refers to a specific type of vulnerability.

As we have explained, there are three layers of vulnerabilities for each security protection object. These layers might interact with each other in certain ways, and the final protection is represented in the form of interaction between the layers.

A, B and C are used to represent three different layers of vulnerabilities.

To calculate the uncertain risk entropy of layers A and B, we will use H(A/B) to represent the risk entropy of existing protections A and B.

If the protection layer A interacts with protection layer B, then we can calculate the effective entropy using the following formula:

$$H\left(\text{A}/\text{B}\right)=\sum_{x=1}^{n}P_x\sum_{j=1}^{m}C_j P_{xj} \ln\frac{1}{1-P_{xj}} \tag{13.7}$$

If A, B and C exist as protection layers but they are not related to each other, then the effective entropy of the three protection layers in terms of protection A will be H(A) as shown in Eq. (13.8):

$$H\left(\text{A}/\text{B}/\text{C}\right)=H\left(\text{A}\right) \tag{13.8}$$

In terms of risk assessment, the entropy weight of the protection index is very important and is usually used to measure the importance of the feature index. The greater the value of the entropy is, the smaller is the entropy weight, and the smaller are the different alternatives in this specific attribute, and the less information the specific attribute provides, the less important this attribute becomes in the decision-making process, as in our example small entropy refers to high protected states.

The entropy value can be calculated using the following formula:

$$W_x = \frac{1 - H(T)_x}{n - \sum_{x=1}^{n} H(T)_x}, \quad W_x \in [0,1], \quad \text{and} \quad \sum_{x=1}^{n} W_x = 1. \tag{13.9}$$

The effective protection weights should be calculated by subtracting the entropy from 1 and dividing the subtractions of the summations of entropies for each row from the number of rows n.

From the above formula, we calculated the weight for all SEs; as you can see, the summations for all entropy weight should be equal to 1.

We used entropy weight method because the weight of the object index represents the important components of the index. We assess the risk of security systems in terms of different security layers which represent our index as we will explain in detail later. In our case study, any security system contains many security entities (SE); Matlab has been used to implement our algorithm and calculate the entropy and the entropy weight. The entropy weight can be obtained by normalising the entropy risk matrix *R*:

$$R = \begin{vmatrix} r_{11} & r_{12} & . & . & r_{1m} \\ r_{21} & r_{22} & . & . & r_{2m} \\ . & . & . & . & . \\ . & . & . & . & . \\ r_{n1} & r_{n2} & . & . & r_{nm} \end{vmatrix}$$

1. Calculate the normalised process.

As each index in the evaluation index system has different dimensions, we need to carry on the normalised process:

$$A_{xy=}(r_{xy})_{mn}$$

$$r_{xy} = \frac{a_{xy} - \min_x \{a_{xy}\}}{\max\{a_{xy}\} - \min\{a_{xy}\}} \quad r_{xy} \in [0,1] \tag{13.10}$$

2. Calculate the entropy: if the index information is smaller, the index provides the bigger information content, so the index of small entropy has important value [7, 4].

$$H(T) = -k\sum_{i-1}^{n} C_i P_i \ln \frac{1}{(1-P_i)} \quad \text{where} \quad k = \frac{1}{\ln n}$$

$$P_{xy} = \frac{r'_{xy}}{\sum_{i=1}^{n} r'_{xy}}, \quad \text{if} \quad P_{xy} = 0 \text{ then } P_{xy} \ln \frac{1}{P_{xy}} = 0.$$

3. Calculate the index entropy weight.

$$W_i = \frac{1 - H(T)_i}{n - \sum_{i=1}^{n} H(T)_i}$$

13.4 Risk Entropy Assessment Model for Security Set

As shown in Fig. 13.3, the attack graph (AG) is using dynamic impact cost-centric metrics, with a total of seven states to represent the dynamic cost-centric attack states for PC1 and PC2 and all possible paths. The main goal is to reduce the visual complexity of an AG using a cost-centric methodology. The presentation of the full-scale scenario for the test pad indicates a dramatic reduction of the complexity of an AG. State zero (s0) represents the attacker's initial stage; an assumption has been made that the attacker can initiate the attack on his/her own PC and he has full privileges on it (ATT, 2). The other state represents the status of PC1 and PC2 according to the vulnerability type. Exploit SE1 means the attacker exploits a vulnerability in PC1 without gaining any privileges and without any authentications (PC 1, 0), while SE3 means the attacker exploits a vulnerability with a User's privilege or the attacker escalates the privileges as a User as a result of the attack. SE5 allows the attackers to gain Root/Administrator privileges because of exploiting a vulnerability.

To go through the entropy weight method to the same prototype, the security entities for the models are defined as follows:

SE1 represents the None privilege vulnerabilities existing in PC1 as shown in Table 13.1.

SE2 represents the User privilege vulnerabilities existing in PC1 as shown in Table 13.1.

SE3 represents the Root/Administrator privilege vulnerabilities existing in PC1 as shown in Table 13.1.

SE4 represents the None privilege vulnerabilities existing in PC2 as shown in Table 13.2.

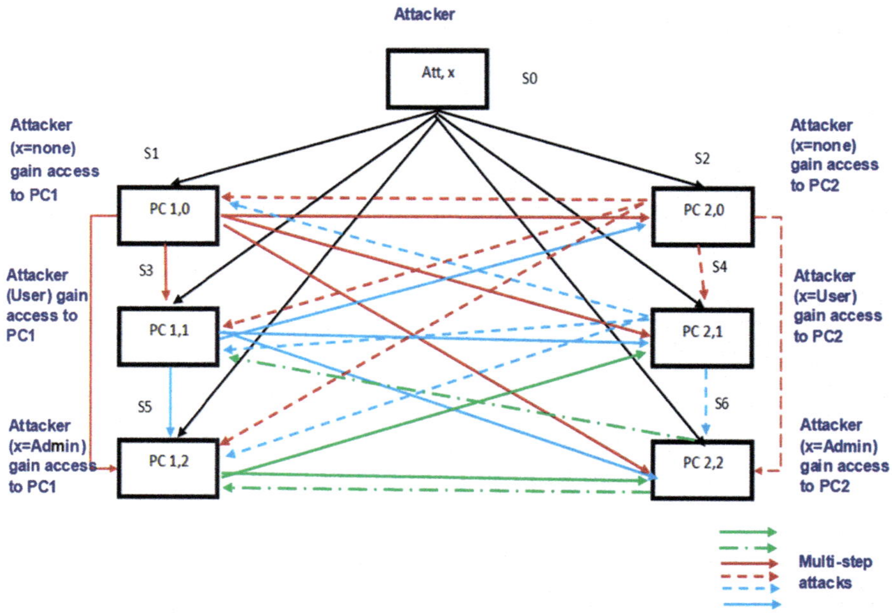

S0	S1	S2	S3	S4	S5	S6
Attacker	PC 1, 0	PC 2, 0	PC 1, 1	PC 2, 1	PC 1, 2	PC 2, 2
	56.31	68.58	70.47	76.47	89.12	91.12

Fig. 13.3 Dynamic cost-centric attack graph (a full-scale scenario)

SE5 represents the User privilege vulnerabilities existing in PC1 as shown in Table 13.1.

SE6 represents the Root/Administrator privilege vulnerabilities existing in PC2 as shown in Table 13.2.

As you can see in Table 13.1, some zero-day vulnerabilities have been added to justify our calculations.

The indirect cost in multistep attack will depend mainly on the destination state as shown in Fig. 13.3. For example, (SE3, SE2) cost = 68.58 as the attacker tries to exploit the vulnerability in PC2 with OLC 2, 0.

In the experiment using dynamic cost-centric AG, there are 10 vulnerabilities in PC1 and 17 vulnerabilities in PC2. The AG representation in Fig. 13.3 represents a full-scale scenario, taking into account all the possibilities of multistep attack.

We will use N_x where ($x = 1, 2, 3, \ldots n$) to refer to the different paths to exploit the SEs.

$$\text{Cost}\left(N_x\right) = W\left(N_x\right) = W\left(SE_1,,,SE_2,,,,\ldots,,,SE_n\right)$$

We presume that the route contains n entities, and we denote it as N_x to refer to the arbitrary entity in the path denoted as Ph (SE_1, SE_2,… SE_n).

And we use Cost(N_x) = Cost(SE_1, SE_2, …, SE_n) to represent the cost of the path.

To find the most vulnerable route, the attacker can use to exploit a specific vulnerability in the victim is the route with highest entropy weights (cost), we can calculate the effective cost in terms of entropy weights for all SEs in the route as shown in the following formula:

$$\begin{aligned}\text{Cost}\left(N_x\right) &= \wedge\left(Ph\left(SE_1\right),,Ph\left(SE_2\right),,\ldots Ph\left(SE_n\right)\right)\\ &= \wedge\left(W1 + W2 + \ldots + Wn\right)\\ &\text{where } \wedge = \min\end{aligned}$$

We could calculate the effective risk in terms of probability of attack and impact as follows:

$$\text{Risk} = P\left(Att\right)^{*} P\left(N_x\right)^{*} C$$

P(Att) is the probability of attack during the time of exploration, which represents the chance that the attacker can compromise a vulnerability in a critical SE and exploit it, which could be calculated using the following formula:

$$P\left(Att\right) = \sum_{x=1}^{n}\left(1 - P_x\right)/100$$

where P is the protection number for each security entity.

As the probability of attack logically is the invert of the protection, in other words, it is the exploitation of unprotected area or weakness. P(Att) represents the probability of successful exploitations of a vulnerability in critical SE in terms of malicious attack. The value of $P(N_x)$ is associated to the protection layers' efficiency, the higher value represents a weak protection and C represents the attack impact.

$$\text{Risk} = \sum_{i=1}^{n} P\left(Att\right)^{*} \frac{1}{e^{W(N_i)}}^{*} C$$

We will use the example network shown in Fig. 13.3 to implement our model of cybersecurity evaluation.

Using our Matlab setting, we constructed the risk evaluation matrix using the values attributed to the SEs as shown in Table 13.3 to formulate the Matrix R, and then we counted on the normalised process applying Eq. 13.10. Afterwards, Eq. 13.6 is utilised to calculate the entropy, and Eq. 13.9 is being applied to compute the index entropy weight.

Table 13.3 The risk weight of different vulnerabilities and entropy weight

	Protection					
	Vul-n1	Vuln-n2	Vuln-n3	Vuln-n4	Vuln-n5	Entropt weight
SE1	57	47	38	44	41	**0.0546**
SE2	52	71	60	44	33	**0.048**
SE3	89	90	87	100	89	**0.009**
SE4	50	64	69	54	51	**0.0424**
SE5	48	77	54	65	66	**0.038**
SE6	92	91	90	82	100	**0.009**

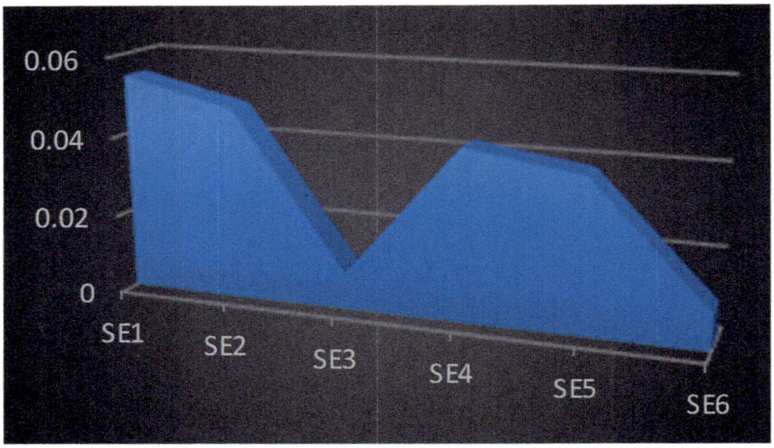

Fig. 13.4 The Matlab results for entropy weights for different SEs

The vulnerability values for each SEs have been used to calculate the entropy and entropy weight using the above presented algorithm. The value of the entropy weight shown in Table 13.3 and Fig. 13.4 reflects the protection classification facts of vulnerabilities, according to exploitations and the escalation of the privileges as explained in Sect. 13.2. The next step classified the security entities according to entropy weight into three main types:

1. Low protected entities when the entropy weight is ≥0.05, shown in white in Fig. 13.5
2. Medium protected entities when the entropy weight is >.03 and less than 0.04
3. High protected entities when the entropy weight is less than 0.04

We assume that the route contains *n* entities, we use Ni to refer to the arbitrary entity in the path denoted as PH (SE1, SE2,…SE*n*) and we use cost (SE1, SE2,… SE*n*) to represent the cost of the path.

As we noted earlier, the most vulnerable route for the attacker to exploit a vulnerability in the victim SE is the route with highest possible protection weight.

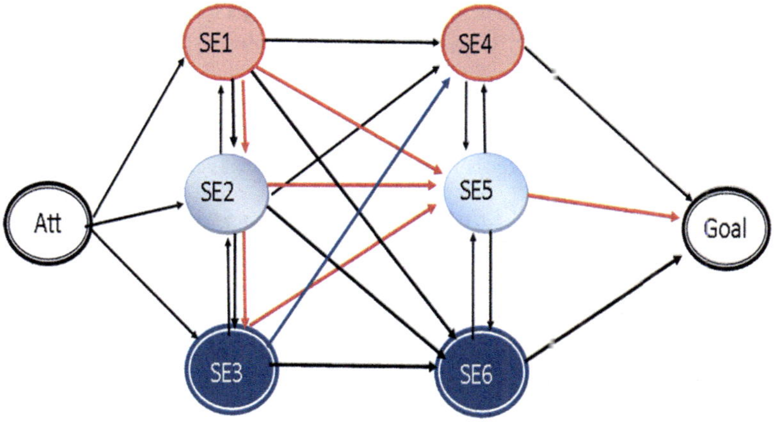

Fig. 13.5 The infrastructures of the SEs according to the network topology and values

In our case study, the critical path for the attacker to turn over the victim is equal to the path that contains the high or medium entropy weight values across the link as shown in Fig. 13.5. We suppose the security entity 1 has a maximum entropy weight value that the attacker can reach with value = 0.0546, and from that point as multistep attack, the attackers can choose the path to security entity 4 with vulnerable entropy value = 0.0424, which has direct access with the victim.

$$\begin{aligned}\text{Cost}(N_x) &= \wedge\left(Ph(SE_1), Ph(SE_4)\right)\\ &= \wedge\left(W_1 + W_4\right)\\ &= 0.0546 + 0.0424 = 0.097\end{aligned}$$

We could calculate the effective risk in terms of probability of attack and impact as follows:

$$\text{Risk} = P(Att)^* P(s)^* C$$

$$\text{Risk} = \sum_{i=1}^{n} P(A)^* \frac{1}{e^{W(i)}}^* C$$

We calculate $P(\text{A})$ as follows:

$$P(A) = \sum_{i=1}^{P} (1 - Pi)/100$$

where P is the number of protections for each security entity, $P(\text{SE1}) = 0.41$.

In the same way, we calculated the probability of attacks of the security entities in the paths from the attacker to the victim. These values should be assessed and rectified by experts:

$P(\text{SE2}) = 0.434$, $P(\text{SE3}) = 0.52$ and $P(\text{SE4}) = 0.556$.

We suggest the Impact C according to the protection level as follows:

$$C_{SE1} = 0.6,\ C_{SE2} = 0.4,\ C_{SE3} = 0.687\ and\ C_{SE4} = 0.66$$

The risk we calculated using our methodology for the attacker to turn over to the victim equals 21.46015608. Employing the best origin and having full understanding of the topology and the infrastructures, the risk will be less. The risk value shown in this example gives clear indications to system administrators to increase the protection on two security entities SE1 and SE4 as the entropy weight for these security entries, showing a value ≥0.5.

We can use our model and methodology for a specific path in the network to protect a valuable asset and detect the vulnerable security entities.

13.5 Security System Risk Assessment

In this section, the algorithm has been established on a relatively small prototype network as shown in Fig. 13.6. This example represents LAN exploit style to commence with the attacker attached to Host-A and needs to exploit a vulnerability at Host-E or Host-F, both of which are protected by a firewall. The firewall rules just permit communications of Host-C or Host-D to Host-E. Router (Ro) connects segments 1 and 2. Additionally, all hosts have a CVE-2003-0818: a vulnerability in the single service called "Microsoft Windows ASN.1 Library Integer." This vulnerability enabled the attackers to gain the system Administrator or Root privilege, using buffer overflow by sending a large packet that contains the arbitrary code in it to

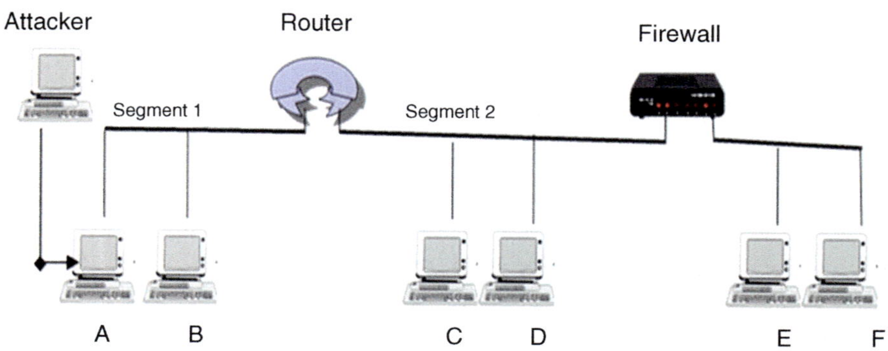

Fig. 13.6 An example network (Modified from Franqueira et al. [8])

Table 13.4 The result obtained from entropy weight algorithm castoff in the experimental approach using the normalised cost matrix

	S1	S2	S3	S4	S5	S6	S7	S8	S9	S10	S11
SE1	0.0375	0.0442	0.0438	0.0352	0.0363	0.0208	0.0368	0.0284	0.0419	0.0301	0.0337
SE2	0.0294	0.0215	0.0252	0.02	0.0488	0.0435	0.0307	0.0583	0.0346	0.0401	0.044
SE3	0.044	0.0169	0.0123	0.0115	0.0488	0.0335	0.0661	0.0461	0.0245	0.0101	0.1074
SE4	0.0188	0.0275	0.0296	0.0156	0.0488	0.0235	0.0361	0.0302	0.0431	0.01	0.0273
SE5	0.0334	0.052	0.0401	0.0386	0.0323	0.0301	0.029	0.0273	0.0225	0.015	0.0559
SE6	0.035	0.0414	0.0203	0.039	0.054	0.0403	0.0292	0.0475	0.0328	0.0154	0.0461
SE7	0.0336	0.0315	0.0205	0.0394	0.0329	0.0405	0.0394	0.0277	0.0531	0.0158	0.0363
SE8	0.034	0.0116	0.0407	0.0398	0.0232	0.0307	0.0496	0.0179	0.0134	0.0262	0.0365
SE9	0.036	0.0317	0.051	0.0402	0.0335	0.061	0.0298	0.0381	0.0237	0.0466	0.0267
SE10	0.0339	0.0218	0.0311	0.062	0.0238	0.0211	0.023	0.0483	0.044	0.037	0.0469
SE11	0.039	0.0319	0.0411	0.041	0.0341	0.0313	0.0102	0.0285	0.0343	0.0374	0.0471
SE12	0.024	0.021	0.0214	0.0413	0.0242	0.0414	0.0403	0.0486	0.0344	0.0376	0.0372

Table 13.5 Experimental results from entropy weight algorithm

S(0)	S(1)	S(2)	S(3)	S(4)	S(5)	S(6)	S(7)	S(8)	S(9)	S(10)	S(11)
	0.044	0.052	0.051	0.062	0.054	0.061	0.0661	0.0583	0.0531	0.0154	0.0559

escalate the privileges, which can cause a comprehensive impact on its availability (A), confidentiality (C) and integrity (I).

Table 13.4 shows the entropy weight obtained when the algorithm castoff in the experimental approach is using the normalised cost matrix.

The final results driven from the weight matrix are presented in Table 13.5, which represents the critical state where the security professionals and systems administrators should give a high priority for countermeasures and mitigations to secure the prototype.

13.6 Conclusions

In this book chapter, we used Shannon entropy to represent the uncertainty of information used to calculate system risks and entropy weight method since the weight of the object index is normally used and points to the significant components of the index. We evaluated the risk of security systems in terms of dynamic risk assessments instead of constant empirical static values used by CVSS, taking into considerations the interaction between vulnerabilities to improve the risk detections and mitigations.

The dynamic impact scoring of vulnerabilities is used to pass on the cost-centric method to specify a unique severity, cost for each host by dividing the loads of the vulnerabilities to three principal degrees of privileges: (1) None; (2) User and (3) Root. The approach then classifies these levels into operational levels to identify and calculate the severity cost of multistep vulnerabilities. These approaches and methods that provide an important step to secure network systems and to assist the security investment decision makers in the process of selecting a proper security solution based on the cost/benefit analysis, as the cost-centric method will be used in building and ranking a novel cost-centric attack graph [9].

The cost of the total weight of all vulnerabilities for each host has been used to develop a unique dynamic severity, using DVSS scores of intrinsic, time-based and ecological metrics by combining related sub-scores and modelling the problem's parameters into a mathematical framework. In the test pad, Nessus scanner 4.x is used to discover known vulnerabilities. Open-source programs have been developed to process and combine different dynamic files to produce dynamic scores. The capability of protections is measured by introducing the concept of protection effectiveness. We wrote an algorithm to normalise the protective matrix and calculate the entropy and the entropy weight, and then we used the weight and paths to evaluate and calculate the total risk in the system and give the system administrator a clear guidance on the vulnerable security entities. We developed a novel approach to evaluate the cybersecurity suitable for the majority of cyber systems by introducing the term security entities; we can use our model and methodology to evaluate the cybersecurity risks for a specific path in the network and to protect a valuable asset and detect the vulnerable security entities, and with different SE and protection layers, we can evaluate the security for cybercrime cybersecurity systems.

References

1. https://www.gov.uk/government/news/more-small-businesses-hit-by-cyber-attacks, last visit 22/05/16.
2. Hicks, C., McGovern, T., & Earl, C. F. (2000). Supply chain management: A strategic issue in engineer to order manufacturing. *International Journal of Production Economics, 65*, 179–190.
3. Smith, C.L. (2004). The development of a security systems research and test laboratory at University. Proceedings of IEEE International Carnahan Conference on Security Technology, pp. 111–115.
4. Dai, J. J., Hu, H. M., & Cai, Q. (2011). Effectiveness evaluation of security system based on entropy theory. *Applied Mechanics and Materials, 40*, 806–811.
5. Xiaohu, Li. (2011). A stochastic model for quantitative security analyses of networked systems.

6. Ammann, P., Wijesekera, D., & Kaushik, S.. (2002) Scalable graph-based network vulnerability analysis. Proceedings of the 9th ACM Conference on Computer and Communications Security. ACM.
7. Balocco, A., & Capone P. Construction site risk analysis based on shannon entropy: A case study application. The First international conference on safety and security engineering, pp. 171–181.
8. Franqueira, V. N. L., & van Keulen, M.. (2008). Analysis of the NIST database towards the composition of vulnerabilities in attack scenarios." Centre for Telematics and Information Technology (CTIT), University of Twente, Enschede, The Netherlands, Tech. Rep. TR-CTIT-08-08.
9. Buchley, J. J., & Chanas, S. (1989). A fast method of ranking alternatives using fuzzy numbers (short communications) [J]. *Fuzzy Sets and Systems, 30*(3), 337–339.

Chapter 14
Designing Health Technology: The Ethical Dimension

Claire Craig and Paul Chamberlain

14.1 Introduction

Global demographics and the rising number of people living with long-term conditions have placed increasing pressure on existing healthcare delivery. Technological innovation in the form of telehealth and telecare has been posited as one possible solution, offering individuals with chronic conditions a set of tools to monitor and better manage their health, reducing pressure on existing services and making possible cost savings through fewer admissions to hospital [30]. These advances are challenging and reshaping existing models of care, revolutionizing how and where healthcare is provided.

However, whilst these technologies offer great potential, they also raise important ethical questions. Some of these relate to the application of technologies: for instance, the use of tracking and surveillance (particularly in the context of people living with dementia). Other questions these advances raise relate to issues regarding access and accessibility to the technology, particularly for older people and individuals from communities with low socio-economic status.

This chapter explores some of these ethical challenges. It begins by considering the broader context in which these health technologies have emerged before identifying the ethical principles that underpin current health and social care provision, most notably autonomy, beneficence, non-maleficence and justice. Drawing on the research undertaken by the authors, this chapter then considers how technology can potentially contravene these principles and the complexity and ethical dilemmas developers, designers and providers of such health technologies can face. Finally it reflects on ways of navigating these tensions.

C. Craig • P. Chamberlain (✉)
Sheffield Hallam University, Sheffield, UK
e-mail: p.m.chamberlain@shu.ac.uk

M. Dastbaz et al. (eds.), *Technology for Smart Futures*,
DOI 10.1007/978-3-319-60137-3_14

14.2 The Broader Context: What Is Health Technology and Why Is It Such a Key Aspect of Current Healthcare?

Globally the population is ageing, and the fastest-growing demographics are individuals ageing 85 and over [28]. Whilst there is much to celebrate, population demographics also pose a set of challenges, as the prevalence of long-term conditions increases with age [26]. Indeed, it is estimated that the burden of disability caused by long-term conditions is three to four times larger for people ageing over 75 years than those ageing 45–55. As a consequence, an individual's age is one of the factors which contribute to the growing expenditure on public health and long-term care. In the United Kingdom, for instance, it is estimated that people living with long-term conditions comprise 30% of the population, and this 30% account for 50% of all GP appointments, 64% of all outpatient appointments and 70% of the total spend of the NHS budget [8].

Current models of healthcare delivery are increasingly untenable, and as a consequence, a paradigm shift has occurred with the emergence of approaches where responsibility is placed on the individual to recognize and manage the symptoms of their condition [16]. The emergence of technologies to support self-monitoring and deliver healthcare remotely played a key role in this strategy, changing both how and where care is delivered [12]. At present, it is estimated that between 1.6 and 1.7 million people in the United Kingdom benefit from technology-focused health services [23], and this number is set to increase. The United Kingdom is not alone, with similar investments made in other European countries including the United4Health project, a large-scale deployment of telecare and telehealth across 14 regions in 10 countries engaging 12,000 patients living with long-term conditions [29]. In America, the most integrated example of telehealth is the Care Coordination and Home Telehealth (CCHT) programme.

Technological developments extend to almost every aspect of care including health informatics, portable diagnostics, home rehabilitation and digital therapeutics, telehealth and telecare. Whilst the importance of ethics relates to all areas of e-health and health technology, for the purpose of this chapter, particular focus will be placed on telehealth, telecare and telemedicine. As health technologies continue to evolve, so does discussion and debate around the terminology used to describe these developments.

14.3 Defining Health Technology

There is a general consensus across health services that telehealth refers to devices concerned with the remote monitoring of physiological data, enabling health professionals to support diagnosis and disease management. Telehealth equipment has been used particularly in supporting people to manage long-term conditions in the community. Individuals are provided with the skills and technology to regularly record a range of physiological data (e.g. blood glucose readings, blood pressure) in the comfort of their own home. This can provide a more comprehensive picture, which can be utilized by clinicians who can gain a fuller understanding of the progression and management of the condition leading to more accurate treatment/management and potential savings in clinician time [25]. Additional applications of this technology has included as a shorter-term diagnostic tool and in the support of individuals discharged from hospital following an acute illness (op.cit).

Telecare on the other hand is defined as the 'use of communication technology to provide health and social care direct to the user (patient)' ([25], p. 193) and comprises of systems of sensors and alarms that can detect possible problems such as smoke or gas and has the ability to alert carers or monitoring control centres should the need arise. This remote monitoring enables individuals who might otherwise be required to move to live in a residential care environment to continue to live at home. Telecare literature is very much distinguished between first-, second- and third-generation devices. First-generation devices are typically alarm buttons that the person wears and presses it during an emergency for help. Second-generation devices are subtler, utilizing a broader range of sensors that are not dependent on the person to press the alarm. These range from smoke and natural gas detectors to infrared sensors that would be activated by movement. These sensors could be deployed in a number of ways. For example, movement sensors located in the bedroom could be connected to the lighting circuit, so when the person gets up in the middle of the night, lights would be turned on to reduce the risk of falls. Sensors installed by the front or back door will activate a sound recording reminding a person with memory problems to remember to take their key.

Third-generation telecare systems can effectively form a type of lifestyle monitoring, tracking a person's usual habits and activating an alarm when someone deviates from usual behaviour, for instance, when activity levels fall or when a person stops cooking. These signs can trigger early support and some form of intervention to prevent deterioration in health and wellbeing.

Finally, telemedicine is defined as 'The delivery of health care services, where distance is a critical factor, by all health care professionals using information and communication technologies for the exchange of valid information for diagnosis, treatment and prevention of disease and injuries, research and evaluation, and for the continuing education of health care providers, all in the interests of advancing the health of individuals and their communities' ([27], pp. 8–9).

Telemedicine offers a medium through which specialist doctors and clinicians can be called on remotely to review a patient's care and recommend best treatment. This technology is used particularly in the area of dermatology and chronic wound management and is invaluable in remote areas where access to such specialist input would be difficult.

14.4 What Does the Evidence Tell Us Are the Potential Benefits of This Approach?

The potential benefits offered by telehealth, telecare and telemedicine are vast and wide-ranging. The World Health Organization [29] highlights the key role these technologies play in providing individuals living with long-term conditions greater control over their health, promoting confidence and independence, mitigating risk and enabling people to live at home independently longer. Telemedicine particularly can increase the access afforded to individuals who may otherwise be excluded from care provision as a consequence of geography or socio-economic status. Increasingly there is recognition that these advances have potential to be used in a more preventative way, saving money and placing less pressure on existing health service resources.

However in spite of the undoubted opportunities that these health technologies offer, the literature also raises a number of concerns. The Whole Systems Demonstrator Action Network ([30], p. 3), for instance, identifies a number of key barriers including 'a lack of robust evidence for the cost effectiveness of telehealth and telecare associated with current high cost of deploying some of the technology at scale'. Questions have also been raised regarding user acceptance and the acceptability of many of the technologies developed with low uptake cited as being particularly problematic [9]. Moreover, the rapid development of these technologies, the speed of deployment and the vulnerability of many of the individuals the technology is aimed at have led to a number of ethical concerns being raised in relation to the use of these technologies which have yet to be fully addressed. This chapter now turns to these ethical dilemmas and questions, firstly considering what these broader ethical questions are and how health technologies can potentially challenge the moral frameworks that underpin much of health and social care.

14.5 What Are Ethics and Why Should We Consider Them?

Ethics are best described as a 'set of moral principles that govern a person's behaviour' (Oxford English Dictionary). Within medicine, 'codes of ethics have been a long-standing element in the professional control of the behavior of doctors and indicate a commitment to act with integrity in extreme circumstances. When patients seek medical care they are not entering an ordinary social relationship; they often feel vulnerable but need to expose and share intimate and important aspects of their lives. Ethical codes of conduct offer some tangible protection to both patients and doctors in these circumstances. The Hippocratic Oath is perhaps the best known code of this kind' ([15], p. 394). Over time, these ethical codes have become a feature of many health and social care professions with codes of conduct and professional suitability being mandatory requirements of registration.

The most commonly accepted framework of medical ethics is that developed by Beauchamp and Childress [3]. Underpinning the framework are four principles [23]: autonomy (the ability of the person to make choices), beneficence (the principle of working for the benefit of the individual), non-maleficence (the principle of doing no harm) and justice (in this context, it relates to fair and equal access to resources).

These principles are directly relevant to the ethics of health technology. For instance, on the one hand, it could be argued that health technology may increase autonomy and independence, decreasing reliance on professional carers. The installation of telecare services such as alarms, for example, can enable individuals to live at home safely and delay admission to residential care settings [23] where freedom and independence may be limited by the environment. Home monitoring as a preventive measure can ensure that any deterioration in a person's condition is identified immediately, promoting wellbeing and benefit for the individual (beneficence). Telecare has the potential to 'increase comfort through environmental sensors and controls' ([23], p. 5) as reflected in the findings of an early study by Beale et al. [2]. In terms of justice, digital health is frequently regarded as one of the mechanisms through which services can be provided to populations in remote areas who would otherwise be denied access to specialist support.

However, this must be balanced against the ethical challenges that technologies can present and potential threats to autonomy, non-maleficence, beneficence and justice. In spite of the potential to generate debate, the broader literature has been relatively inconsistent in its discussion and consideration of the ethical and moral issues that health technologies present. Korhonen et al.'s [14] review of technology and ethics within the nursing and medical literature 2000–2013 concluded that 'the ethics of technology had not been established as a guiding principle' (p. 561). The authors found that many of the studies reviewed neglected to address ethical issues, and where ethics were discussed, they were framed either as the benefits of technology or focused on making broad generalizations relating to the challenges people faced when accessing health technologies.

This is also reflected at policy level: the quantity and speed of technological innovation in the area of health has meant that policy frameworks, standards and guidance that govern other medical devices have been slow to materialize. The World Health Organization [29], for example, recently highlighted in their report on technology and e-health that whilst 'most countries in the WHO European Region have policies or strategies outlining eHealth goals, measures and implementation objectives and achievements…there is still a need for adequate governance and legislation to address the advances that ehealth is bringing and for strengthened monitoring…developers and users of e-health have to navigate many legal grey areas and adequate guidance and support are still limited' [29].

This situation is not confined to Europe: Hall and McGraw [11], for example, have highlighted that in North America 'no federal agency currently has authority to enact privacy and security requirements to cover the telehealth ecosystem' (p. 216) and that without this 'the success of telehealth could be undermined if serious security risks are not addressed' (op.cit). The authors provide two examples where the ethical principle of non-maleficence may be compromised. The first relates to instances where sensors to detect medical emergencies are installed in a person's home. The sensors monitor patterns of behaviour (in order to detect falls), but in doing so, these sensors also potentially collect sensitive information about household activities including when the person is not at home and the house is empty. The second example they cite is where routine transmissions from a medical device are collected and stored by the device or app manufacturer and not just the healthcare provider. This can be the case where apps are financed by sharing this potentially sensitive data with other organisations [11].

Where reference is made in the broader literature to ethics in relation to health technology, this tends to be in the context of people living with cognitive impairments [5, 17]. The Social Care Institute of Excellence [23] and the Nuffield Council on Bioethics [18], for example, have both written about ethical issues in the use of telecare with a particular focus on people with dementia. The documents describe potential ethical issues that can arise from the point of pre-installation making close reference to the ethical framework proposed by Beauchamp and Childress [3]. The main challenge identified by the SCIE and the Nuffield Council in relation to ethics and technology is that of informed consent. According to Beauchamp and Childress [3], consent relates to the principle of respect for autonomy and the 'non-subjection of autonomous actions to controlling constraints by others; and respectful treatment in disclosing information, probing for and ensuring understanding and voluntariness and fostering autonomous decision making' [23].

A potential scenario described by the Social Care Institute for Excellence where autonomy could be compromised is in the fitting of exit sensors to a person living with dementia's home. The exit sensors are intended to alert the family or health professionals in order to ensure that the person does not leave the house alone and puts themselves in danger when crossing the road or becoming lost. However, in doing so, the sensors could be seen to act as a threat and a restriction to the individual's liberty. Numerous similar scenarios exist including the use of 'tagging technology' so that a person living with a cognitive impairment can be tracked in case they become lost. The Nuffield Council on Bioethics in Dementia [18] concludes that 'at the heart of the issue is balancing safety with freedom, deciding what is in the best interests of the person with dementia and recognizing that the needs of the person with dementia may sometimes conflict with the needs of others' (xvii). Whilst the council highlights the value of technology for people living with long-term neurological conditions such as dementia in increasing quality of life, promoting choice and enabling individuals to live more independently at home longer, it also recognizes that there are '*possible detrimental effects such as the intrusion of privacy, stigma (particularly with reference to tracking devices) and the risk of reduced human contact)*" ([26], p. 61).

14.6 Ethical Questions: Beyond Cognitive Impairment

However the issues raised by the Social Care Institute for Excellence should not be confined to individuals with cognitive impairment. Underpinning the principles of informed consent is being able to make a clear judgement based on an understanding of the information provided. Yet telecare and telehealth are likely to be unfamiliar to many potential recipients and individuals without cognitive impairment who are unacquainted with particular technologies who may not understand the privacy or risk implications of sharing sensitive health data with others. Similarly the stigma (challenge to the ethical principle of non-maleficence) that can arise from wearing a pendant alarm or having monitoring equipment in the home can pose a threat to identity for individuals who are cognitively well. Health technologies are not neutral objects but are embodiments of ourselves and cultural values [10], and a growing body of literature is emerging that suggests that the medicalization of the home environment brought about by residential technologies can impact on social relationships ([20], Arras and Neveloff-Dubler in [7], p. 112). Authors such as Oudshoorn [20] have suggested that rather than increasing autonomy by enabling people to live in the home environment for longer telehealth and telecare may restrict people's movements, making people more likely to be confined to the home. These concerns have been echoed by research showing that for individuals who have little or no family or friendships, telehealth and care technologies can remove the little human contact they receive from direct staff contact, further increasing levels of social isolation [23].

14.7 Insights into Telehealth and Care Technologies

Whilst the literature is patchy, the issues highlighted by this chapter to date suggest that in terms of building clearer understanding of the ethical implications of current telehealth and care technologies, further research is required. One of the challenges is that where literature does exist regarding the moral and ethical imperatives that these technologies raise, much is written from the perspective of clinicians, developers and providers of the technology. Less is known regarding what end users of these technologies identify as their ethical concerns. Finding ways to access these views is important. Firstly if individuals are to be supported to engage with and use these technologies, designers and providers of the products and systems need to understand people's concerns and factors that will prevent or promote uptake and use. It is also important to engage end users so that they become active participants and collaborators in care. This is an ethical precedent in itself and one that reflects the paradigm of self-management and increased personal responsibility for health in which current health technologies are situated. This very much reflects the Department of Health's mandate that 'the importance and value of self-care and patient education are truly understood and where shared decision making and supportive self care are seen as integral elements of Long Term Conditions management' ([8], p. 21 Assisted living innovation platform, p. 62).

Lab4Living is a transdisciplinary research cluster within Sheffield Hallam University, bringing together researchers and practitioners in design, creative practice, health, engineering and people using products and services with the aim of designing product services and interventions to promote dignity and wellbeing.

It embodies the Living-Labs model which offers a 'real-life test and experimentation environment where users and producers co-create innovations' (ENOLL).

According to the European Commission, Living Labs are involved in four main activities: *co-creation, codesign by users and producers; exploration, discovering emerging users' behaviours and market opportunities; experimentation, implementing the scenarios within communities of users and evaluation, assessment of concepts, products and services according to socio-ergonomic, socio-cognitive and socio-economic criteria* (www.openlivinglabs.eu/aboutus).

Insights into Telehealth and Care Technologies (InTacT) is a research project that has been undertaken by Lab4Living which very much draws on the value of adopting a critical artefact methodology as a way of understanding factors that end users identify as being important in the design of digital health devices. The present study is supported by the Collaboration for Leadership in Applied Health Research and Care/Yorkshire and Humberside (CLAHRC/YH) funded by the National Institute for Health Research (NIHR). The overall aim of the research is to explore the inequalities in telehealth and care technologies and identify and creatively challenge cultural (e.g. language, rituals, socio-economic) barriers to adoption.

The study explores participatory methods and approaches to engage people who are frequently under-represented in telehealth/telecare research by virtue of their age, ethnicity or socio-economic status, in meaningful ways. The focus of the first phase of the research has been to build understanding of end users' attitudes to technology in everyday life, identifying potential ethical concerns and how these technologies might most appropriately be adopted to support their personal healthcare.

Initial research was located in the North of England. Ethical approval was obtained for the study, and workshops were held with individuals attending voluntary and third sector organisations.

To date, 56 socially and ethnically diverse community living individuals have been recruited to the research. Individuals were invited to attend one of six workshops that were held in community venues and were facilitated by the research team. Each workshop lasted on average for 2 h. The workshops began with a general introduction from the research team and an invitation for participants to share (verbally or through drawing) the images and associations that came to mind when they heard the word technology. A critical artefact methodology was then employed [6] using objects to engage participants and to scaffold conversation. Written consent was obtained to video and audio record the session, and these were transcribed following the groups. This data was analysed using framework analysis [21]. This enabled the development of a matrix of themes and related subtopics from the data as well as identification of the links across themes, different participants and venues.

Four themes relating to the ethics of health technology emerged from the data. Three focused around ideas and concepts of trust and trustworthiness: trust from the perspective of what happens to the data collected, trust in relation to the reliability of the technology and of the accuracy of the readings given, and the final *theme related to social exclusion.*

14.8 Trust: What Happens to the Data

Across all the workshops held, participants expressed questions and a concern regarding what happens to the data. This in part related to who could access the information. In the words of one participant:

> Trust is interesting one from us because I am interested in medical products and whilst I am happy to share my health information with my family and friends I am not so happy for the information to be shared with my health insurance company.. .

Sharing of data with third parties was seen by participants as a way of increasing health insurance, of reducing access to state benefits and in one instance where the individual was living in a refuge following an attempt on their life from their partner as a threat to their safety and of the safety of their children. These concerns were fuelled mainly by anecdotal accounts from friends and from stories appearing in the popular press rather than by direct experience.

> It does worry people…in the news here we have had some situations where people have left CDs on the train of things like information on everyone who is claiming child benefit, do you remember that? He left it on the train, and it had everyone's bank info on? People could get hold of all sorts of information on you.

> That's the same I thought about, you have the watches that share the data about your pulse, and it can be uploaded to the computer. Just like the insurance company, they can read that and say ok, that's not ok.

A number of participants had reflected on the tensions and dilemmas surrounding what should happen to data. As one person expressed:

> I think a lot of trouble is about lack of trust, for example computers getting a lot of information about people but there is in fact a conflict which you have to make your own mind up about it – there is for example criminal activities as opposed to privacy – if you were to discover something electronically that saves lives …I am just asking you to think about that – the detection of criminal activities against your own privacy

14.9 Trust and the Reliability of the Technology

Trust could also relate to the reliability of the technology both in terms of the accuracy of reporting and also whether it could be depended on not to break down as exemplified by the following comment:

> There is almost a built-in obsolescence…

Participants commented that many examples of technological products in the home had a life-span determined by the manufacturer,

> There is just one other point that has popped into my head, what you said about technology – if you buy a printer nowadays, it is already made that it will only print for about

> 5 years, so then you will have to get another one. So there are really lots of tricks behind technology…

This led to concerns regarding an over-dependence on such innovation, particularly if it was to be trusted with key medical aspects of care,

> So lots there – security…and still insecurity.

14.10 Social Exclusion

As well as issues of trust and privacy, the fourth theme to emerge from the study related very much to the ethical principles of beneficence and non-maleficence. The speed of change and the challenges of mastering new and unfamiliar technologies meant that a number of participants described how technology could lead to feelings of inadequacy and loss of confidence. This was particularly the case in relation to the experience of older participants who were less familiar with using computers. One individual commented:

> I should like to be able to learn a new technology it's just that I don't understand it….

And another:

> I don't think I am capable any longer of using the technology to talk to someone. It's terrible. I'm sorry but it's true.

For some participants, this could be stress provoking when individuals were also attempting to come to terms with loss of the ability to recall information or learn new skills and could be an unwelcome reminder of failing capacities:

> I used to enjoy it and know what was happening but now something has happened to my brain and I've spent ages getting something ready to send to someone and then I've waved my hand over the key and I've lost it – everything and I am finding it extremely frustrating.

At the extreme, this could lead to feelings of disenfranchisement and of no longer being a part of society.

Clearly this research is still a work in progress, but finding ways to hear the voices of end users is important. Presently work is being undertaken by the authors exploring access to health technologies amongst groups with high socio-economic need. According to the Office of National Statistics [19], 6 million people in the United Kingdom have never used the Internet, and 10 million people lack basic digital skills, with the majority of these individuals belonging to social class DE and 31% living with a disability. This picture is not confined to the United Kingdom with the European Commission citing figures of 47% of the European population lacking in digital skills and 65% of these individuals coming from disadvantaged populations. These figures raise important questions when considering the principle of justice and equitable access to health technology.

14.11 Conclusion and Recommendations

As this chapter has highlighted, health technology has the potential to completely reshape how and where care is delivered. This innovation is regarded globally by governments as policymakers as offering a way to meet the increasing demands placed on healthcare systems as a consequence of global ageing and of the rising number of people living with long-term conditions. Considerable government investment continues to be made into extending and mainstreaming technology-based health services [23, 29]. However whilst such initiatives potentially offer individuals the opportunity for increased choice and autonomy in relation to being able to live at home longer, to identify and monitor the illness trajectory this chapter have also shown that these same technologies can also increase 'threats to individual choice and privacy through more extensive use of monitoring' ([23], p. 2), particularly for individuals who are emotionally or cognitively frail. For the true potential of this technology to be realised, it is necessary to continue to build understanding of the ethical questions it raises and for end users, developers and healthcare practitioners, service designers and policymakers to engage work together to begin to consider, challenge and to address these bigger questions. This chapter ends with a series of reflections and recommendations of the direction these discussions and this research may take.

14.11.1 Recommendations

First and foremost, before further huge investments, it is a moral imperative to undertake further research to build understanding of the efficacy of health technologies and to determine the value of the innovation and the contexts where it may be used most effectively. As the Technology Strategy Board [26] highlights 'evidence on the positive outcomes for older individuals using telecare devices is inconsistent' (p. 51). Whilst the results of the UK Whole demonstrator trial were promising, a number of authors have called for caution, citing the inclusion of people at low risk, the level of extra support and the short follow-up period [24]. Botsis and Hartvigsen [4] in their review of international papers on health technology in chronic disease have highlighted a lack of evidence to show a direct relationship between health technology and improvement of health outcomes and quality of life.

Work is required to build understanding of end user experience. Health technologies are not used in a vacuum, and if stigma is to be avoided, it is important for designers and developers of technologies to better understand the broader physical and social environments in which these technologies will operate and how they relate to the contexts of the lives of end users. This is necessary because according to Greenhalgh et al. [10], the things we use and make (technologies) are not neutral objects but embodiments of ourselves and cultural values. Where a disconnect between these values emerge this impacts on the individuals relationship with the world.

There is an increased need for transdisciplinary research so that dialogues between designers and developers of technology and the healthcare practitioners who will assess and delivery these technologies take place. There is a necessity to understand different perspectives and also the moral and ethical frameworks and codes in which different professionals operate.

Clearer national and international guidance is required. Legal questions regarding accountability where health technology fails and leads to the death of a patient has yet to be resolved. The standards and guidance that govern other medical devices needs to be developed to relate to health technology products.

References and Bibliography

1. Anderson, J. (2007). Ethical and legal barriers to e-health. *International Journal of Medical Informatics, 76*, 480–483.
2. Beale, S., Sanderson, D. S., & Kruger, J. (2009). *Evaluation of the telecare development programme: Final report*. York: Health Economics Consortium.
3. Beauchamp, T. L., & Childress, J. F. (1989). *Principles of biomedical ethics* (3rd ed.). Oxford: Oxford University Press.
4. Botsis, T., & Hartvigsen, G. (2008). Current status and future perspectives in telecare for elderly people suffering from chronic diseases. *Journal of Telemedicine and Telecare, 14*(4), 195–203.
5. Bjorneby, S., Topo, P., Cahill, S., Begley, E., Jones, K., Hagen, I., Macijauskiene, J., & Holthe, T. (2004). Ethical considerations in the ENABLE Project. *Dementia, 3*(3), 297–312.
6. Chamberlain, P., & Craig, C. (2013). Engagingdesign, methods for collective creativity. In *Human-computer interaction. Human-centred design approaches, methods, tools, and environments, Series lecture notes in computer science* (Vol. 8004, pp. 22–31). Berlin/Heidelberg: Springer.
7. Demiris, G., & Hensel, B. (2008). Technologies for an ageing society: A systematic review of "smart home" applications. *Methods of Information in Medicine, 47*(Suppl. 1), 33–40.
8. Department of Health. (2012). Long term conditions compendium of information. In DH (Ed.). London: HMSO.
9. Foster, A., Horspool, K. A., Edwards, L., Thomas, C. L., Salisbury, C., Montgomery, A., & O'Cathain, A. (2015). *Who does not participate in telehealth trials and why?* A cross-sectional survey. *Trials, 16*(258), 1–10.
10. Greenhalgh, T., Wherton, J., Sugarhood, P., Hinder, S., Procter, R. N., & Stones, R. (2013). *What matters to older people with assisted living needs?* A phenomenological analysis of the use and non-use of telehealth and telecare. *Social Science & Medicine, 93*, 86–94.
11. Hall, J. L., & McGraw, D. (2014). For telehealth to succeed privacy and security risks must be identified and addressed. *Health Affairs, 33*(2), 216–221.
12. Huang, H.-T., Chang, C.-M., Liu, L.-F., Lin, H.-S., & Chen, C.-H. (2013). Trajectories and predictors of functional decline of hospitalised older patients. *Journal of Clinical Nursing, 22*, 1322–1331.
13. Kluge, E. (2011). Ethical and legal challenges for health telematics in a global world: Telehealth and the technological imperative. *International Journal of Medical Informatics, 80*, e1–e5.
14. Korhonen, E., Nordman, T., & Eriksson, K. (2015). Technology and its ethics in nursing and caring journals: An integrative literature review. *Nursing Ethics, 22*(5), 561–576.
15. Limentani, A. E. (1999). The role of ethical principles in health care and the implications for ethical codes. *Journal of Medical Ethics, 25*, 394–398.

16. Lorig, K., & Holman, H. R. (2003). Self-management education: Definition, outcomes and mechanisms. *Annals of Behavioural Medicine, 26*(1), 1–7.
17. Marshall, M. (Ed.). (2000). *ASTRID: A guide to using technology within dementia care*. London: Hawker Publications.
18. Nuffield Council on Bioethics. (2009). *Dementia: Ethical issues*. London: Nuffield Council on Bioethics.
19. Office of National Statistics. (2015). *Internet users 2015*. London: ONS.
20. Oudshoorn, N. (2012). How places matter: Telecare technologies and the changing spatial dimensions of healthcare. *Social Studies of Science, 42*(1), 121–142.
21. Ritchie, J., & Spencer, L. (1994). Qualitative data analysis for applied policy research. In A. Bryman & R. G. Burgess (Eds.), *Analyzing qualitative research* (pp. 173–194). London: Routledge.
22. Rutenberg, C., & Oberle, K. (2008). Ethics in telehealth nursing practice. *Home health care management and practice, 20*(4), 342–348.
23. Social Care Institute of Excellence. (2010). *Ethical Issues in the use of telecare*. London: SCIE.
24. Steventon, A., et al. (2012). Effect of telehealth on use of secondary care and mortality: Findings from the Whole System Demonstrator cluster randomized trial (2012). *British Medical Journal, 344*, e3874.
25. Stowe, S., & Harding, S. (2010). Telecare, telehealth and telemedicine. *European Geriatric Medicine, 1*, 193–197.
26. Technology Strategy Board. (2013). *Assisted living innovation platform*. London: Technology Strategy Board.
27. World Health Organisation. (2009). *Telemedicine: Opportunities and developments in Member States: Report on the second global survey on ehealth*. Geneva: WHO.
28. World Health Organisation. (2011). *Global health and ageing*. Geneva: WHO.
29. World Health Organisation. (2016). *From innovation to implementation: ehealth in the WHO European Region*. World Health Organisation: Copenhagen.
30. Whole Systems Demonstrator Action Network. (2010). *Sustaining innovation in telehealth and telecare*. London: Kings Fund.

Chapter 15
Testing the Comprehensive Digital Forensic Investigation Process Model (the CDFIPM)

Reza Montasari

15.1 Introduction

Nowadays, the nature of evidence presented in courts of law is less likely to be paper-based due to the ubiquitous nature of information technology. Evidence of computer crime differs from that related to traditional crimes for which there are well established standards and procedures [3, 41, 42]. In order for digital evidence to be admissible, investigators need to demonstrate that they have specialised knowledge and that the evidence was acquired using reliable principles and methods [22]. As with other types of evidence, digital evidence is not assumed to be valid and reliable: empirical testing in relation to the theories and techniques of its production is required [3, 28]. Careful notice is taken in court of the manner in which the digital investigative process has been carried out [14, 22, 28]. A digital forensic investigator might discover significant and incriminating evidence, but if they cannot present the evidence in a coherent and understandable way to the lay audience (such as judge and jury), the case may be lost [40]. The complexity of tools and methodologies used to perform a digital investigative process requires investigators to be able to explain the process in a manner that a judge and jury can understand it [22]. Such tools and methodologies must also adhere to some standards of practice and be accepted by other investigators operating in the field [3, 7, 22].

Nevertheless, the field of digital forensics still lacks a 'formal' process model that courts can employ to determine the reliability of the digital evidence presented to them [5, 23, 29, 44, 46]. A further issue with the existing models is their tendency to

---

The original version of this chapter was revised. An erratum to this chapter can be found at DOI 10.1007/978-3-319-60137-3_18

R. Montasari (✉)
School of Computing, Engineering and the Built Environment, Birmingham City University, Millennium Point, Curzon Street, Birmingham B47XG, UK
e-mail: Reza.Montasari@bcu.ac.uk

M. Dastbaz et al. (eds.), *Technology for Smart Futures*,
DOI 10.1007/978-3-319-60137-3_15

focus on one specific area of digital forensics, neglecting other environments [29, 30]. Unlike in other domains of forensic practice, digital forensic investigators operate in various fields [4, 9, 13, 17]. Therefore, as Carrier and Spafford [11] argue, 'A model must be abstract and apply to law enforcement investigations, corporate investigations, and incident response'. However, instead of being generic, previous models have often focused on only one specific area, such as law enforcement, commerce or incident responses [3, 34], therefore failing to consider the requirements of those operating in different domains. The result has been the hindering of the development of a generic model that can be applied in the various fields of digital forensics. A third significant concern associated with the previous models is that they are not comprehensive, failing to cover the entire investigative process. The models often focus on the 'middle part' of the investigative process, that being 'Identification, Acquisition and Examination Processes,' excluding other essential stages. Beebe and Clark argue that a more comprehensive and generally accepted framework is needed to enhance scientific rigour and facilitate education, application and research [9].

The remainder of the paper is structured as follows: Sect. 15.2 presents a background to the existing digital forensic investigation process models (DFIPMs). Section 15.3 provides an overview of the proposed model, while Sect. 15.4 presents the research methodology. In Sect. 15.5, the CDFIPM is applied into a case study, and a walkthrough of the model is performed. Finally, the paper is concluded, and future work is discussed in Sect. 15.6.

15.2 Background to the Existing DFIPMs

Prior to designing and developing the proposed model, presented in Montasari [31], a critical review of the existing models was carried out, and the results of this review was presented in Montasari [30] and Montasari and Peltola [29]. Since the latter two papers discuss such a review in detail, this section provides only a summary of the findings of this critical analysis. The review of the existing models revealed that these models have often been developed by digital forensic practitioners based on their own personal experience on an ad hoc basis without consideration to establish standardisation within the field [46]. This has prevented the establishment of formal processes that are urgently needed by courts of law [4, 29]. As Table 15.1 clearly demonstrates, existing DFIPMs display significant disparities in terms of the number of phases, scope and the specific domains that they have been developed for. As a result, these models have often been criticised for being too specific [11, 35], too high level [9], too broad [36], too technical [45] and too complex [39]. Due to such shortcomings, many researchers are increasingly calling for scientific approaches and formal methods for describing the digital investigation processes [10, 15, 18, 25, 34]. Therefore, as discussed in Sect. 15.1, the Comprehensive Digital Forensic Investigation Process Model (the CDFIPM) was proposed in Montasari [31] to address the stated issues in relation to the existing DFIPMs. By implementing the CDFIPM, this model will be of immediate value to both digital forensic investigators (DFIs) operating within the stated fields and courts of law.

Table 15.1 The comparative summary of the existing DFIPMs

The Comparative Summary of the Existing Digital Forensic Investigation Process Models (DFIPMs)																				
Existing DFIPMs	Palmer (2001)	Ashcroft (2001)	Reith et al. (2002)	Carrier and Spafford (2003)	Baryamureeba and Tushabe (2004)	Ciardhuáin (2004)	Rogers (2004)	Beebe and Clark (2005)	Kent et al. (2006)	Kohn et al. (2006)	Rogers et al. (2006)	Freiling and Schwittay (2007)	Khatir et al. (2008)	Selamat et al. (2008)	Cohen (2009)	Yusoff et al. (2011)	Agarwal et al. (2011)	Valjarevic and Venter (2012)	Kohn et al. (2013)	Adams et al. (2014)
Readiness				✔	✔		✔	✔				✔							✔	
Deployment				✔			✔	✔											✔	
Policy/ Procedure										✔		✔							✔	
Operational Readiness				✔			✔					✔							✔	
Infrastructure Readiness				✔			✔					✔							✔	
Incident Detection (Awareness)			✔	✔	✔	✔						✔						✔	✔	
Report Incident (Notification)				✔	✔	✔		✔		✔				✔					✔	
Assess Incident.								✔					✔						✔	
Confirm Incident.				✔	✔		✔						✔	✔					✔	
Authorisation				✔	✔	✔	✔			✔		✔	✔	✔			✔	✔	✔	✔
Incident Response				✔			✔	✔	✔			✔						✔		
Planning (Approach Strategy)			✔			✔		✔		✔	✔	✔	✔	✔			✔	✔	✔	✔
Understand Task Requirements																				✔
Determine Overall Picture																				✔
Determine Required Outcomes																				✔
Determine Parameters																				✔
Consider Physical Constraint																				✔
Consider Timing Constraint																				✔
Consider Data Constraint																				✔
Plan Logistics																				✔
Create Outline Plan																	✔			✔
Preparation			✔					✔		✔							✔	✔		
Attend Site											✔						✔			✔
Securing the Scene			✔					✔									✔			
Address Safety Issues																	✔			✔
Communication Shielding																	✔			
Triage											✔									
Examine User Usage Profiles											✔									
Examine Chronology Timeline											✔									
Examine Browsing Activities									✔		✔									
Case Specifics											✔									
Carry Out Preliminary Survey																				✔
Documentation of Scene				✔		✔		✔									✔	✔		✔
Update Outline Plan																	✔			✔
Search				✔	✔	✔	✔	✔		✔							✔		✔	
Survey				✔	✔		✔	✔									✔			
Identification	✔					✔		✔	✔	✔					✔		✔	✔		
Preservation	✔		✔	✔	✔		✔	✔							✔	✔	✔		✔	
Collection	✔	✔	✔	✔	✔	✔	✔	✔	✔	✔		✔	✔		✔	✔	✔	✔	✔	✔
Volatile Evidence Collection									✔			✔					✔			
To Be Continued ...																				

Table 15.1 (continued)

The Comparative Summary of the Existing Digital Forensic Investigation Process Models (DFIPMs)

Existing DFIPMs	Palmer (2001)	Ashcroft (2001)	Reith et al. (2002)	Carrier and Spafford (2003)	Baryamureeba and Tushabe (2004)	Ciardhuáin (2004)	Rogers (2004)	Beebe and Clark (2005)	Kent et al. (2006)	Kohn et al. (2006)	Rogers et al. (2006)	Freiling and Schwittay (2007)	Khatir et al. (2008)	Selamat et al. (2008)	Cohen (2009)	Yusoff et al. (2011)	Agarwal et al. (2011)	Valjarevic and Venter (2012)	Kohn et al. (2013)	Adams et al. (2014)
Non-Volatile Evidence Collection									✔			✔								
Authenticate																			✔	
Seizure								✔											✔	✔
Package																	✔			✔
Transport						✔				✔					✔		✔	✔	✔	✔
Storage						✔				✔					✔		✔	✔	✔	✔
Examination	✔		✔	✔		✔		✔	✔	✔			✔		✔		✔		✔	
Harvest								✔				✔							✔	
Reduce								✔				✔							✔	
Identify				✔															✔	
Classify																			✔	
Organise												✔							✔	
Compare																			✔	
Analysis	✔		✔	✔				✔	✔	✔			✔		✔	✔	✔	✔	✔	
Attribute															✔				✔	
Evaluate																			✔	
Hypothesis						✔													✔	
Interpretation															✔				✔	
Reconstruction				✔	✔		✔	✔				✔			✔				✔	
Reporting		✔		✔				✔	✔			✔		✔			✔			
Presentation	✔		✔	✔	✔	✔	✔	✔		✔			✔		✔	✔	✔	✔	✔	
Proof / Defence						✔		✔		✔										
Decision	✔							✔											✔	
Review				✔	✔		✔	✔									✔		✔	
Dissemination						✔		✔											✔	
Returning Evidence			✔					✔					✔							✔
Digital Crime Scene Investigation				✔	✔		✔	✔												
Physical Crime Scene Investigation				✔	✔		✔											✔	✔	
Documentation				✔	✔	✔	✔	✔					✔					✔	✔	✔
Preserving Chain of Custody						✔	✔					✔								✔
Preserving Digital Evidence						✔		✔										✔		
Information Flow						✔												✔		
Case Management						✔							✔							

End of the Table

15.3 Overview of the Proposed Model

The soundness of a digital forensic investigation process model is a function of usability and acceptability [9, 35]. To achieve the soundness, classes, processes and principles were incorporated into the CDFIPM. A class is the highest level (first layer) in the CDFIPM; it is the main group containing one lower layer, namely, processes. A process is the next level down from a class and the second layer in the CDFIPM. Processes are obvious and individually separate steps; they can sometimes be a function of time and therefore can be sequential or sometimes iterative. In contrast, the action principles included in the concurrent process class are those processes that are not confined to a single point in time during an investigation. Instead, they have to be maintained concurrently throughout the whole or parts of the other processes in the CDFIPM. As Fig. 15.1 illustrates, there are six classes in the CDFIPM with each class containing a certain number of processes. There is also a total number of eight action principles contained in the concurrent process class. As the CDFIPM has already been discussed in detail in Montasari [31], the following subsections provide only a brief overview of the model's classes together with the various processes included in each class.

15.3.1 The Readiness Process Class

Forensic readiness is the ability of an organisation to maximise the collection of credible digital evidence from an incident environment and minimise the cost of a forensic incident response [38]. Organisations should focus their efforts on establishing two components including operational readiness and infrastructure readiness.

15.3.2 The Initialisation Process Class

The initialisation process class initiates the digital investigation and contains five processes described as follows.

15.3.2.1 Incident Detection Process

Incident detection is the first step in a digital investigation where an incident is detected by either internal events such as an intrusion detection system or external events such as crime being reported to the police.

15.3.2.2 First Response Process

During this process, the first responders must secure the crime scene in order to ensure preservation of digital device(s) suspected of containing potential digital evidence. Preservation should include disconnecting digital device(s) from a networked environment and detecting the corrupted data. The network must also be monitored to detect incoming calls and IP addresses.

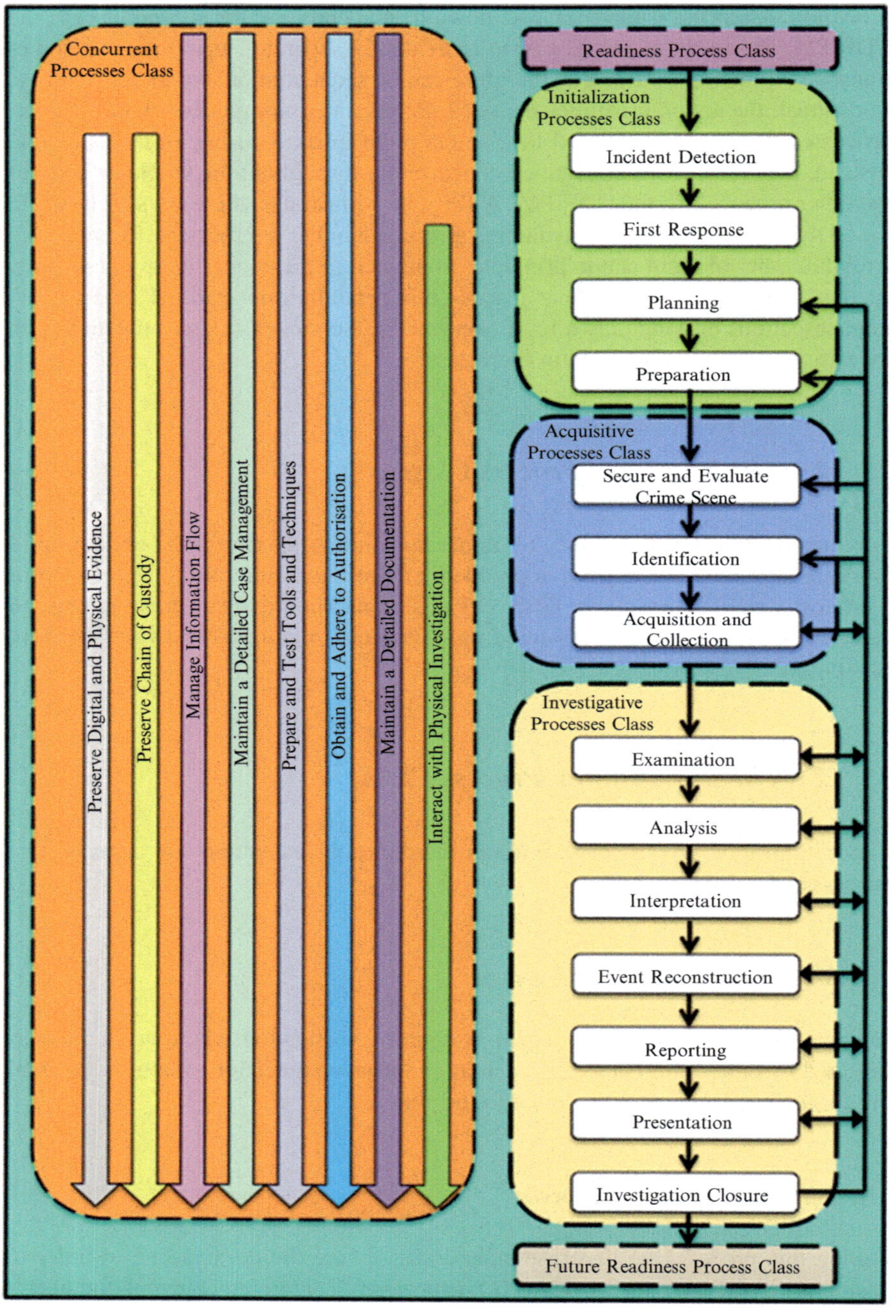

Fig. 15.1 The proposed model (the CDFIPM)

15.3.2.3 Planning Process

During this process, investigators must plan and develop proper procedures and define methodologies, the choice of tools to use and the appropriate human resources that should be involved in the investigation. At this stage, investigators are not expected to produce anything beyond a rough outline of a plan [3].

15.3.2.4 Preparation Process

The preparation process involves implementing those planning made in the previous process. This includes, but is not limited to, preparing appropriate equipment (both hardware and software), infrastructure, human resources, raising awareness, training and documentation.

15.3.3 The Acquisitive Process Class

The acquisitive process class pertains to the processes that are concerned with the acquisition of evidence; this class includes five processes described as follows.

15.3.3.1 Secure and Evaluate the Crime Scene Process

During this process, investigators must enforce a lock down of the entire crime scene to preserve the integrity of both digital device and the digital evidence [12]. Investigators must also carry out a preliminary survey of the physical crime scene to obtain an idea about how to process the physical crime scene. They should aim to identify the obvious pieces of physical evidence, identify any technical issue, determine the mixture of laboratory and on-site data acquisition and finally develop an initial theory about the crime [3].

15.3.3.2 Identification Process

The identification process enables investigators to fill the gap in relation to the location, size and format of the digital device suspected of containing the potential digital evidence [3]. During this phase, investigators must perform a methodical search for digital evidence which can be both in physical and logical form.

15.3.3.3 Acquisition and Collection Process

This process requires investigators to determine whether to carry out a live or a static data acquisition. Various factors will influence such a decision including the type of authorisation, the type of environment in which the device is operating and also the type and size of storage media (ISO/IEC 27043 49). Investigators will need to determine the most appropriate methods of acquiring digital evidence in line with the common practice suggested by ISO/IEC 27037 [21] and ACPO [1]. During the data acquisition, master copy and working copy of the raw data must be acquired by creating verifiable image of all the bits and bytes contained within the digital device. The original source and the digital evidence copies should then be verified with a proven function such as MD5 or SHA1 so that the extracted data can attain legal validity as genuine. In certain circumstances, it is not practical or permissible to acquire a digital evidence copy of the entire evidence source due to its large storage size. In such circumstances, investigators should perform a logical acquisition that targets only specific data types, directories or locations.

15.3.4 The Investigative Process Class

The investigative process class contains those processes that pertain to investigating the incident or crime that has been the reason for the digital forensic investigation. This class includes six processes described as follows.

15.3.4.1 Examination Process

During the examination process, investigators must survey the digital crime scene preferably in a forensic laboratory on the image of the system. In circumstances where this activity must be performed on a live system, investigators must ensure to perform field searches by booting the system into a trusted environment in order to prevent the modification of the digital evidence. Investigators must also identify and locate potential evidence possibly within unconventional locations. A large number of techniques might be performed to find obfuscated data which might have been deleted or hidden, and there might be large volumes of data to be examined. Therefore, automated techniques should be employed using tools such as FTK or EnCase in order to support the investigators. The data should then be harvested by giving a logical structure to the entire data set. The result of the harvesting activity is a logical structured data set in which the extracted raw data becomes structured information. This denotes that the harvested information can now be mounted and read by the original file system such as FAT or NTFS. The data also needs to be reduced to expedite the examination process due to the fact that there can be very large amount of data. Identifying known elements can enable investigators to reduce the data. Investigators will need to use the metadata and unique identifiers, such as

MD5, in order to remove known system files and other application data. The data that will remain will be modified data or data that could be uniquely attributed to the users of a specific computer system. Digital evidence with similar identifying patterns should also be classified based on the types of investigation.

15.3.4.2 Analysis Process

This process involves investigators reconstructing fragments of data based on their significance and determining a possible root cause of the incident. Based upon the results of the examination process, investigators must now be able to define what the exact characteristics of the incident are and who is to be held accountable for the incident. Investigators must be able to formulate a hypothesis of how the incident took place by reconstructing a sequence of events which have resulted in the current state of the system under investigation. Investigators must thoroughly examine and test the data that was organised in the examination process against the hypothesis that was formulated in the previous activity. Moreover, the investigators must also question the legal validity of the possible digital evidence by considering issues such as relevance, admissibility and weight. This will enable them to test the hypothesis by identifying the best possible evidence. Digital evidence should then be linked and attributed to a specific user or the event which is the root cause of the incident or crime. Finally, under this process, investigators must evaluate their findings in order to ensure that the hypothesis they have developed holds true. Backtracking from the analysis process to the examination process is often to be expected as the investigators acquire a better understanding of the events which resulted in the investigation in the first place.

15.3.4.3 Interpretation Process

Investigators must interpret digital evidence to produce meaningful statements in the legal context. After interpreting the analysis results, investigators will need to classify the interpreted evidence according to relevance by organising the evidence in a way that they can differentiate which digital evidence items are more important than the others. Event reconstruction is another activity through which investigators should be able to reconstruct a possible event sequence reflecting the incident results by using the series of events known to them that they have deduced from the digital evidence. Investigators should use this process to explain how the incident might have occurred, prior to assessing the review results against the original hypothesis formulated in the analysis process. This will be to determine whether they have obtained all the evidence required to support the original hypothesis. If all the evidence has not been obtained, investigators will then need to iterate to the analysis process, in which the hypothesis development activity will form a cycle that needs to be repeated until investigators can explain the incident. If there is no need to backtrack to the analysis process, any areas of improvement will need to be identified to address those required improvements.

15.3.4.4 Reporting Process

This process requires investigators to compile an accurate report based on the their findings constructed in an opinion to be presented to a relevant audience. This report must contain conclusions that can be reproduced by independent third parties. Also, since an investigation might produce many incriminating digital evidence items, investigators must ensure that all digital evidence items are listed in the report so that no valuable item of evidence is left out. The report must be in a simple language and be well-defined, concise and unambiguous in order for the lay person (such as judge and jury) to be able to understand it.

15.3.4.5 Presentation Process

This process involves presenting the output of the reporting process, which should be a well-written report, to a wide variety of audience such as courts of law, law enforcement and management in an organisation. Presenting the report can be carried out in the form of the report itself or can be accompanied by other formats such as multimedia presentation and expert witness, etc.

15.3.4.6 Investigation Closure Process

This process involves reviewing the existing policies and procedures of the victim organisation based on the outcome of the investigation. Lessons from the incident must be identified and learnt in order to enable the organisation to apply the findings and be better prepared for the future incidents. A decision must also be made regarding whether to return, cleanse and reuse or destroy the evidence. In certain circumstances, the evidence might need to be stored for a certain period of time before any of the three possibilities can be applied. The decision made concerning the investigation must be recorded ideally on a database for the future reference. Relevant information regarding the entire investigation must also be disseminated and communicated to all stakeholders. This includes communicating the need to return to a previous process, deciding on the acceptance or rejection of the hypothesis or providing any reports or documents from the presentation process.

15.3.5 The Future Readiness Process Class

This aim of this class is to enable victim organisations to prepare for and mitigate the risks of potential future incidents. This class involves victim organisations applying the lessons learnt and also improving their existing policies and procedures based on the review of the outcome of the case from the preceding process.

Ideally case studies should also be developed for the future reference to enable both victim organisations as well as other corporates to learn from the incident which has been investigated.

15.3.6 The Concurrent Process Class

Concurrent process class comprises of nine overriding principles or action principles that are applicable to other processes in the model. These principles are objectives that need to be achieved in a given digital investigation and should be performed concurrently throughout the whole or parts of the other processes in the CDFIPM. Maintaining these principles in a digital investigation ensures the admissibility of digital evidence in a court.

15.3.6.1 Preserve Digital and Physical Evidence

This principle refers to protecting both the physical and digital evidence against damage or alteration. In order to enable the investigators to preserve the evidence in a forensically sound manner, organisations and law enforcement agencies will need to establish and maintain certain strict procedures, effective quality systems such as standard operating procedures (SOPs) or procedural workflows.

15.3.6.2 Preserve Chain of Custody

In order to preserve chain of custody, investigators must adhere to all legal requirements and must properly document each given process within the CDFIPM. Chain of custody is of extreme importance; cases where the chain of custody has not been properly preserved can be easily challenged in courts. An example of preserving chain of custody is when evidence copies are required to be shared with other experts in other locations.

15.3.6.3 Manage Information Flow

A defined information flow should exist between each given process in a digital investigation so that it can be protected and supported technologically. An example of the information flow can be the exchange of digital evidence between two investigators involved in the same investigation. This information flow can be protected, for example, through the use of trusted public key infrastructures (PKI) and time stamping to identify the different investigators, protect the evidence integrity and also protect the confidentiality of the evidence through PKI-based encryption [13].

15.3.6.4 Maintain a Detailed Case Management

This overriding principle applies to the role of managers who often lead a team of investigators during an investigation. Managers will need to undertake certain tasks including guiding investigators in the right direction, creating an overall picture of the investigation, determining the cost of investigation, identifying team members for each given process, etc.

15.3.6.5 Prepare and Test Tools and Techniques

It is vital that investigators prepare an appropriate set of tools and techniques during the course of an investigation so that each process of the investigative process can be carried out effectively. Cases where untested tools have been used to carry out digital investigations are easily challenged in courts. Therefore, investigators must select tools that are court-approved such as EnCase, AccessData FTK and ProDiscover.

15.3.6.6 Obtain and Adhere to Authorisation

Any digital investigation that is commissioned to be carried out necessitates proper authorisation, whether it is an internal or an external authorisation. This principle ensures that the rights of the system owners, custodians, principles or users are not infringed and that no law is violated.

15.3.6.7 Maintain a Detailed Documentation

It is extremely important to document all activities carried out throughout the entire investigative process in order to enable other investigators to authenticate the process and results. This principle involves recording all information applicable or produced during the investigative process to support decision making and the legal, administrative processing of those decisions.

15.3.6.8 Interact with Physical Investigation

A digital investigation and a physical investigation are often interrelated and dependent on one another. In cases where a physical investigation requires an assistance from a digital investigation, an example can be to use a digital forensic investigation to reveal communications between terror suspects via computers, mobile phones, online social network activities, email communication, communication via chat rooms and forums, etc. An example of digital investigation being dependent on a physical investigation is when a suspect is interviewed to provide a password to a system under investigation. Defining the relationship between a digital investigation and a physical investigation is required to preserve chain of custody, preserve the integrity of the digital evidence, protect the digital evidence from damage and ensure an efficient investigation [46].

15.4 Research Methodology

The design science research (DSR), widely adopted in the domain of information systems (IS) [19, 20, 26, 32, 33, 47], has been selected as the methodology to conduct the research presented both in this paper and also in Montasari [31]. The DSR involves the design of novel or innovative artefacts and the analysis of the performance or use of such artefacts [24, 48]. The development and evaluation of artefacts form an important part in the DSR [20, 27]. Artefacts include, amongst others, models, methods, constructs, instantiations and design theories [26, 27], social innovations and new or previously unknown properties of technical, social or informational resources [27]. The artefact related to the research presented in Montasari [31] and this paper is a new model, the Comprehensive Digital Forensic Investigation Process Model (CDFIPM), that encompasses the entire digital investigative process. The organisational context associated with this research is that of law enforcement, corporates and incident response. This research also addresses the hitherto unsolved problem that there does not exist a comprehensive model encompassing the entire digital investigative process that is both formal, in that it synthesises, harmonises and extends the existing models and, in generic, in that it can be employed in the different fields of law enforcement, incident response and commerce. The selection of the DSR over other alternative methodologies (such as Requirements Engineering: RE) is justified as it is particularly suited to the task of creating a new process model (an IT artefact) [3]. Armstrong and Armstrong [7], as cited by Adams [3], state that with the DSR's focus on designing solutions, it is ideal when approaching the problem domain of digital forensics. Various researchers both within and outside the IS domain have provided guidance to define the DSR and have described what goals should be followed in its production [2, 6, 16, 19, 20, 26, 32, 33, 37, 43, 47, 49]. These researchers have often proposed various methods, processes or theoretical frameworks to rationalise the DSR studies. However, the design science research process (DSRP) model proposed by Peffers et al. [33] has been selected as the appropriate DSR to conduct the research in this paper and Montasari [31]. The rationale for doing so is due to the fact that it provides a graphical representation of the conceptual process for both carrying out and presenting the DSR. Such a mental model facilitates the application of the DSR and can also assist the author in producing and presenting a high-quality DSR that would be accepted as valuable, rigorous and publishable within the field of digital forensics science.

The Peffers et al.'s [33] DSRP consists of seven components as shown in its graphical representation in Fig. 15.2.

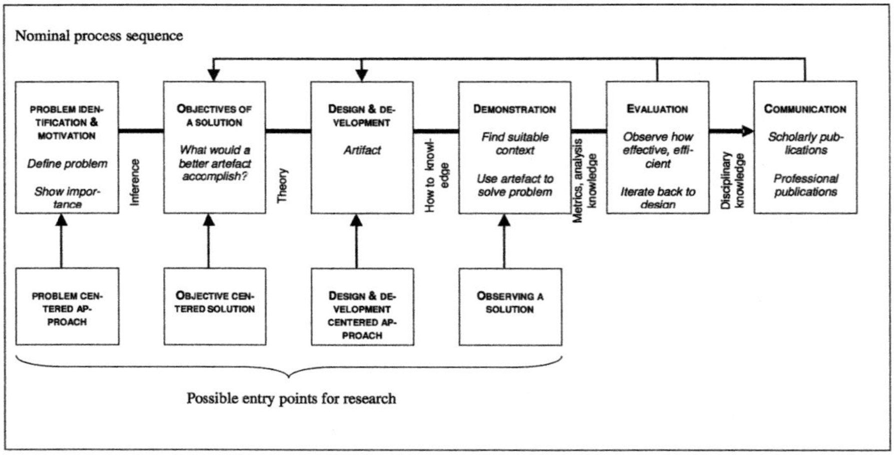

Fig. 15.2 The design science research process (DSRP) model after Peffers et al. [33]

The first three components of the DSRP, namely, problem identification and motivation, objectives of a solution and design and development, have already been covered in the previous research paper [31]. This paper covers the two out of the three remaining components, namely, the demonstration and communication activities, with the remaining component, namely, the evaluation activity, being covered in the future work (see Sect. 15.6). Nevertheless, the following three subsections including Sects. 15.3.1, 15.3.2, and 15.3.3 describe how the first three components of the Peffers et al.'s [33] DSRP were applied to the research presented in Montasari [31]. The remaining three subsections including Sects. 15.3.4, 15.3.5, and 15.3.6 then discuss how the two out of the three remaining components have been applied to the research in this paper and how the remaining component will be applied to the future work.

15.4.1 Problem Identification and Motivation

The research presented in Montasari [31] addressed the following problem:

> *That there does not exist a comprehensive model encompassing the entire digital investigative process that is formal in that it synthesises, harmonises and extends the existing models and that is generic in that it can be applied in the different fields of law enforcement, commerce and incident response*

A profound knowledge of the problem addressed by the research was acquired through a detailed analysis and assessment of the literature related to previous DFIPMs in Montasari [30] and Montasari and Peltola [29].

15.4.2 Objectives of a Solution

The aim of the research was as follows:

To develop a comprehensive model encompassing the entire digital investigative process that is formal in that it synthesises, harmonises and extends the exiting models and that is generic in that it can be applied in the different fields of law enforcement, commerce and incident response

The research aim was formulated according to the definition of the problem. In order to formulate the stated aim, a deep knowledge was acquired of the state of the problem and its current solution in the form of previously proposed DFIPMs and their efficacy.

15.4.3 Design and Development

Prior to designing and developing the CDFIPM, all previously proposed models were analysed in order to identify which could contribute to the new model. Law enforcement, commerce and incident response were the three environments on which the research in Montasari [31] focused. Therefore, the existing models within those three domains were considered for their possible contributions to the new model. Once the most reliable models were identified, their specific key contributions were determined for inclusion in the new model. Following this, the essential components necessary for the new model were identified from the specific key contributions. These formed the basic structure of the CDFIPM. The prevailing models were then built upon by the construction of a new set of domain-specific components. Contribution of the previous models in the form of identified components as well as the new set of constructed components were used to develop the new model. The CDFIPM was graphically represented in the form of UML Activity Diagram.

15.4.4 Demonstration

Following the design and development of the CDFIPM, its implementation needed to be demonstrated within an appropriate environment. Peffers et al. [33] as well as various other researchers in the field [4, 9, 11–13, 46] suggest the use of case studies as such an appropriate environment. Therefore, the CDFIPM was applied to a case study (see Sect. 15.5) in order to demonstrate its application and effectiveness within the three stated domains.

15.4.5 Evaluation

Based on the DSRP's requirements, the CDFIPM will also need to be evaluated to determine how well it supports the solution to the stated problem. The evaluation activity, which is not covered in this paper but in the future work (see Sect. 15.6), will aim to compare the CDFIPM's application and effectiveness with the characteristics set out in its research aim. Evaluation activity will involve the submission of the CDFIPM to digital forensic practitioners within the three domains that are the focus of this research and also to judges, barristers and researchers in academia, those being experts within two other domains to which the model has relevance. The aim of such an approach will be to enable the author to acquire insightful and reliable feedback as to the effectiveness of the CDFIPM from authoritative external reviewers (an approach also undertaken by other researchers such as those in [3, 32, 36]. Once the evaluation has been carried out, the author will have been able to judge whether to repeat the design and development phase of the CDFIPM in order to make improvements. Any such amendments will be subsequently introduced to the design and implementation stages of the CDFIPM.

15.4.6 Communication

Again in accordance with the DSRP's requirements, the problem addressed by this research and its importance, its solution (the CDFIPM), its utility and novelty, the rigour of its design and implementation and its effectiveness all needed to be communicated to the intended user community. Therefore, the communication activity of the DSRP in relation to this research was achieved through publications such as Montasari [30, 31] and Montasari and Peltola [29] in well-known and peer-reviewed journals and conferences. In addition, there was direct interaction with a wide variety of experts, including digital forensic practitioners, legal practitioners and experts in academia.

15.5 Testing the CDFIPM

This section follows the demonstration activity of the Peffers et al.'s [33] DSRP, used in this research. The demonstration activity of the DSRP requires a researcher to apply the artefact in an appropriate environment such as 'experimentation' and 'case study' to solve the stated problem [3, 8, 19, 33]. Resources needed for the demonstration activity include effective knowledge of how the artefact should be applied to solve the stated problem. Therefore, in order to assess how the CDFIPM addresses the stated research problem, the CDFIPM is applied into a case study and a walkthrough of the model is performed. The case study presented is based on an actual situation and is intended to demonstrate the potential deployment of the CDFIPM. The following case study, which is modelled after Ciardhuáin [13], relates to the exploitation of a vulnerability found in an online service operated by a bank:

15.5.1 Case Study

The following case study is modelled after Ciardhuáin's [13] paper. This investigation started when Bank X in London (England) received an email claiming to have found a vulnerability in an online service operated by the bank. The email offered to provide details of the vulnerability in exchange for payment. On checking their logs, Bank X concluded that an unauthorised access had been made to their web server. The bank received further emails threatening to reveal the vulnerability to the press and public, including a link to a website which the suspect intended to use to disclose the vulnerability. Bank X reported the issue to the police in London who initiated the investigation. It became obvious that the compromised web server was located in Manchester (England) from Bank X's headquarters and that the source of the emails was in Cardiff (Wales). Therefore, another police force, namely, the South Wales Police (SWP), took up the case to start the investigation. Throughout the following stages of the digital investigative process, investigator R and investigator P adhere to all the eight overriding principles of the CDFIPM (see Sect. 15.2).

15.5.1.1 Readiness Process

Bank X has already implemented both operational and infrastructure readiness capabilities and has an in-house incident response team and procedures for forensic readiness and incident detection implemented. The bank also has its own standard operating procedures (SOPs).

15.5.1.2 Incident Detection Process

The first step in this investigation is the incident detection and the creation of awareness that the investigation is needed. In this case, the incident has been reported by the suspect himself to the Bank X. After the incident has been detected, the bank requests its senior IT Administrator, Mr. Thompson, to look into the issue to confirm or refute the validity of the incident as this might be a hoax. Mr. Thompson contacts the head of the incident response team for assistance in this matter. To validate and assess the incident, Mr. Thompson and the incident response team examine the emails and log files and confirm that the system's security has been compromised. Bank X then reports the incident to the London MPS (Metropolitan Police Service), who initiates the investigation of its own. It becomes clear that the compromised web server is based in Manchester and that the suspect is located in Cardiff; therefore, the investigation is passed to SWP (the South Wales Police). Up to this point, the reporting of the incident has taken place three times: when Bank X receives the emails, when the bank reports it to the MPS and when the investigation is passed to the second police force, SWP. During this process, both internal and external authorisations are needed. The internal authorisation is obtained when Bank X instructs its senior IT administrator to conduct the investigation. The external authorisation is acquired when the MPS realises that SWP are the police force who are authorised to carry out the investigation. The search warrant is the example of this authorisation. Moreover, during this process, detailed and contemporaneous documentation is made.

15.5.1.3 First Response Process

Due to the fact that this incident involves law enforcement and is investigated externally, this process is only partly applicable to this case scenario. The application of this process is when the incident response team assists the senior IT administrator in examining the log files.

15.5.1.4 Planning Process

The Planning activity is conducted by both Bank X and the two police forces. This activity takes place in the bank's investigation when they perform an examination of the logs and decide to involve the police based on what they have found. The Planning activity also takes place in the two police forces investigations where they plan their own respective approaches to be undertaken to identify the suspect and collect the needed evidence. Under the planning stage, the two police forces consider data constraint, timing constraint, physical constraint and authorisation, as well as performing risk assessment, planning logistics and creating their own outline plans.

15.5.1.5 Preparation Process

Under the preparation activity, Bank X and the two polices forces simply implement the plans that they have drawn in the Planning stage.

15.5.1.6 Secure and Evaluate the Crime Scene Process

This activity takes place when the SWP police officers raid the premises of the suspect's place of employment. The first step they take is to address the safety issues such as the safety of the officers and employees, followed by preserving the crime scene. Since the suspect is at the crime scene, he is detained and briefly interviewed as the authorisation allows the questioning of the suspect. Investigators then survey the crime scene in order to determine the location of digital device(s) and establish the combination of on-site and off-site data acquisition. Throughout the entire process, detailed contemporaneous notes of all activities are maintained.

15.5.1.7 Identification Process

Identification initially takes place when Bank X identifies their log files to determine what has occurred. Both police forces, MPS and SWP, later carry out the same activity to locate the sources of the emails. Moreover, SWP conducts a physical search which results from the information obtained from the previous searches. Secure and evaluate crime scene process and identification processes both overlap as each process requires searching the physical crime scene.

15.5.1.8 Acquisition and Collection Process

This process takes place when the search of the employer's premises in the previous process led to the seizure of a computer. Since the suspect's computer system is not a mission-critical system and the authorisation permits its seizure, the investigators decide to size the system and conduct an off-site data acquisition in the police forensic laboratory. However, since the system is running, officers decide to conduct a live acquisition of volatile data first prior to shutting down the system in case the RAM might contain valuable information which might be lost after powering down the system. Using FTK, the officers perform a live acquisition of the volatile data and duplicate the master copy of the captured image of RAM. Both copies are then verified using MD5 and SHA1 checksums. Since the data on the system is stable, the officers remove the power source directly from the suspect's computer. They then record, remove and secure connections prior to labelling and packaging the system. The transport phase takes place when the system is seized and physically transferred to the police. This phase also occurs in three other occasions including when the captured image of the RAM is taken to the police, when log files are transferred from server to the police for later examination and analysis and when the emails are transferred from the bank to the police. The storage phase occurs when the police retain the seized computer, the captured images of both hard drive and RAM, log files as well as emails in a secure storage facility. In the forensic laboratory, the investigators image the hard drive of the system and verify it using MD5 and SHA1 checksums. They also duplicate the master copy to become the working copy on which the subsequent Examination and Analysis will be performed.

15.5.1.9 Examination Process

This activity initially occurs when the bank examines their log files. It also occurs when the police examine log files, emails and the working copy of both hard drive and RAM images of the suspect's system in the forensic laboratory. During the examination process, investigators process the deleted and hidden data to ensure that emails and log files are recognised from the evidence. Investigators also harvest data to provide structure to data which they are interested in so that it can be mounted on the investigating machine. Since the suspect's system contains large amount of data such as known systems files, investigators use metadata and unique identifiers to reduce the data by removing known system files and different other application data. The investigators are now left with the log files and the emails that the suspect sent to Bank X. Now, the emails can be uniquely attributed to the suspect who has been the user of that specific system.

15.5.1.10 Analysis Process

This process initially takes place when Bank X's system administrator, Mr. Thompson, and the incident response team conclude from the log files that an unauthorised access has been made to their web server. Later on, this process is conducted in a forensic laboratory by investigators who develop the initial hypothesis for the identity of the suspect and for the manner in which the incident has taken place.

15.5.1.11 Interpretation Process

This process occurs after investigators evaluate their findings in the analysis process and determine that their formulated hypothesis is true. During this process, investigators interpret digital evidence to produce meaningful statements for later reporting and presentation regarding how the suspect made an unauthorised access to Bank X's web serve. As part of this process, investigators classify and organise the interpreted evidence such as log files and emails according to their relevance in order to distinguish which digital evidence items are more important than the others.

15.5.1.12 Event Reconstruction Process

This process occurs when investigators reconstruct the events which led to the identification of the suspect and the subsequent seizure of the suspect's computer. This entails the investigators iterating in the CDFIPM and results in a more detailed hypothesis. Through this process, investigators are able to explain how the suspect has carried out the intrusion to Bank X's web server. Investigators consolidate and review their findings prior to assessing the results of their review against the original hypothesis that they have formulated. Through this assessment, investigators ensure that they have gathered all relevant evidence related to that attack to support their hypothesis.

15.5.1.13 Reporting Process

This process takes place when investigators compile a report based on their findings to be presented in a court.

15.5.1.14 Presentation Process

This process occurs five times during the entire investigation. This includes when the bank's IT administrator presents the evidence to the management within the bank, when the bank approaches the MPS police and present their evidence to investigators, when the MPS police pass the investigation to SWP, when evidence is presented to acquire a search warrant and when investigators present the evidence in the court. The formality of the evidence increases as the investigation proceeds. Prior to presenting the findings to the court, investigators meet with the legal team to understand the presentation requirements. Through the meeting, the target audience in the court are determined. Investigators also carry out the necessary preparation prior to attending the court such as preparing expert testimony, exhibits and appropriate presentation aids. During the presentation in the court, investigators are able to assist the judge and jury in understanding the technical points made by avoiding complex arguments and delivering their conclusion in a logical and structured manner. The presentation contains factual data that investigators have deduced. Moreover, investigators use the CDFIPM to enable the judge to comprehend the processes that they have followed during the investigation. In the court, the investigators have to prove and to defend the validity of the hypothesis as it is challenged by the court and the defence lawyers. Since investigators have followed a formal model which enabled them to carry out the investigation in a forensically sound manner, the opposite hypothesis is refuted. The court decides that the suspect has made an unauthorised access to Bank X's web server and sentences him to prison.

15.5.1.15 Investigation Closure Process

This process takes place after a formal decision is reached by the court concerning the incident. During this process, based on the outcome of the investigation, Bank X reviews its existing policies and procedures concerning its IT security. As there is no need to backtrack to the previous stages in the investigation, bank management decide to accept the hypothesis. During this process, the bank identifies the lessons learnt and the suspect's system is returned to his employer's company. The result of the case is recorded on the database for the future readiness. The final phase of this activity is disseminate investigation results, where relevant information concerning this incident and its outcome are communicated to all stakeholders. The initial communication is carried out prior to the completion of the trial in order to remove the sensitive data from the disseminated information. The final phase, the review phase, discusses how the incident could have been handled better. One outcome is to instal the central log server sooner than planned. The review phase includes a review of the investigation and identifies some new analysis techniques that are employed. The techniques are added to the official analysis procedures. During the analysis, new suspect files are identified, and they are added to the hash database so that they can be quickly found in the future investigations.

15.5.1.16 Future Readiness Process

This activity takes place both in Bank X and also the two police forces. Bank X starts applying the lessons that they have learnt from this incident; a set of recommendations are made on how they can improve in terms of securing their digital data assets. As a result, they improve their existing forensic readiness procedures and incident detection systems. The two police forces also develop case studies based on this particular investigation for future training of other officers.

15.5.2 Case Study Discussion

This section followed the demonstration activity of Peffers et al.'s [33] DSRP and discussed the method employed to demonstrate the potential deployment of the CDFIPM. In order to assess how the CDFIPM addressed the stated research problem, the model was applied to a case study and a 'walkthrough' of the model was performed. The walkthrough of the CDFIPM employing the case study successfully mapped the entire processes of the model to the corresponding activities carried out by digital forensic investigators. This method of testing the CDFIPM using the case study clearly demonstrated that the proposed model is very efficient in enabling investigators to account for every investigative process carried out through the iterative structure of the CDFIPM. Notice that it would have been possible to investigate this case study using a different model other than the CDFIPM. However, it is argued that the application of the CDFIPM to any digital investigation within the three stated domains covered by the research scope would be more effective as the proposed model has inherited all the benefits of the previous models by rigorously synthesising, harmonising and building upon them. Having completed the demonstration activity, the next stage in the Peffers et al.'s [33] DSRP is the evaluation activity, which will need to be conducted by a number of digital forensic investigators, legal practitioners, experts and researchers in the field of digital forensics. The evaluation activity, however, will be the subject of future work.

15.6 Conclusion and Future Work

A new and Comprehensive Digital Forensic Investigation Process Model has now been described in Montasari [31] and tested in this paper. Synthesising, harmonising and building upon the previous models as well as including the overriding principles in the new model have made the CDFIPM much more comprehensive than previous models. It is argued that the CDFIPM provides a foundation for the development of techniques and especially tools to support the work of investigators. Although the CDFIPM is mainly aimed at the UK jurisdiction, it could easily be adapted to other jurisdictions. Without such modification, the model already has relevance to those jurisdictions which employ a similar legal basis for evaluating the digital investigative process. As the future work, to determine the CDFIPM's usability and utility further, an independent evaluation of the CDFIPM will need to be carried out by the high-tech crime units (HTCUs) of different police forces, experts in corporate and incident response environments, legal practitioners and researchers in academia. Similarly, although the case study to which the CDFIPM was applied represents each of the three fields of digital forensics to which the model is relevant, the CDFIPM would benefit from further case studies to identify task hierarchies. This would improve task development efforts and facilitate scenario development, assisting its users, researchers and tool developers in understanding how to take advantage of and apply the model. Finally, as with the future work, the CDFIPM must be validated through the method validation under ISO 17025 accreditation, which will be published in the UK within the next 18 months (as of July 2016), to determine its compliance.

References

1. ACPO. (2012). ACPO good practice guide for digital evidence. U.K. Association of Chief Police Officers.
2. Adams, L., & Courtney, J. (2004). Achieving relevance in IS research via the DAGS framework. *37th Hawaii International Conference on System Sciences* (pp. 1–10). Big Island, HI, USA.
3. Adams, R. (2012). The Advanced Data Acquisition Model (ADAM): A process model for digital forensic practice. PhD thesis. Murdoch University.
4. Adams, R., Hobbs, V., & Mann, G. (2014). The advanced data acquisition model (ADAM): A process model for digital forensic practice. *Journal of Digital Forensics, Security and Law, 8*(4), 25–48.
5. Agarwal, A., Gupta, M., Gupta, S., & Gupta, C. (2011). Systematic digital forensic investigation model. *International Journal of Computer Science and Security, 5*(1), 118–130.
6. Archer, L. (1984). *Systematic method for designers*. London: Wiley.
7. Armstrong, C., & Armstrong, H. (2010). Modeling forensic evidence systems using design science. *IFIP WG International Working Conference* (pp. 282–300).
8. Balci, O. (2004). Quality assessment, verification, and validation of modeling and simulation applications. *Proceedings of the 2004 Winter Simulation Conference* (pp. 1–8). Washington DC.

9. Beebe, N., & Clark, J. (2005). A hierarchical, objectives-based framework for the digital investigations process. *Digital Investigation, 2*(2), 147–167.
10. Carlton, H., & Worthley, R. (2009). An evaluation of agreement and conflict among computer forensic experts. *42nd Hawaii International Conference on System Sciences* (pp. 1–10). Washington DC.
11. Carrier, B., & Spafford, E. (2003). Getting physical with the digital investigation process. *International Journal of Digital Evidence, 2*(2), 1–20.
12. Casey, E. (2011). *Digital evidence and computer crime: Forensic science, computers and the internet* (3rd ed.). New York: Elsevier Academic Press.
13. Ciardhuáin, O. (2004). An extended model of cybercrime investigations. *International Journal of Digital Evidence, 3*(1), 1–22.
14. Cohen, F. (2011). Putting the science in digital forensics. *Journal of Digital Forensics, Security and Law, 6*(1), 7–14.
15. Cohen, F. (2012). Update on the state of the science of digital evidence examination. *Proceedings of the Conference on Digital Forensics, Security, and Law* (pp. 7–18). Richmond, USA.
16. Eekels, J., & Roozenburg, N. (1991). A methodological comparison of the structures of scientific research and engineering design: Their similarities and differences. *Design Studies, 12*(4), 197–203.
17. Freiling, C., & Schwittay, B. (2007). A common process model for incident response and computer forensics. *3rd International Conference on IT-Incident Management & IT-Forensics* (pp. 19–40). Stuttgart, Germany.
18. Garfinkel, S., Farrell, P., Roussev, V., & Dinolt, G. (2009). Bringing science to digital forensics with standardized forensic corpora. *Digital Investigation, 6*, 2–11.
19. Hevner, A., & Chatterjee, S. (2010). *Design research in information systems*. New York: Springer.
20. Hevner, A., March, S., Park, J., & Ram, S. (2004). Design science in information systems research. *MIS Quarterly, 28*(1), 75–105.
21. International Organisation for Standardization. (2012). *ISO/IEC 27037:2012. Information technology–Security techniques–Guidelines for identification, collection, acquisition and preservation of digital evidence*. Geneva: International Organization for Standardization.
22. Kessler, C. (2010). Judges' awareness, understanding, and application of digital evidence. PhD thesis, Nova Southeastern University.
23. Kohn, M., Eloff, M., & Eloff, J. (2013). Integrated digital forensic process model. *Computers & Security, 38*, 103–115.
24. Kuechler, B., & Vaishnavi, V. (2008). On theory development in design science research: anatomy of a research project. *European Journal of Information Systems, 17*(5), 489–504.
25. Leigland, L., & Krings, A. (2004). A formalization of digital forensics. *International Journal of Digital Evidence, 3*(2), 1–32.
26. March, S., & Smith, G. (1995). Design and natural science research on information technology. *Decision Support Systems, 15*(4), 251–266.
27. March, S., & Storey, V. (2008). Design science in the information systems discipline: An introduction to the special issue on design science research. *MIS Quarterly, 32*(4), 725–730.
28. Mason, S. (2007). *Electronic evidence: Disclosure, discovery and admissibility*. London: LexisNexis Butterworths.
29. Montasari, R., & Peltola, P. (2015). Computer forensic analysis of private browsing modes. In *Proceedings of 10th international conference on global security, safety and sustainability: Tomorrow's challenges of cyber security* (pp. 96–109). London: Springer International Publishing.
30. Montasari, R. (2016). An Ad Hoc detailed review of digital forensic investigation process models. *International Journal of Electronic Security and Digital Forensics, 8*(3), 203–223.
31. Montasari, R. (2016). A comprehensive digital forensic investigation process model. *International Journal of Electronic Security and Digital Forensics (IJESDF), 8*(4), 285–301.

32. Nunamaker, J., Chen, M., & Purdin, T. (1990). Systems development in information systems research. *Journal of Management Information Systems, 7*(3), 89–106.
33. Peffers, K., Tuunanen, T., Gengler, C., Rossi, M., Hui, W., Virtanen, V. & Bragge, J. (2006). The design science research process: A model for producing and presenting information systems research. *1st International Conference on Design Science Research in Information Systems and Technology* (pp. 83–106). USA.
34. Pollitt, M. (2009). The good, the bad, the unaddressed. *Journal of Digital Forensic Practice, 2*(4), 172–174.
35. Reith, M., Carr, C., & Gunsch, G. (2002). An examination of digital forensic models. *International Journal of Digital Evidence, 1*(3), 1–12.
36. Rogers, M., Goldman, J., Mislan, R., Wedge, T. & Debrota, S. (2006). Computer forensics field triage process model. *Conference on Digital Forensics, Security and Law* (pp. 27–40). Las Vegas, USA.
37. Rossi, M., & Sein, M. (2003). Design research workshop: A proactive research approach. *26th Information Systems Research Seminar in Scandinavia* (pp. 9–12). Haikko, Finland.
38. Rowlingson, R. (2004). A ten step process for forensic readiness. *International Journal of Digital Evidence, 2*(3), 1–28.
39. Selamat, S., Yusof, R., & Sahib, S. (2008). Mapping process of digital forensic investigation framework. *International Journal of Computer Science and Network Security, 8*(10), 163–169.
40. Sherman, S. (2006). A digital forensic practitioner's guide to giving evidence in a court of law. *Proceedings of the 4th Australian Digital Forensics Conference* (pp. 1–7). Perth Western, Australia.
41. Smith, R., Grabosky, P., & Urbas, G. (2011). *Cyber criminals on trial*. Cambridge: Cambridge University Press.
42. Stanfield, A. (2009). *Computer forensics, electronic discovery and electronic evidence*. Chatswood: LexisNexis Butterworths.
43. Takeda, H., Veerkamp, P., Tomiyama, T., & Yoshikawa, H. (1990). Modeling design processes. *AI Magazine, 11*(4), 37–48.
44. US-CERT. (2012). Computer forensics. U.S. Department of Homeland Security. Available at: https://www.us-cert.gov/security-publications/computer-forensics. Accessed 17 June 2016.
45. Venter, J. (2006). Process flow for cyber forensics training and operations. Available at: http://researchspace.csir.co.za/dspace/handle/10204/1073. Accessed 17 June 2015.
46. Valjarevic, A., & Venter, H. (2015). A comprehensive and harmonized digital forensic investigation process model. *Journal of Forensic Sciences, 60*(6), 1467–1483.
47. Walls, J., Widmeyer, G., & El Sawy, O. (1992). Building an information system design theory for vigilant EIS. *Information Systems Research, 3*(1), 36–59.
48. Watts, S., Shankaranarayanan, G., & Even, A. (2009). Data quality assessment in context: A cognitive perspective. *Decision Support Systems, 48*(1), 202–211.
49. Wieringa, R. (2009). Design science as nested problem solving. *4th International Conference on Design Science Research in Information Systems and Technology* (pp. 8–19). Philadelphia, USA.
50. International Organisation for Standardization. (2015). *ISO/IEC 27043:2015. Information technology–Security techniques–Incident investigation principles and processes*. Geneva: International Organization for Standardization.

Chapter 16
Systematic Literature Review of Using Knowledge Management Systems and Processes in Green ICT and ICT for Greening

Alexandra Klimova

16.1 Introduction

Our world is facing complex challenges including climate change; water, air, and soil pollution; and resource depletion. Those problems require urgent attention. Therefore, environmental protection and green energy became top priorities in many countries, and various principles of sustainable development have been integrated into national policies and programs [3, 37, 38, 60].

One of the most widely adopted definitions of sustainability is so-called Brundtland definition, which was given by the World Commission on Environment and Development [69]. According to it, sustainability means "development that meets the needs of the present without compromising the ability of future generations to meet their needs."

One of the providers of better opportunities in a journey toward sustainability in the knowledge economy is knowledge management (KM). It is a "trans-disciplinary approach to improving organisational outcomes and learning, through maximising the use of knowledge. It involves the design, implementation and review of social and technological activities and processes to improve the creating, sharing, and applying or using of knowledge" [4].

Knowledge could be tacit and explicit [45]. "Explicit knowledge typically refers to knowledge that has been expressed into words and numbers. Such knowledge can be shared formally and systematically in the form of data, specifications, manuals, drawings, audio and videotapes, computer programs, patents, and the like" [6]. "Tacit knowledge includes insights, intuitions, and hunches. It is difficult to express and formalize, and therefore difficult to share." [6]. Managing both tacit and explicit knowledge is the challenge of KM.

A. Klimova (✉)
ITMO University, Kronverksky pr 49, Saint Petersburg 197101, Russia
e-mail: alexandra.kgsu@gmail.com

M. Dastbaz et al. (eds.), *Technology for Smart Futures*,
DOI 10.1007/978-3-319-60137-3_16

The KM concept was introduced in the beginning of 1990. Since then, knowledge has been considered as the key of a sustainable competitive advantage in the new economy and an essential player for the improvement and achievement of performance excellence [51]. Therefore, many scientists tried to evaluate the importance of KM for sustainable development and assess its impact on the subject. According to Wong [68], "KM and sustainable development has been coined for slightly more than decade now, as an essential solution to the global crisis." Results of the empirical assessment of KM application to sustainability [39] showed that KM is very important for innovation and sustainable development, especially for efficient use of resources. The same author mentioned that information and communication technology (ICT) is a very effective tool for knowledge mobilization in such a knowledge-intensive field as sustainable development. Furthermore, Faucheux and Nicolaï [15] claimed that ICT has the same influence on the knowledge economy as electricity had on the industrial revolution.

The article focuses on investigation of the potential contribution of KM to green ICT and ICT for greening. The paper is organized as follows. In the next section, relationships between KM, ICT, and sustainability are described. The section entitled "Research Method" provides the methodology of selection, extraction, and analysis of the literature. Research results provide an analysis of the body of literature, summary of the papers, description of knowledge management tools, and examples of using KM tools and technologies for green ICT and ICT for greening.

16.2 Sustainability, KM, and ICT: Relationships Between Three Aspects

Relations between KM, ICT, and sustainable development can exist in different combinations and proportions. Figure 16.1, which was adapted from Mohamed et al. [39] with modifications, shows possible combinations.

16.2.1 KM and ICT

More than 20 years ago, spoke about incorporation of ICT practices into KM "as a combination of the capabilities of technology and the generic features of KM, for example considering the Internet as a knowledge repository." Today ICT plays a vital role for implementation of KM applications [6], by supporting processes of knowledge discovering, sharing, and capturing an application within the organizations. Even though social aspects of organizations, such as communities of practice (CoP), are also important, web-based technologies, artificial intelligence, expert-based and decision support systems, object-oriented and relational databases, semantic networks, etc. continue to transform the field of KM. Garsia-Alvarez [19], who presented the analysis of effects of ICT in KM and innovation, emphasized importance of ICT for both tacit and explicit knowledge. It is obvious that the use of cutting-age ICT enables significant improvement in KM.

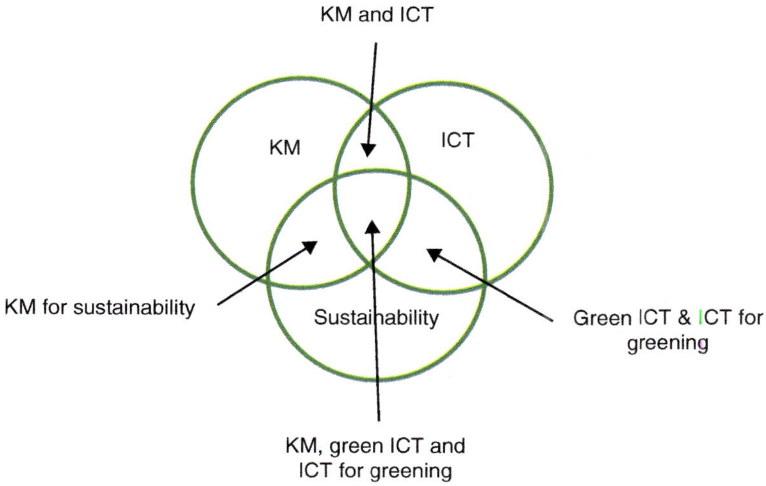

Fig. 16.1 Relations between KM, ICT, and sustainability (Source: Adapted from Mohamed et al. [39])

16.2.2 KM for Sustainability

As well known, sustainable development consists of three components: ecological, social, and economical. Presented research is mostly concentrated on the ecological component, in particular, on the connections between KM and sustainable ICT. The description of "sustainable ICT" will be given later. However, it is worth mentioning that a large number of papers are dedicated to influence of knowledge management to economical and social aspects of sustainable development ([20, 36, 51, 68], etc.). These papers emphasize the role of KM for organizational change and development of sustainable competitive advantages. KM systems and practices improve core business competencies and organizational commitment, foster innovation development, and contribute to decision-making processes. McNeil [36] also emphasized increases in opportunities for innovation and operational excellence while presenting KM benefits for the business, for the community, and for the individual.

16.2.3 Green ICT and ICT for Greening

Since the 2000s, ICT has become an essential player in the journey toward a low-carbon economy. The Smart 2020 report [62] recommended the intensive deployment of ICT both for enhancing the monitoring of the environment and human activities (in industry, building, transport, etc.) and for distributing smart ICT systems to mitigate pollution; waste, food quality, and supply problems; energy constraints; etc. Certain conceptual frameworks have been developed for the classification of ICT effects on the environment and sustainability [22].

Murugesan [42] defined "green IT" as "the study and practice of designing, manufacturing, using, and disposing of computers, servers, and associated subsystems [...] efficiently and effectively with minimal or no impact on the environment." Other definitions are "the using of IT resources in an energy-efficient and cost-effective manner" [8] and "an initiative to encourage individuals, groups, and organisations engaged in the use of ICT to consider environmental problems and find solutions to them" [11]. Green ICT includes such areas as green software development; data center design, layout, and location; power management; server virtualization; recycling; green metrics; etc. [42].

The term "green ICT" should be distinguished from the term "ICT for greening," or "greening by ICT." Indeed, green ICT is concerned with how to make ICT goods and services more sustainable over their whole life cycle, mainly by reducing the energy and material flows they invoke [22], whereas ICT for greening describes how ICT can be used to make other sectors more environmental friendly.

There are numerous papers, which study KM and ICT (e.g., [9, 26, 43]), KM for sustainability (e.g., [20, 36, 51, 68]), ICT and sustainability (e.g., [22, 46, 52]). However, there is a lack of literature, which examines the relationship between all three concepts.

16.3 Research Method

In the study, the systematic literature review methodology was applied for data selection, extraction, analysis, and syntheses [56, 64].

The main objective of that work is to offer researchers a comprehensive review of previous papers related to applying KM to green ICT and greening by ICT.

The process of data selection includes a systematic periodic search for articles related to sustainability and knowledge management. As a search engine, the following databases were used: Science Direct, Emerald, and Scopus databases. The following journals were also explored: ACM Transactions of Knowledge Discovery from Data, Data Mining and Knowledge Discovery, IEEE Transactions of Knowledge and Data Engineering, Journal of Cleaner Production, Knowledge and Information Systems, Knowledge Management for Development, Knowledge-Based Systems Journal, and Knowledge and Process management.

For the research the following keywords were used: "knowledge management" and "knowledge" along with "green ICT," "sustainability," and "greening by ICT."

In this research were considered only articles published in peer-reviewed journals for the period of 12 years, starting from 2005. Books, book chapters, and conference proceedings were also included.

Based on the abstracts, those papers were excluded from the selection, which contained information about societal sustainability or concentrated only on green ICT and ICT for greening. Published papers, which contained information only on sustainability or sustainability and ICT, regardless of knowledge management, were also excluded.

In order to retrieve the full publications for the research, Science Direct was used. The final sample of the study consists of 41 papers.

16.4 Research Results

Table 16.1 gives a list of the 41 publications sampled for this study. The information includes the following: the type of the sector (green ICT or ICT for greening), aim of the paper, and type of ICT tool or technology to support sustainability.

Figure 16.2 presents results of distribution publications by the field of technology. As it is stated, only 7% of the sample dedicated for implementation of KM to green ICT. Majewski et al. [35] described a combination of KM with business process simulation and its influence in green ICT. Two other publications are dedicated to green software development [1] and green datacenters [17], where an integrated mechanism to identify energy-saving opportunities within data centers was presented.

Below key KM applications to ICT for greening are presented.

16.4.1 Green Constructions

The majority of papers from the sample (24%) are dedicated to the use of KM and ICT in green constructions. The US Environmental Protection Agency defines green construction as "the practice of creating structures and the using of processes that are environmentally responsible and resource-efficient throughout a building's life-cycle: from siting to design, construction, operation, maintenance, renovation, and demolition" (https://www3.epa.gov/). Green constructions include efficient use of energy and water, protection of the occupant's health, reduction in waste, and pollution and environmental degradation.

The fact that the amount of papers dedicated to green constructions is so high is not surprising. The building sector is considered as one of the major consumers of energy. Indeed, according to International Energy Agency [25], heat represented 47% of the final energy consumption in 2009 with the building sector accounting for more than 40% of the final energy consumption in the EU. As a result, building energy efficiency attracted increasing research efforts in recent years. At the same time, benefits of green buildings are obvious [44]: air and water quality improvement, waste streams reduction, natural resources conservation and restoration, minimizing global warming, etc. Therefore, as noted by Shelbourne et al. [55], the construction industry is having a tendency to include sustainability aspects into its daily practices.

Nilashi et al. [44] described a knowledge-based expert system for assessing the performance level of green buildings from the environmental, social, and economical perspectives. They presented a tool, which operated on the basis of fuzzy logic, and analyze the effect of different factors in sustainable building development.

Hercheui [21] analyzed behavior change under the influence of knowledge management tools. For the illustration they use Microsoft Holm, free-of-charge Internet tool, which helps householders to better manage energy consumption. Data obtained from the best management practices were compared with household data, and based on that, householders have received instructions on how to improve energy consumption.

Table 16.1 Summary of the papers

№	Authors (year)	Type of the sector	Aim of the paper	ICT tool or technology
1	Abdullah et al. [1]	Green software development (GSD)	To assesses literature on GSD in regard to the evolution of green computing and discuss how KM comes in to assist in managing the knowledge of GSD	Warehouse
2	Ali and Avdic [2]	Sustainable rural development	To provide a framework for knowledge management in sustainable rural development and an inventory of existing frameworks for that	Internet/intranet web portal
3	Awasthi et al. [5]	Sustainable transport	To present a multi-criteria decision-making approach (MDMA) for selecting sustainability transportation systems under uncertainty	MDMA; Fuzzy TOPSIS
4	Belkadi et al. [7]	Sustainable manufacturing	To explore the capabilities of knowledge-based frameworks combined to PLM approaches as a backbone for supporting resources optimization	Global KB-PLM framework
5	Cash et al. [10]	Environmental monitoring	To develop a framework for understanding the effectiveness of systems that link knowledge to action for sustainability	Not clear
6	Cinar and Kayakutlu [12]	Energy sector	To propose a decision model that will support the researchers in forecasting and scenario analysis fields	Bayesian network (BN) models
7	Dong and Hussain [13]	Healthcare	Framework for a semantic service matchmaker that takes into account the ambiguous, heterogeneous nature of service information in digital health ecosystems	Semantic service matchmaking (Semantic Health Service Search Engine)
8	Dorasamy et al. [14]	Disasters management	Systematic review of papers pertaining to the application of knowledge-driven systems in support of emergency management	Knowledge-based emergency management information systems (EMIS)
9	Fan et al. [16]	Green buildings	To present a time series data mining methodology for temporal knowledge discovery in big building automation system data	Data mining
10	Ferreira and Pernici [17]	Green datacenters	To explore the characteristics of knowledge-based agents to discover energy-saving opportunities within these dynamic systems	Service-based applications
11	Gamarra et al. [18]	Sustainable manufacturing	To review technical literature and introduce an innovative approach to microgrid planning	Data mining
12	Hercheui [21]	Green buildings	To show the relevance of designing green ICT solutions, which cope with tacit and explicit knowledge, and reduce the complexity in managing information on sustainability	Internet tool
13	Howland et al. [23]	Natural resource management	Design of strategies for farmer engagement in the knowledge-sharing online platform of the AES-CE (acronym in Spanish for Sharing Experiences for Site Specific Agriculture)	Knowledge-sharing online platform
14	Kivits and Furneaux [27]	Sustainable construction	To outline the benefits, enablers, and barriers associated with BIM and make suggestions about how these issues may be addressed	Building Information Modeling

15	Kontopoulos et al. [28]	Green buildings	To present an ontology-driven decision support system for facilitating the selection of domestic solar hot water systems	Ontology-based decision support tool
16	Linger et al. [30]	Natural resource management	Describes an action-oriented TbKM framework aimed at building capability for policy work in order to address the challenges of a complex policy environment	Task-based KM (TbKM) framework
17	Li et al. [29]	Energy sector	To determine the appropriate spread parameter in using the GRNN for power load forecasting	Generalized regression neural network
18	Liu et al. [31]	Sustainable transport	To improve utilization efficiency of public transportation services, according to people's real demand for public transportation	Not clear
19	Liu et al. [32]	Disaster management	To contribute a reference in decision-making for prevention of grassland fire disaster and for stockbreeding sustainable development planning	GIS and information diffusion-based methodology
20	Lwoga [33]	Agricultural indigenous	To present application of KM approaches for the management of IK and its integration with other knowledge systems for agricultural development in developing countries, including Tanzania	Not clear
21	Majewski et al. [35]	Green ICT	To present a literature review and theoretical investigation combining the areas of business process simulation (BPS) and knowledge management (KM)	Not clear
22	Majchrzak et al. [34]	Healthcare	To present a business rules-based decision support system for the allocation of traumatized patients	Decision support system
23	Mohamed et al. [39]	KM for sustainability	Empirically assess the importance of KM to sustainable development	Not clear
24	Mohamed et al. [40]	ICT, KM, and sustainability	To quantitatively evaluate the importance of ICTs for sustainable development	Knowledge-oriented ICT infrastructure
25	Morik et al. [41]	Data mining for sustainability	To present data mining techniques that explore and analyze environmental spatiotemporal data and help operate better sustainable systems	Data mining
26	Nilashi et al. [44]	Green buildings	Performance assessment tool that analyzes the effect of factors in developing the sustainable building	Knowledge-based expert system: fuzzy inferences system and green buildings rating system
27	Pietrosemoli and Monroy [47]	Sustainable construction	To present references about the positive relationship existing among KM, sustainable construction, and the global sustainability goals	Lessons-learned systems
28	Pinto et al. [48]	Electricity market	To complement the ALBidS strategies by combining them and taking advantage of their different perspectives through the use of the "six thinking hats" group decision technique	Adaptive Learning strategic Bidding System: Decision support system

(continued)

Table 16.1 (continued)

№	Authors (year)	Type of the sector	Aim of the paper	ICT tool or technology
29	Reed et al. [49]	Sustainable land management	To present a hybrid methodological framework	Not clear
30	Shaheen et al. [53]	Energy development	To provide method for classifying a nation's hydrocarbon development into one of five classes: futuristic, conforming, sustainable, unsustainable, and critical	Data mining
31	Shaluf and Ahamadun [54]	Disaster management	To provide with background on the technological emergencies, expert system, and technological emergencies expert system development	Expert system (decision support)
32	Shelbourne et al. [55]	Sustainable construction	To describe the development of the C-SanD tool and supporting web portal implementation, evaluation, and take-up in the project's industry partners	Process protocol method Web portal
33	Smirnov et al. [57]	Disaster management	To propose an approach to emergency situation response that benefits from the ubiquitous computing	Decision support system
34	Srivastava [58]	Sustainable transport	To present an anomaly detection method based on virtual sensors to help detect overconsumption of fuel in aircraft	Virtual sensors (anomaly detection method)
35	Stojanova et al. [59]	Disaster management	To build predictive models of estimating the risk of fire outbreaks in Slovenia by using state-of-the-art data mining techniques	Data mining, decision-making system
36	Tan et al. [61]	Sustainable construction	To understand and describe the types of sustainability knowledge of importance to CPO organizations	Web-based KMS
37	Togawa et al. [63]	Energy circulation systems	To introduce an innovative information network system integrating a regional geographical information system and ICT application and evaluate its performance and future outlooks	Innovative information network system
38	Trotta [65]	Sustainable product development	To analyze important tools used by PLM to formalize knowledge for sustainable new products development	Centralized interfacing for complex databases; integrated platforms
39	Venters et al. [66]	Sustainable construction	To introduce SSM as a method for conceptualizing the industry's knowledge environment and increasing sustainability in construction industry practice	Soft system methodology (SSM)
40	Vikhorev et al. [67]	Manufacturing	Framework for energy monitoring and management in the factory	Decision support system
41	Xu et al. [70]	Green building	To consider a bi-level multi-objective benefit trade-off problem between a green building developer and a contractor under a fuzzy environment	Interactive fuzzy programming technique

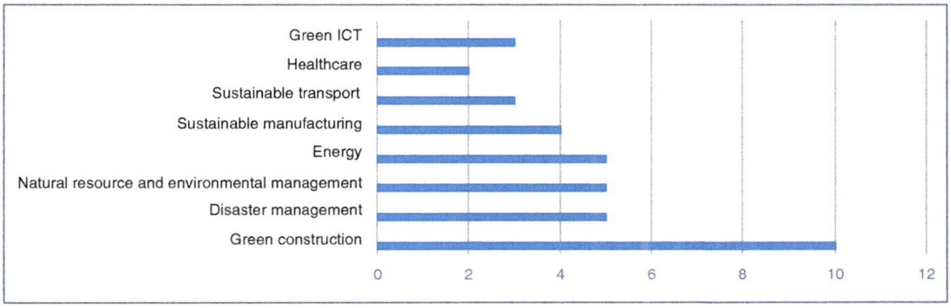

Fig. 16.2 Distribution of ICT for greening

Venters et al. [66] used soft system methodology as a way to increase in construction industry practice. Tan et al. [61] presented web-based knowledge management system to support higher education institutions in the course of sustainable construction. Shelbourne et al. [55] described tools, methods, and architectures for creation, sustaining, and disseminating knowledge for sustainable construction; they have presented C-SanD portal, an integration tool.

Kontopoulos et al. presented an ontology-based decision support tool for optimizing domestic solar hot water system selection. Fan et al. [16] described data mining methodology for knowledge discovery in big building automation system data and suggested method for the efficient post-processing of knowledge discovered. Method, proposed in that paper, helps not only reduce energy consumption of the building but also decrease the required computational costs. However, Pietrosemoli and Monroy [47] conclude that limitations in the mechanisms to promote knowledge processes in green construction industry still exist.

16.4.2 Disaster or Emergency Management

Disasters have a complex and dynamic environment; therefore, KM systems must operate along with a high degree of uncertainties. Dorasamy et al. [14] presented a literature survey dedicated to KM systems in support of disasters management, in particular to knowledge-based emergency management information systems. According to the paper, simulation tools, DDS, and sensor network are the most popular technologies for disaster management.

Stojanova et al. [59] presented improved models that predict the risk of fire outbreaks in Slovenia by using state-of-the-art data mining techniques. Another research on managing fire was made by Liu et al. [32], who described geographical information systems and information diffusion-based methodology for spatiotemporal risk analysis of grassland fire disaster to livestock production in the grassland area of the Northern China.

16.4.3 Environmental Monitoring and Natural Resource Management

Environmental monitoring describes the processes and activities that need to take place to characterize and monitor the quality of the environment. Cash et al. [10] described a framework for understanding the effectiveness of systems that link knowledge to action for sustainability. They provided several case studies and presented applications of KM for environmental monitoring, in particular, for managing ocean fisheries, reduction of transboundary air pollution, El Nino forecasts, and enhancing agricultural productivity. Agricultural productivity is a main theme of the paper, written by Lwoga [33]. The paper described the application of KM for improved farming activities in Tanzania. Linger et al. [] suggested task-based knowledge management (TbKM) approach for application to the case of natural resource management in Indonesia in the context of economic development and response to climate change.

16.4.4 Sustainable Manufacturing and Product Development

The US Department of Commerce's Sustainable Manufacturing Initiative defines sustainable manufacturing as "the creation of manufactured products that use processes that minimize negative environmental impacts, conserve energy and natural resources, are safe for employees, communities, and consumers and are economically sound" (http://trade.gov/competitiveness/sustainablemanufacturing/).

The optimization of resources and energy consumption of production systems is one of the fields in which industries should make relevant improvements in order to successfully integrate sustainability aspects in the factory of the future [7]. Sustainable manufacturing today is a business imperative. Indeed, according to Vikhorev et al. [67], in 2008, the global industrial sector consumed around 98 EJ of energy, and this figure was expected to increase by 44% between 2006 and 2030. Twenty-eight percent of European energy use was consumed by the industrial sector. Manufacturing industry is also responsible for 38% of global CO2 emissions [24]. These figures indicate the necessity of using KM processes and systems in order to provide energy optimization and reduce CO2 emissions.

Vikhorev et al. [67] presented the framework for energy monitoring and management in the factory. Belkadi et al. [7] described knowledge-based and product lifecycle management facilities for sustainability perspective in manufacturing. The paper presents guideline that helps to apply the knowledge model in a specific manufacturing domain. Trotta [65] presents different tools in order to optimize product lifecycle management. Gamarra et al. [18] describes knowledge discovery method that allows minimizing environmental impact of manufacturing processes and makes it more sustainable.

16.4.5 *Sustainable Transportation and Supply Chain Management*

Sustainable transportation can be defined as one that is able to meet transportation needs without compromising the ability of future generations to meet their transportation needs [50]. It is affordable, operates efficiently, limits emissions and waste within the planet's ability to absorb them, minimizes consumption of nonrenewable resources, and reuses and recycles its components.

Awasthi (2011) described multi-criteria decision-making system that helps to evaluate sustainable transportation. The system identified the criteria for sustainability, and then experts provide "linguistic ratings to the potential alternative against the selection criteria," selected the best possibility, and determined "an influence of criteria weight on the decision making process." Srivastava [58] presented method based on virtual sensor technique that helps to detect overconsumption of fuel in aircraft and, therefore, reduce the environmental impact of the aviation. Liu et al. [31] provided method of intelligent bus routing that helps to improve the efficiency of public transportation services, based on people's actual demand for public transportation.

16.5 Discussion and Conclusions

The paper is focused on current research status of using knowledge management systems and processes in green ICT and ICT for greening.

Figure 16.3 reflects different technological tools that were applied in order to support KM processes. They included decision support systems (DSS), knowledge-based systems, data mining technique, web portals and Internet tools, simulation tools, sensors, GIS, service-based applications, etc. As stated in Fig. 16.3, the most frequently used technologies, enlisted in the papers, are DSS, data mining, web portals, and knowledge-based systems.

DSS are mentioned in eight papers, for example, by Awasthi et al. [5], Majchrzak et al. [34], Pinto et al. [48], Smirnov et al. [57], Stojanova et al. [59], and Vikhorev et al. [67]. DSS are a commonly used technology for knowledge application systems and processes that "support the process through which some individuals utilize knowledge possessed by other individuals without actually acquiring that knowledge" [6]. As it follows from Table 16.1, DSS are widely employed by disaster management, where decision-making process, decision quality, and timeliness are highly important.

DSS are followed by data mining (six mentioning) and web portals/Internet tools (five mentioning). Data mining enables knowledge creation systems, and it is utilized for the knowledge discovery process. Examples of such applications are listed by Fan et al. [16], Morik et al. [41], Shaheen et al. [53], etc. They have been applied for such areas as green construction or energy sector for temporal knowledge discovery in big building automation system data. Web portals is one of the most popular

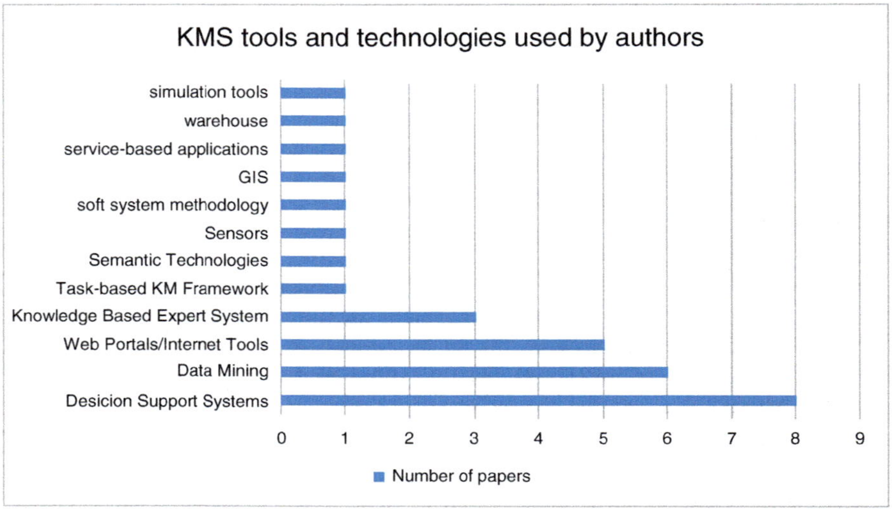

Fig. 16.3 KMS tools and technologies used by authors

tools for knowledge sharing, aimed for exchange knowledge and information as platform of the AES-CE (Sharing Experiences for Site Specific Agriculture) described by Howland et al. [23].

Therefore, analyses of the literature have proved the existence of many different strategies for application of KM to ICT for greening fields, such as green building and constructions, disaster management, natural resource management and environmental monitoring, sustainable manufacturing, sustainable transport, etc. However, review results also show that an urgent need still exists in investigation of application of KM systems and processes to the field of green ICT. Apart from green datacenters and green software development, there are many other fields such as green networking, green hardware, pervasive computing, etc. Therefore, much more work could be done in that field.

Acknowledgment The author would like to show her gratitude and thank all the partner institutions and researchers of PERCCOM program.

References

1. Abdullah, R., Abdullah, S., Din, J., & Tee, M. (2015). A systematic literature review of green software development in collaborative knowledge management environment. *International Journal of Advanced Computer Technology, 9*, 136.
2. Ali, L., & Avdic, A. (2015). A knowledge management framework for sustainable rural development: The case of Gilgit-Baltistan, Pakistan. *The Electronic Journal of Knowledge Management, 13*(2), 103–165.

3. Australian Government. (2010). Australian government ICT sustainability plan 2010–2015. Retrieved from http://www.environment.gov.au/system/files/resources/0478ab3c-032f-4bcd-8e7d-2df70fe1734f/files/ict-plan.pdf
4. Australian Standard AS 5037-2005.
5. Awasthi, A., Chauhan, C. C., & Omranic, H. (2011). Application of fuzzy TOPSIS in evaluating sustainable transportation systems. *Expert Systems with Applications*. doi:10.1016/j.eswa.2011.04.005.
6. Becerra-Fernandez, I., & Sabherwal, R. (2010). *Knowledge management: Systems and processes*. New York: ME Sharpe. ISBN: 076562351x.
7. Belkadi, F., Bernard, A., & Laroche, F. (2015). Knowledge based and PLM facilities for sustainability perspective in manufacturing: A global approach. *Proceedings of the 22nd CIRP conference on Life Cycle Engineering*. doi:10.1016/j.procir.2015.01.065.
8. Bose, R., & Luo, X. (2011). Integrative framework for assessing firms' potential to undertake green IT initiatives via virtualization – A theoretical perspective. *Journal of Strategic Information Systems, 20*(1), 38–54.
9. Butler, T., & Murphi, C. (2007). Understanding the design of information technologies for knowledge management in organizations: A pragmatic perspective. *Information Systems Journal, 17*(2), 143–163.
10. Cash, D. W., Clark, W. C., Alcock, F., Dickson, N. M., Eckley, N., Guston, D. H., Jager, J., & Mitchell, R. B. (2003). Knowledge systems for sustainable development. *Proceedings of the National Academy of Sciences of the United States of America, 100*, 8086–8091.
11. Chai-Arayalert, S., & Nakata, K. (2011). The evolution of green ICT practice: UK Higher Education Institutions Case Study. *Proceedings of the International Conference on Green Computing and Communications* (pp. 220–225). IEEE Computer Society. doi:10.1109/GreenCom.2011.45.
12. Cinar, D., & Kayakutlu, D. (2010). Scenario analysis using Bayesian networks: A case study in energy sector. *Knowledge-Based Systems, 23*(3), 267–276. doi:10.1016/j.knosys.2010.01.009.
13. Dong, H., & Hussain, F. K. (2011). Semantic service matchmaking for Digital Health Ecosystems. *Knowledge-Based Systems, 24*(6), 761–774.
14. Dorasamy, M., Raman, M., & Kaliannan, M. (2013). Knowledge management systems in support of disasters management: A two decade review. *Technological Forecasting and Social Change, 80*(9), 1834–1853.
15. Faucheux, S., & Nicolaï, I. (2011). IT for green and green IT: A proposed typology of eco-innovation. *Ecological Economics, 70*(11), 2020–2027. doi:10.1016/j.ecolecon.2011.05.019.
16. Fan, C., Xiao, F., Madsen, H., & Wang, D. (2015). Temporal knowledge discovery in big BAS data for building energy management. Energy and Buildings, 109, 75–89. http://dx.doi.org/10.1016/j.enbuild.2015.09.060
17. Ferreira, A. M., & Pernici, B. (2013). Using intelligent agents to discover energy saving opportunities within datacenters, in:R E4SuSy@RE, 2013.
18. Gamarra, C., Guerrero, J. M., & Montero, E. (2016). A knowledge discovery in databases approach for industrial microgrid planning. *Renewable and Sustainable Energy Reviews, 60*, 615–630. doi:10.1016/j.rser.2016.01.091.
19. García-Álvarez, M. T. (2014). Analysis of the effects of ICTs in knowledge management and innovation: The case of Zara Group. *Computers in Human Behavior, 51*, 994–1002.
20. Halawi, L., Aronson, J., & McCarthy, R. (2005). Resource-based view of knowledge management for competitive advantage. *The Electronic Journal of Knowledge Management, 3*(2), 75–86. Available online at www.ejkm.com
21. Hercheui, M. D. (2011). Using knowledge management tools in fostering green ICT related behavior change. In *Handbook of research on green ICT: Technology, business and social perspectives*. IGI Global. doi:10.4018/978-1-61692-834-6.ch020.
22. Hilty, L. M., & Lohmann, W. (2013). An annotated bibliography of conceptual frameworks in ICT for sustainability, in: *ICT4S*, 288–300. doi:10.5167/uzh-84887.

23. Howland, F., Muñoz, L. A., Staiger-Rivas, S., Cock, J., & Alvarez, S. (2015). Data sharing and use of ICTs in agriculture: Working with small farmer groups in Colombia. *Knowledge Management for Development Journal, 11*(2), 44–63.
24. IEA. (2007). Tracking industrial energy efficiency and CO2 emissions. Tech. rep., The International Energy Agency (IEA), http://www.iea.org/textbase/nppdf/free/2007/tracking_emissions.pdf
25. IEA. (2011). *Key world energy statistics*. The International Energy Agency (IEA). http://iea-gia.org/wpcontent/uploads/2012/08/key_world_energy_stats-2011-27Dec11.pdf.
26. Jain, P. (2014). Indigenous knowledge management in Botswana using ICT applications. In K. J. Bwalya, N. M. Mnjama, & P. M. M. Sebina (Eds.), *Concepts and advances in information knowledge management. Studies from developing and emerging economies* (pp. 167–191). Chandos Publishing. doi:10.1533/9781780634357.2.167.
27. Kivits, R. A., & Furneaux, C. (2013). BIM: Enabling sustainability and asset management through knowledge management. *The Scientific World Journal, 2013*, 1–14. http://dx.doi.org/10.1155/2013/983721.
28. Kontopoulos, E., Martinopoulos, G., Lazarou, D., & Bassiliades, N. (2016). An ontology-based decision support tool for optimizing domestic solar hot water system selection. *Journal of Cleaner Production, 112*(5), 4636–4646.
29. Li, H. -Z., Guo, S., Li, C. -J., & Sun J. -Q. (2013). A hybrid annual power load forecasting model based on generalized regression neural network with fruit fly optimization algorithm. Knowledge-Based Systems, 37(2013), 378–387. http://dx.doi.org/10.1016/j.knosys.2012.08.015.
30. Linger, H., Aarons, J., McShane, P., & Burstein, F. (2013). A Knowledge Management Framework for Sustainable Development: A Case of Natural Resource Management Policy Work in Indonesia. *PACIS 2013 Proceedings*. 93. http://aisel.aisnet.org/pacis2013/93
31. Liu, Y., Liu, C., Yuan, N. J., Duan, L., Fu, Y., Xiong, H., Xu, S., & Wu, J. (2016). Intelligent bus routing with heterogeneous human mobility patterns. *Knowledge and Information Systems*. doi:10.1007/s10115-016-0948-6.
32. Liu, X., Zhang, J., Cai, W., & Tong, Z. (2010). Information diffusion-based spatio-temporal risk analysis of grassland fire disaster in northern China. *Knowledge-Based Systems, 23*(1), 53–60.
33. Lwoga, E. T. (2011) "Knowledge management approaches in managing agricultural indigenous and exogenous knowledge in Tanzania", Journal of Documentation, 67 3, 407–430. doi:http://dx.doi.org/10.1108/00220411111124523.
34. Majchrzak, T., Noack, O., Neuhaus, P., & Uckert, F. (2011). Supporting the allocation of traumatized patients with a decision support system. *International Journal of Information Systems for Crisis Response and Management*. doi:10.4018/jiscrm.2011070103.
35. Majewski, G., Usoro, A., & Su, P. (2014). Chapter 17. Simulation of knowledge intensive process. In E. Ariwa (Ed.) *Green technology applications for enterprise and academic innovation*. doi:10.40187978-1-4666-5166-1.cb017.
36. McNeil, R. (2011). Application of knowledge management for sustainable development in institutions of higher education. *Dalhousie Journal of Interdisciplinary Management, 7*, 1–13.
37. METI. (2008). Green IT initiative in Japan. Available at: http://www.meti.go.jp/english/policy/GreenITInitiativeInJapan.pdf. Accessed 21 Dec 2015, vol. 2013.
38. Ministry of Science, Technology and Innovation of Denmark. (2008). Action plan for green IT in Denmark. Retrieved from: https://www.apc.org/en/system/files/Action_Plan_for_Green_IT_in_Denmark.pdf
39. Mohamed, M., Stankosky, M., & Mohamed, M. (2009). An empirical assessment of knowledge management criticality for sustainable development. *Journal of Knowledge Management*. doi:10.1108/13673270910988105. Source: DBLP.
40. Mohamed, M., Murray, A., & Mohamed, M. (2010). The role of information and communication technology (ICT) in mobilization of sustainable development knowledge: A quantitative evaluation. *Journal of Knowledge Management, 14*(5), 2010.
41. Morik, K., Bhaduri, K., & Kargupta, H. (2012). Introduction to data mining for sustainability. *Data Mining and Knowledge Discovery, 24*, 311–324. doi:10.1007/s10618-011-0239-5.

42. Murugesan, S. (2008). Harnessing Green IT: Principles and Practices. *IT Professional, 10*(1), 24–32.
43. Nguyen, T., & Burgess, S. (2014). A case analysis of ICT for knowledge transfer in small businesses in Vietnam. *International Journal of Information Management, 34*(3), 416–421. doi:10.1016/j.ijinfomgt.2014.02.009.
44. Nilashi, M., Zakaria, R., Ibrahim, O., Majid, M. Z. A., Zin, R. M., Muhammad, W. C., Abidin, N. I. Z., Sahamir, S. R., & Yakubu, D. A. (2005). A knowledge-based expert system for assessing the performance level of green buildings. *Knowledge-Based Systems, 86*(2015), 194–209. http://dx.doi.org/10.1016/j.knosys.2015.06.009.
45. Nonaka, I., & Takeuchi, H. (1995). *The knowledge creating company: How Japanese companies create the dynamics of innovation*. New York: Oxford University Press.
46. Osseyran, A. (2013). Green IT current developments—A strategic view on ICT changing the global warming trend. In J. H. Appelman (Ed.), *Green ICT and energy: From smart to wise strategies* (pp. 29–42). CRC Press Book. ISBN 9780415620963.
47. Pietrosemoli, L., & Monroy, C. R. (2013). The impact of sustainable construction and knowledge management on sustainability goals. A review of the Venezuelan renewable energy sector. *Renewable and Sustainable Energy Reviews, 27*(2013), 683–691. http://dx.doi.org/10.1016/j.rser.2013.07.056.
48. Pinto, T., Barreto, J., Praça, I., Sousa, T. M., Vale, Z., & Pires, E. J. S. (2015). Six thinking hats: A novel metalearner for intelligent decision support in electricity markets. *Journal of Decision Support Systems*. doi:10.1016/j.dss.2015.07.011.
49. Reed, M. S., Buenemann, M., Atlhopheng, J., Akhtar-Schuster, M., Bachmann, F., Bastin, G., et al. (2011). Cross-scale monitoring and assessment of land degradation and sustainable land management: A methodological framework for knowledge management. *Land Degradation and Development, 22*, 261–271.
50. Richardson, B. C. (2005). Sustainable transport: Analysis framework. *Journal of Transport Geography, 13*(1), 29–39. doi:10.1016/j.jtrangeo.2004.11.005.
51. Roblek, V., Meško, M., Bach, M. P., & Bertocelj, A. (2014). Impact of the knowledge management for sustainable development in innovative economy. *Refereed Proceedings of the Business Systems Laboratory - 2nd International Symposium "SYSTEMS THINKING FOR A SUSTAINABLE ECONOMY. Advancements in Economic and Managerial Theory and Practice"*. January 23–24, 2014 Universitas Mercatorum, Rome (Italy). doi:10.13140/2.1.1981.9522.
52. Rondeau, E., Lepage, F., Georges, J. P., & Morel, G. (2015). Measurements and sustainability, Chapter 3. In M. Dastbaz, C. Pattinson, & B. Akhgar (Eds.), *Green information technology, 1st Edition, a sustainable approach* (pp. 29–59). Morgan Kaufmann. doi:10.1016/B978-0-12-801379-3.00003-6.
53. Shaheen, M., Shahbaz, M., Guergachi, A., & Rehman, Z. (2011). Mining sustainability indicators to classify hydrocarbon development. *Knowledge-Based Systems, 24*(2011), 1159–1168. doi:10.1016/j.knosys.2011.04.016.
54. Shaluf, I. M., & Ahamadun, F. R. (2006). Technological Emergencies Expert System (TEES). *Disaster Prevention and Management, 15*(3), 414–424.
55. Shelbourne, M., Bouchlaghem, D., Anumba, C., Carrillo, P., Khalfan, M., & Glass, J. (2006). Managing knowledge in the context of sustainable construction. *Itcon: Journal of Information Technology in Construction, 11*, 57–71.
56. Silvius, A. J. G., & Schipper, R. P. J. (2014). Sustainability in project management: A literature review and impact analysis. *SOCIAL BUSINESS, 2014, 4*(1), 63–96. http://dx.doi.org/10.1362/204440814X13948909253866
57. Smirnov, A., Levashova, T., Shilov, N., & Kashevnik, A. (2011). Ubiquitous computing for personalized decision support in emergency. *International Journal of Information Systems for Crisis Response and Management, 3*, 55–72.
58. Srivastava, A. N. (2011). Greener aviation with virtual sensors: A case study. *Data Mining and Knowledge Discovery (2012), 24*, 443–471. doi:10.1007/s10618-011-0240-z.

59. Stojanova, D., Kobler, A., Ogrinc, P., Ženko, B., & Džeroski, S. (2012). Estimating the risk of fire outbreaks in the natural environment. *Data mining and knowledge discovery*. doi:10.1007/s10618-011-0213-2.
60. Suryawanshi, K., & Narkhede, S. (2013). Evolution of green ICT implementation in education sector: A study of developed and developing country. *International Journal of Management IAEME*, 91–98. doi:10.1109/MITE.2013.6756344.
61. Tan, X., Lin, K.-Y., Anumba, C. J., Tan, H.-C., & Li, H.-M. (2012). *Exploratory Study of Sustainability Knowledge Management in Capital Project Offices*. West Lafayette, IN: Construction Research Congress 2012.
62. The climate group (2008). SMART 2020: Enabling the low carbon economy in the information age. Retrieved from GeSI. http://www.smart2020.org/_assets/files/02_smart2020Report.pdf. Accessed 21 Sept 2015.
63. Togawa, T., Fujita, T., Dong, L., Ohnishi, S., & Fujii, M. (2016). Integrating GIS databases and ICT applications for the design of energy circulation systems. *Journal of Cleaner Production, 114*, 224–232. doi:10.1016/j.jclepro.2015.07.020.
64. Tranfield, D., Denyer, D., & Smart, P. (2003). Towards a Methodology for Developing Evidence-Informed Management Knowledge by Means of Systematic Review. *British. Journal of Management, 14*(3), 207–222. doi:10.1111/1467-8551.00375.
65. Trotta, M. G. (2010). Product lifecycle management: Sustainability and knowledge management as keys in a complex system of product development. *Journal of Industrial Engineering and Management*. doi:10.3926/jiem.2010.v3n2.p309-322.
66. Venters, W., Cornford, T., & Cushman, M. (2005). Knowledge about sustainability: SSM as a method for conceptualising the UK construction industry's knowledge environment. *CIT Journal of Computing and Information Technology, 13*(2), 137–148. Morgan kaufmann doi:10.1016/B978-0-12-801379-3.00003-6.
67. Vikhorev, K., Greenough, R., & Brown, N. (2013). An advanced energy management framework to promote energy awareness. *Journal of Cleaner Production, 43*, 103–112. doi:10.1016/j.jclepro.2012.12.012.
68. Wong, D. M. L. (2010). Knowledge management catalyst for sustainable development. *ITSim, 2010*(3), 1444–1449.
69. World Commission on Environment and Development (1987). Report of the world commission on environment and development: Our common future. Available at: http://www.un-documents.net/our-common-future
70. Xu, J., Wang, Y., & Tao, Z. (2013). Rough approximation based strategy model between a green building developer and a contractor under a fuzzy environment. *Knowledge-Based Systems, 46*, 54–68.

Chapter 17
PCA-Based Neural Network Model for Identification of Loss of Coolant Accidents in Nuclear Power Plants

T.V. Santosh, Gopika Vinod, P.K. Vijayan, and Jiamei Deng

17.1 Introduction

Safety is an important aspect to be considered in a nuclear reactor. Maximum care is exercised to keep the likelihood of potential risks to a very low value. However, in the event of such an unlikely occurrence, the operator has to take necessary actions relatively fast, which involves complex judgments, making trade-offs between partly incompatible demands, and requires an expert's opinion. Timely and correct decisions in these situations could either prevent an incident from developing into a severe accident or to mitigate the undesired consequences of an accident. The objective of this study is to develop a system, which assists the operator in identifying an accident quickly using ANNs that diagnose the accidents based on reactor process parameters and continuously displays the status of the nuclear reactor. A large database of transient data of reactor process parameters has been generated for reactor core, containment, environmental dispersion, and radiological dose to train the ANNs. These data have been generated using various codes, e.g., RELAP5 – thermal hydraulics code for the core. The present version of this system is capable of identifying large break LOCA scenarios of 220 MWe Indian pressurized heavy water reactors (PHWRs). The system has been designed to provide the necessary information to the operator to handle emergency situations when the reactor is operating. The diagnostic results obtained from ANNs study are satisfactory. The previous work has been able to identify smaller breaks of inlet headers and

T.V. Santosh • G. Vinod • P.K. Vijayan
Reactor Safety Division, Bhabha Atomic Research Centre, Trombay, Mumbai, India

J. Deng (✉)
Reactor Safety Division, Bhabha Atomic Research Centre, Trombay, Mumbai, India

Leeds Beckett University, School of the Built Environment and Engineering, Leeds LS2 9EN, UK
e-mail: j.deng6@hotmail.com

M. Dastbaz et al. (eds.), *Technology for Smart Futures*,
DOI 10.1007/978-3-319-60137-3_17

location well based on NNs but have had problems to identify large breaks [6]. This is one of the major obstacles for a successful tool, because the large break transients are so rapid and many process parameters change almost simultaneously [2–6, 10]. This study attempted to get a better performance and was extended to apply the PCA to optimize the inputs of ANNs. The results of comparison between the classical and PCA-based ANN have been presented in this paper. The simplified ANN model based on PCA is in relatively good agreement with the classical ANN model.

17.2 The Indian PHWR

The primary heat transport (PHT) system of 220 MWe Indian PHWRs is shown in Fig. 17.1 [11]. There are 306 horizontal coolant channels in 220 MWe Indian PHWRs. Each channel consists of 12 fuel bundles contained in a pressure tube, which is surrounded by a calandria tube. The calandria tube is submerged in the relatively cold heavy water moderator. The primary coolant flows inside the pressure tube containing the 19-rod fuel bundle.

Auxiliary systems, like feed and bleed systems, help to maintain the system inventory and the pressure and are achieved with the help of a PHT storage tank, bleed condenser, and pressurizing pump. The reactor is equipped with ECCS with heavy water and light water hydro accumulators along with a long-term pumped recirculation system. Indian PHWR has two shutdown systems, namely, primary shutdown system and secondary shutdown system to bring the reactor to a shutdown state. ECCS is designed to limit the consequences of events such as LOCA.

All the instrumentation and control parameters are continuously displayed on a computerized operator information system (COIS), which is located in the main reactor control room. These process parameters can be utilized to identify the plant state using computational intelligence methods.

17.3 Transient Analysis and ANN Modeling

To study the system behavior under large break LOCA condition, a wide range of break sizes is considered. A spectrum of break sizes have been analyzed for the core and containment. Following are some of the typical process parameters' profiles obtained from RELAP5 simulation for a 200% break in RIH with the availability of ECCS. For example, Fig. 17.2 shows the header pressure in the broken and unbroken pass and Fig. 17.3 shows the D_2O inventory in the PHT system [9].

The events are modeled based on the relevant reactor process parameters' time-dependent data as available in the COIS. A large amount of time-dependent data has been generated in order to train and test the neural networks for the selected scenarios so that these scenarios can be identified during reactor operations [1, 8].

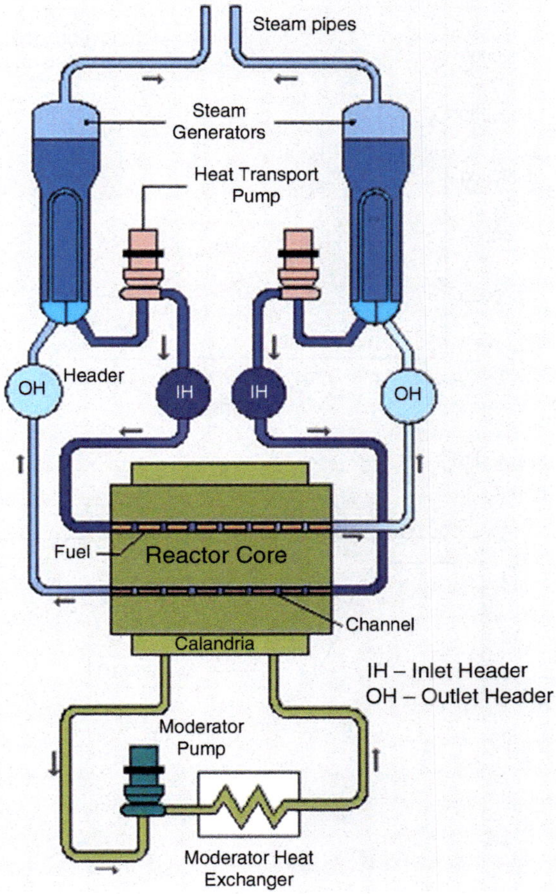

Fig. 17.1 PHT system of Indian PHWR

The input parameter list consists of 35 analog and 2 digital parameters derived from two analog parameters such as containment pressure and high log rate. The analog parameters will have a time-dependent transient data, whereas the digital parameters indicate the status of certain process states such as reactor trip and pump room pressure high. This type of problem is modeled as a pattern recognition problem in the artificial neural networks, wherein a set of input values (known as pattern) with respect to time represents a class of output. Thus, the events are classified into several classes based on the input patterns. In order to illustrate the developed methodology, large break LOCA in RIH with ECCS is selected. The break sizes considered for this event are 20%, 60%, 75%, 100%, 120%, 160%, and 200% of double-ended break in RIH. In addition to these break cases, the normal operating condition of the

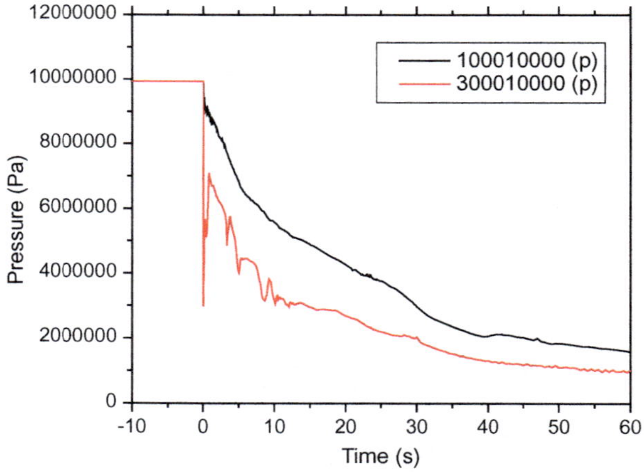

Fig. 17.2 RIH pressure (Pa)

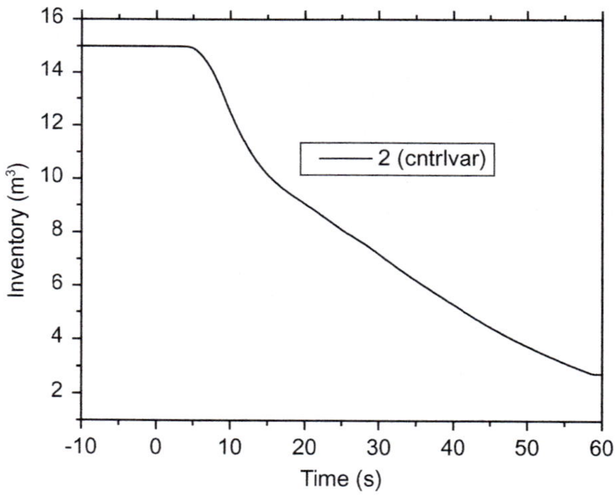

Fig. 17.3 D_2O inventory in PHT system (m^3)

reactor is simulated. A 60 s transient duration is considered in this case under the assumption that this time duration is sufficient to identify the large break LOCA.

To accomplish this task, a neural network consisting of 37 input neurons and 3 output neurons has been selected. A 3-neuron output pattern represents as follows: the first parameter of the output represents the size of the break (percent of cross-sectional area), the second parameter representing the location of the break (i.e., 0 for RIH and 1 for ROH), and the last parameter representing the status of ECCS (i.e., 0 for without the availability of ECCS and 1 for with the availability of ECCS). Table 17.1 shows the representation of output neurons for a 20% break case. A similar representation was followed for other break scenarios.

Table 17.1 Output representation

Output pattern	Description
20.0, 0.0, 0.0	20% break in RIH with ECCS
20.0, 0.0, 1.0	20% break in RIH without ECCS
20.0, 1.0, 0.0	20% break in ROH with ECCS
20.0, 1.0, 1.0	20% break in ROH without ECCS

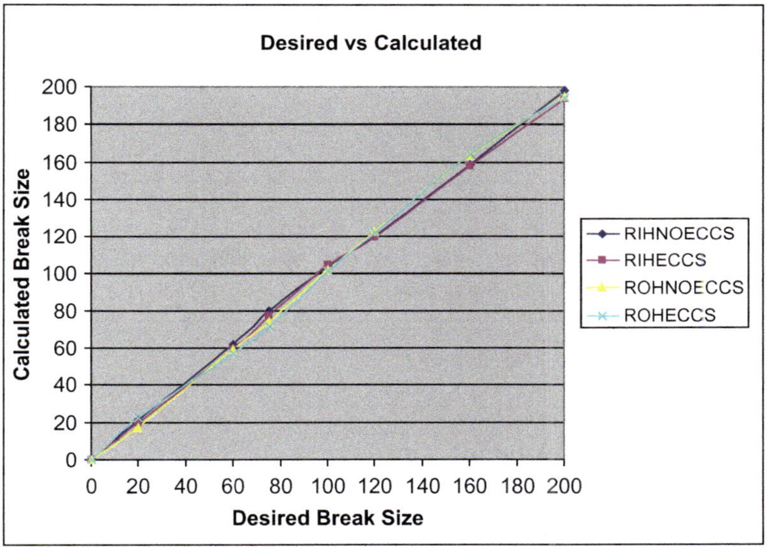

Fig. 17.4 Desired vs. calculated output of ANN

The number of hidden layers and neurons in each layer was selected based on the study presented in [11]. The first-order resilient back propagation algorithm with batch mode of training was employed. Unipolar sigmoid activation function and sum-squared error function were used. Training of ANN was carried out on a typical Pentium IV processor with 1.5 GHz and 512 MB of RAM. The total CPU time the simulator took was approximately 24 h to converge to a minimum error of 1.42E-02 in about 76,000 epochs. A few case studies were carried out with the transients of some selected event scenarios which were not used for training and testing of the ANN. Figure 17.4 shows the plot of desired versus calculated breaks, and Fig. 17.5 shows the predicted break versus the RMS error in percent.

It can be seen from Fig. 17.4 that the predicted breaks are in good agreement. From Fig. 17.5, it is clear that the RMS error for smaller breaks is less compared to larger breaks. This is because of the fact that the large break transients are so rapid that many process parameters come almost simultaneously and ANN may not be distinguished accurately due to lack of uniqueness in the patterns.

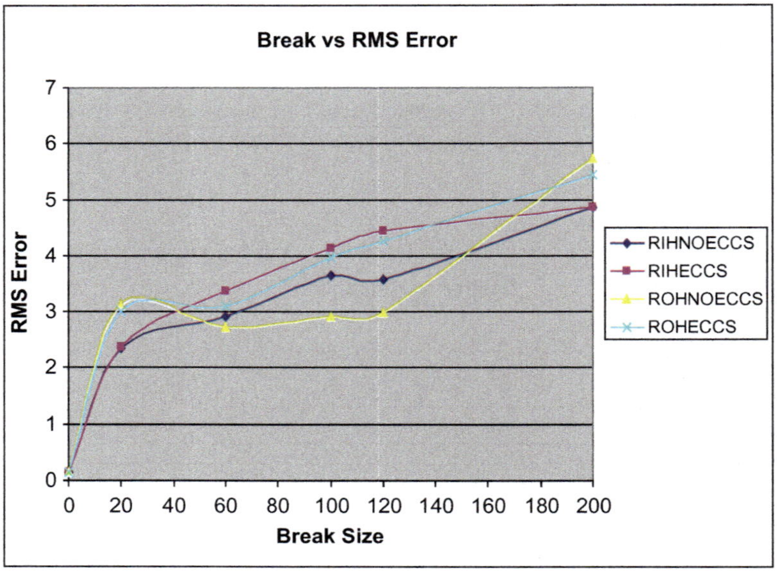

Fig. 17.5 Predicted breaks vs. RMS error (%)

17.4 Principal Component Analysis

A common problem in statistical pattern recognition is that of feature selection or feature extraction. *Feature selection* refers to a process, whereby a *data space* is transformed into a *feature space* that, in theory, has exactly the same dimension as the original data space. However, the transformation is designed in such a way that the data set may be represented by a reduced number of "effective" features and yet retain most of the intrinsic information content of the data; in other words, the data set undergoes a *dimensionality reduction.* To be specific, suppose we have an m-dimensional vector x and wish to transmit it using I numbers, where $I < m$. If we simply truncate the vector x, it will cause a mean square error equal to the sum of the variances of the elements eliminated from x. Clearly, the transformation T should have the property that some of its components have low variance. Principal component analysis maximizes the rate of decrease of variance and is therefore the right choice.

17.5 PCA-Based ANN

The input data for identification of LOCA scenarios has 37 parameters. This requires a substantially high amount of computation during the ANN processing. To reduce this data set, PCA was applied on 37 parameters, thus giving 19 principal components which accounted for 99% variation in the input data. The scree plot of the PCA is shown in Fig. 17.6.

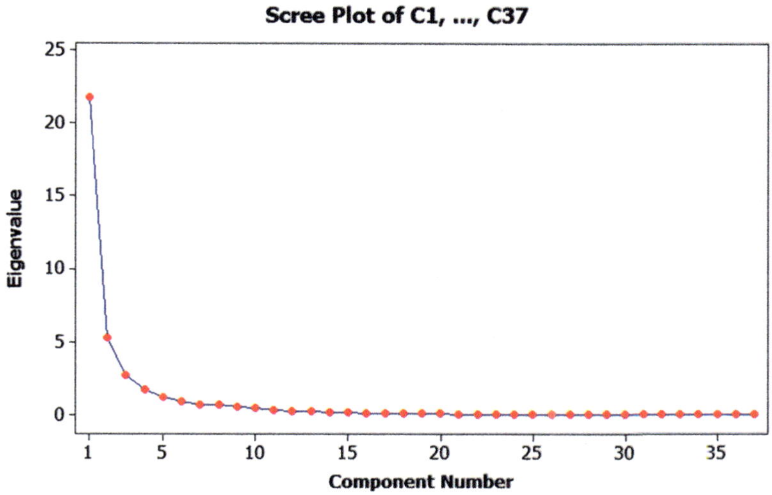

Fig. 17.6 Scree plot of PCA

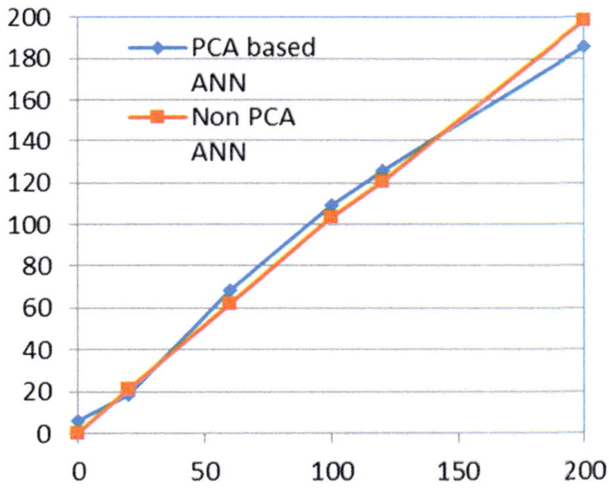

Fig. 17.7 ANN predictions

The architecture used for transient prediction is the feed-forward artificial neural network with resilient back propagation learning algorithm. The learning rate was kept 0.7 and the momentum factor was kept 0.5. Initial weights were kept random between −0.5 and +0.5. The input layer consisted of 19 neurons. Initially the data consisted of 37 parameters of input having 6223 observations. Using principal component analysis, the dimension of the data was reduced to 19 parameters. Hence, the input layer now consisted of 19 neurons. The hidden layer and output layers had ten and three neurons, respectively. 100,000 iterations were run on the data to train the network. The predictions of ANN with and without PCA are shown in Fig. 17.7 for break in inlet header without the availability of ECCS.

17.6 Diagnostic System

Diagnostic system is an artificial intelligence-based operator support system for accident management in 220 MWe Indian PHWRs. It has been developed to monitor the status of the nuclear reactor based on various process parameters, to detect deviations from normal operating conditions, and to determine the significance of situation and recommend an appropriate response in a short time. It performs the abovementioned tasks by operating on a large knowledge base which was generated from RELAP and other popular simulation codes. Currently, diagnostic system can harness the advantage presented by the efficient logging of important plant parameters in various computer-based systems such as computerized operator information system (COIS), radiation data acquisition system (RADAS), etc. Diagnostic system has been set up with a high-speed distributed computing servers for fast processing and real-time response and has been demonstrated successfully for real-time identification of accident scenarios in NPPs. The setup of diagnostic system commissioned at BARC by RSD is shown in Fig. 17.8. The system has the self-diagnostics which monitors the internal servers and various remote tasks to ensure the round-the-clock operation for real-time diagnosis of transients in NPPs. The transient prediction and plume dispersion by diagnostic system are shown in Figs. 17.9 and 17.10, respectively.

17.7 Conclusions

In this paper, the results of an efficient neural network model based on PCA for transient identification have been presented. The prediction obtained from the simplified ANN model based on PCA is in relatively good agreement with the classical ANN model for a typical LOCA scenario. It can be said that the PCA-based ANN

Fig. 17.8 Diagnostic system at BARC

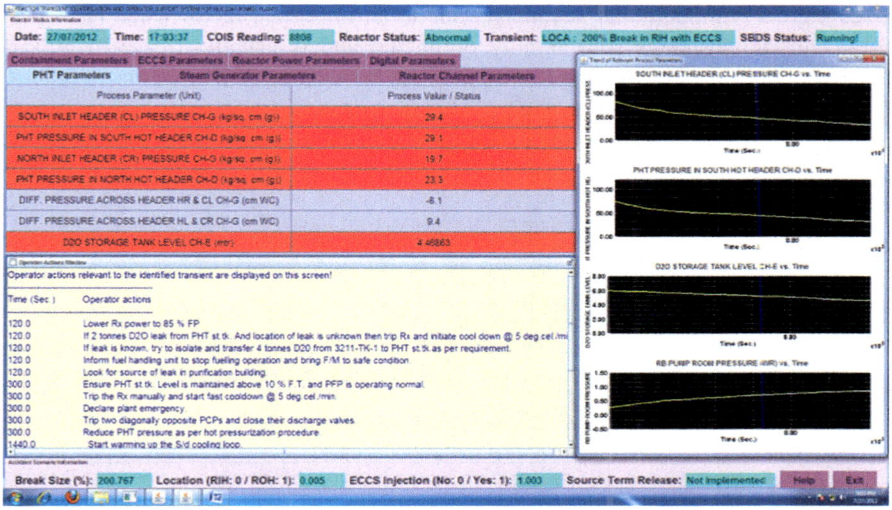

Fig. 17.9 200% LOCA prediction in RIH with ECCS

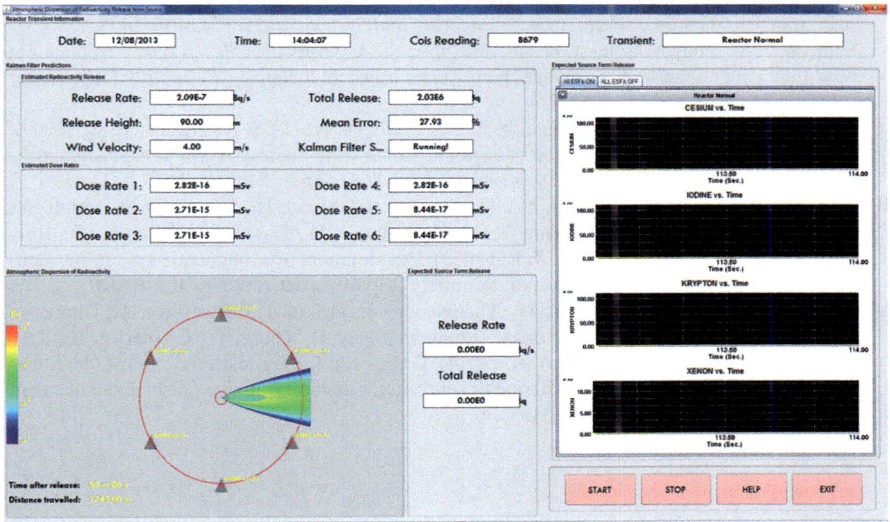

Fig. 17.10 Radioactivity plume dispersion

gives a great computational advantage, due to an important factor when the input parameter dimension is substantially optimized and is usually a case in NPPs. However, there is a scope of improvement in the PCA-based ANN in terms of reduction of error, and this could be achieved by incorporating more of variance during dimension reduction by PCA and also applying different architectures of ANN.

References

1. Computerized Operator Information System (COIS), Kaiga 3 & 4, NPCIL.
2. Coble, J. B., Hashemian, H. M., Meyer, R. M., & Shumaker, B. D. (2012). A review of sensor calibration monitoring for calibration interval extension in nuclear power plants. Instrument: US report, August.
3. Kim, K., & Bartlett, E. B. (2006). Nuclear power plant fault diagnosis using neural networks. *IEEE Transactions on Nuclear Sciences, 43*(4), 2373–2388.
4. Famtoni, F., & Mazzola, A. (1996). A pattern recognition artificial neural networks based model for signal validation in nuclear power plants. *Annual Nuclear Energy, 23*(13), 1069–1076.
5. Renders, J. M., Goosens, A., de Viron, F., & Vlaminck, M. D. (1995). A prototype neural network to perform early warning in nuclear power plant. *Fuzzy Sets and Systems, 74*, 139–151.
6. Santosh, T. V., Vinod, G., Babar, A. K., Kushwaha, H. S., & Raj, V. V. (2002). Symptom based diagnostic system for nuclear power plant operations using artificial neural networks. *Reliability Engineering and System Safety, 82*, 33–40.
7. Santosh, T. V., & Vinod, G. (2007). Application of artificial neural networks to nuclear power plant transient diagnosis. *Reliability Engineering and System Safety, 92*, 1468–1472.
8. Santosh, T. V., Vinod, G., Saraf, R. K., Ghosh, A. K., & Kushwaha, H. S. (2007). Application of artificial neural networks to nuclear power plant transient diagnosis. *Reliability Engineering and System Safety, 92*, 1486–1472.
9. Santosh, T. V., Shrivastava, A., Rao, V. V. S. S., Ghosh, A. K., & Kushwaha, H. S. (2009). Diagnostic system for identification of accident scenarios in nuclear power plants using artificial neural networks. *Reliability Engineering and System Safety, 54*, 759–762.
10. Santhosh, T. V., Gera, B., Kumar, M., Thangamani, I., Prasad, H., Srivastava, A., Dutta, A., Sharma, P. K., Majumdar, P., Verma, V., Mukhopadhyay, D., Ganju, S., Chatterjee, B., Rao, V. V. S. S., Lele, H. G., & Ghosh, A. K. (2013). Development of a diagnostic system for identifying accident conditions in a reactor. Summary Report, BARC/2009/E/013, BARC.
11. Santhosh, T. V., Gera, B., Kumar, M., Thangamani, I., Prasad, H., Srivastava, A., Dutta, A., Sharma, P. K., Majumdar, P., Verma, V., Mukhopadhyay, D., Ganju, S., Chatterjee, B., Rao, V. V. S. S., Lele, H. G., & Ghosh, A. K. (2014). Diagnostic & prognostic system for identification of accident scenarios and prediction of source term in nuclear power plants under accident conditions. BARC Newsletter No. 338.

Erratum to: Testing the Comprehensive Digital Forensic Investigation Process Model (the CDFIPM)

Reza Montasari

Erratum to:
Chapter 15 in: M. Dastbaz et al. (eds.), *Technology for Smart Futures*, DOI 10.1007/978-3-319-60137-3_15

The country name was incorrect in the previous affiliation. It should be UK and not USA. Below is the correct affiliation:

R. Montasari (✉)
School of Computing, Engineering and the Built Environment,
Birmingham City University,
Millennium Point, Curzon Street,
Birmingham B47XG, UK
e-mail: Reza.Montasari@bcu.ac.uk

The online version of the original chapter can be found under
https://dx.doi.org/10.1007/978-3-319-60137-3_15

M. Dastbaz et al. (eds.), *Technology for Smart Futures*,
DOI 10.1007/978-3-319-60137-3_18

Index

M. Dastbaz et al. (eds.), *Technology for Smart Futures*,
DOI 10.1007/978-3-319-60137-3

D

I

K

L

M